HARLEY-DAVIDSON
ALL THE MOTORCYCLES 1903-1983

HARLEY-DAVIDSON
ALL THE MOTORCYCLES 1903-1983

MATTHIAS GERST

Haynes Publishing

© Matthias Gerst 2009

All rights reserved. No part of this publication may be reproduced or stored in a retrieval system or transmitted, in any form or by any means, electronic, mechanical, photocopying, recording or otherwise, without prior permission in writing from Haynes Publishing.

Published in July 2009

A catalogue record for this book is available from the British Library

ISBN 978 1 84425 676 1

Published by Haynes Publishing,
Sparkford, Yeovil, Somerset BA22 7JJ, UK

Tel: 01963 442030 Fax: 01963 440001
Int. tel: +44 1963 442030
Int. fax: +44 1963 440001
E-mail: sales@haynes.co.uk
Website: www.haynes.co.uk

Haynes North America, Inc., 861 Lawrence Drive,
Newbury Park, California 91320, USA

Photo credits: All illustrations are from the archives of Dieter Rebmann, Wolfgang Wiesner and Matthias Gerst

Design and layout by Dominic Stickland
Translation by Jeremy Churchill

Printed and bound in the UK

Contents

Introduction. 100 years of Harley-Davidson: 1903 to 2003	7
Glossary of terms	8
The story up to 1945	10
The early singles, 1903 to 1922	20
The first Big Twins:	37
Model D series, 1909 to 1912	
Model E series, 1912 to 1918	
Model F series, 1913 to 1929	
Model F (Two Cam) series, 1928 to 1929	
Model H series, 1915	
Model J series, 1915 to 1929	
Model J (Two Cam) series, 1928 to 1929	
The Model G forecars, 1913 to 1916	85
The VCR outfit, 1931	87
Servi-Car. The Model G range, 1932 to 1973	88
The Model TA military three-wheeler, 1940/1	100
Sport Twin. Model W, 1919 to 1923	102
The Model A and B singles, 1926 to 1934	106
The Model C singles, 1929 to 1934/7	112
The Forty-Fives: Model D, 1929 to 1931	114
The Forty-Fives: Model R, 1931 to 1936	117
The Forty-Fives: Model W, 1937 to 1951	123
The Forty-Fives and Fifty-Fours: Models K and KH, 1952 to 1956	135
The Model V Big Twins, 1930 to 1936	140
The Model U Big Twins, 1937 to 1948	158
The story from 1945	168
The post-war two-stroke singles, 1948 to 1965	176
Topper scooters, 1960 to 1965	189
Knucklehead: Models E and F Big Twins, 1936 to 1947	194
Panhead & Hydra-Glide: Models E and F Big Twins, 1948 to 1957	210
Duo-Glide. The Model F Big Twins, 1958 to 1964	223
Electra-Glide. The Model F Big Twins, 1965 to 1984	232
The Aermacchi singles, 1961 to 1978	253
Sportster. The XL models, 1957 to 1985	285
Super Glide. The FX models, 1971 to 1986	316
Military machines, 1910 to 1973	346
Appendix 1: Harley-Davidson's non-motorcycle products	369
Appendix 2: 80 years of Harley-Davidson: Model range year-by-year listing	378
Appendix 3: Harley-Davidson production totals, 1903 to 1983	400

100 years of Harley-Davidson

1903 to 2003

When it comes to Harley-Davidson, the motorcycle manufacturer that in 2003 could look back on a century of existence, hasn't everything already been said that there is to be said? Is there anything new to say, something that has not yet been disseminated everywhere? The publisher and the authors are of the opinion that yes, there most certainly is.

This work, set out in two volumes, concentrates exclusively on the motorcycles from Milwaukee. Anyone who hopes to find out about the biker lifestyle and the Harley scene had better close the book now and put it back on the shelf. Because this book is all about rubber, chrome and steel, about model years and technical modifications, about details and about dates, data and still more data. That's exactly what this book is made of.

Despite the abundance of material already published, this mammoth undertaking required several years of research. Around two hundredweight of documents, including much that was very difficult to get hold of, as well as over 20 years experience with Harley-Davidson motorcycles, form the basis of this book. Without the wholehearted support of TÜV SÜD, this book could never have seen the light of day.

While writing this book, the author went to great lengths to ensure that accurate data is given and that authentic illustrations are used. Unfortunately the technical data of the pre-1945 models, even direct from the factory, are not particularly accurate. Where possible, factory documents were always used as primary sources, but even the official data in sales literature and workshop manuals did not always help, or worse, they contradicted each other. Therefore it is essential to ask for the reader's understanding in the event that, despite all the care that has been taken, an error should have crept in here or there.

Should there be specialists in individual models or series, who can contribute to a richer or better coverage of the motorcycles presented in this book and who would like to do so, they are cordially requested to pass on their observations on this publication to the author so that these aspects can be considered for any future edition.

Because the Harley-Davidson Motor Co. of the USA would not give permission for the use of the original pictures that were intended to be used in this book – despite the efforts of the German Harley-Davidson importers, who are expressly thanked here for their efforts – one or two gaps in the pictorial coverage could not, regrettably, be filled. Every cloud, though, has a silver lining; the loss of well-known factory photographs was more than offset by the fact that in almost every case photos from the work of Dieter Rebmann could replace the missing item. This was a real stroke of luck, as he is considered, not without good reason, as one of the world's best Harley-Davidson photographers. Without his material this book could have never have assumed its current form.

The sheer number of Harley-Davidson motorcycles necessitated a separation of the coverage. This first volume covers all Harley-Davidson products up to 1983, the year before the introduction of the new generation of Evolution models. Since throughout their history Harley-Davidson have never introduced changes across the board for all their models at once, but have rather encouraged a flowing and measured process of conversion, the same approach has been used for this book. So the Tour Glide and the Softail (which came on to the market in 1980 and 1984, respectively) could not yet be covered in this book, nor could the five-speed Electra-Glide. However the 61 cu. in. (997 cc.) Sportsters and the four-speed Electra-Glides (up to 1985 and 1984, respectively) are covered. All police models, irrespective of year of production, will be covered in the second volume.

Anyway, enough of the prologue; prepare for the voyage of discovery that is the history of the first 80 years of Harley-Davidson motorcycles. One thing's for sure; it won't be boring . . .

Matthias Gerst, Dieter Rebmann

Glossary of terms

Atmospheric (intake) valve
An early form of valve operation inherited from the original de Dion-Bouton design. The intake valve, located in the top of the cylinder head, opened automatically under atmospheric pressure as the piston descended on the induction stroke, its movement being controlled by an induction stop cast into the cylinder head. As pressure within the cylinder equalized and then started to exceed atmospheric pressure as the piston reached the bottom of its stroke and then rose again, the valve closed, this process being aided by a light return spring. Also known as 'automatic', or 'suction', intake valves, these were used on Harley singles up to 1911-1913 and on the early V-twins built between 1906 and 1910 and are easily recognized by the lack of valvegear on the outside of the engine. So unreliable that riders used to carry spare valve assemblies with them and thought nothing of replacing a defective valve at the roadside, and a severe limitation on engine operating speeds and power outputs, atmospheric intake valves were supplanted in Harley-Davidson's range by the much more reliable 'ioe' system of operation.

Automatic intake valve
See 'Atmospheric intake valve'.

Dow metal
Pistons made of a magnesium alloy created by the Dow Corporation.

Dryer and Cleaner
Specially-equipped sidecar body for commercial sales to laundries, etc.

F-Head
See 'ioe' and 'pocket valve'.

Flathead
Not exclusively a Harley-Davidson expression (any side-valve engine is a 'flathead', due to the flat nature of a cylinder head that is almost totally devoid of porting, arrangements for valvegear, etc.). However, while many Harley-Davidson models have been fitted with side-valve engines, the classic 'flathead' Harley-Davidson models would be the 74- and 80-cubic inch (1200- and 1300 cc.) Models V and U manufactured between 1930 and 1948, and the equally-famous 45 cubic inch (742 cc.) motorcycle and Servi-car models produced from 1929 to 1973.

ioe
Intake-over-exhaust valve layout; i.e., overhead intake valve, side exhaust valve (N.B. Not to be confused with 'eoi' or overhead exhaust, side intake layouts, which were also used in the early years by some manufacturers). A single camshaft with two lobes per cylinder controlled both valves; the side exhaust by means of a cam follower and tappet, while the overhead intake was opened by a rocker via an externally-mounted pushrod; a distinctive feature making identification of this type of engine very easy. Although the overhead valve could be (and was, in other engine designs) located anywhere in the cylinder head, those Harley-Davidson engines using this layout featured the overhead intake valve located directly above the side exhaust valve, both valves opening into a chamber or pocket formed to one side of the main combustion chamber (see illustration on p. 14); this engine configuration thus became known as the 'pocket valve'.

Iron alloy
Pistons made of a low-expansion iron alloy.

Knucklehead
Harley-Davidson's first overhead-valve V-Twin engine of 60 cubic inches (989 cc.), later 74 cu. in. (1207 cc.), fitted to the Model Es built between 1936 and 1947 and so named because of the distinctive shape – thought to resemble knuckles – of the round knobs on the rocker housings.

L-Head
See 'sv'.

Lynite
Pistons made of a light alloy.

Model years.
It must be remembered at all times that Harley-Davidson's model year, like that of most car and motorcycle manufacturers, begins in the autumn of the previous year, when advantage has been taken of the shutting-down of the factory for the annual holiday to re-tool for any changes required by the new models. Thus, for example, a Model VLD of the 1934 model year would have been built at any time from August or September of 1933 to July 1934 and may have been sold at any time from the autumn of 1933 onwards. This is the date that is used in the first two characters of a Harley's engine number (see photo of 1927 Model JD on p. 50), the only means of identifying any Harley up to 1970, when the Company started putting frame numbers on their machines as well. All dates in this book, particularly in the tables of data, are model years. The exceptions, usually specific dates given in the text, will be evident from the context.

N./Av.
Not Available.

Newspaper Side Van
Specially-equipped sidecar body for commercial sales to newspaper-sellers, etc.

ohv
OverHead Valves. Both intake and exhaust valves located in the cylinder head above the combustion chamber and controlled from the camshaft by long pushrods and rockers.

Package Truck
Specially-equipped sidecar body for commercial sales to delivery companies.

Panhead
Harley-Davidson's second overhead-valve V-Twin engine, replacing the 'Knucklehead' in 1948 and manufactured until 1965 – thus being fitted to both E and F models – when it was replaced by the 'Shovelhead'. So named because of the distinctive shape of the valve/rocker covers at the top of the engine. Made in 60- and 74-cubic inch (989 and 1207 cc.)

capacities, the 'Panhead', essentially, was a new top end, featuring aluminium-alloy cylinder heads and hydraulic tappets for cooler running and reduced maintenance, on the existing 'Knucklehead' bottom end.

Pocket valve
Strictly speaking, this term refers only to those engines in which the overhead intake valve was controlled by a lobe on the camshaft, the side exhaust valve being controlled by the other camshaft lobe (see 'ioe'). It is occasionally used also of the early engines with atmospheric type intake valve (see above), purely because of their similar location in the engine. The 'pocket valve' name arose because both intake and exhaust valves opened into a chamber or pocket formed to one side of the main combustion chamber (see illustration on p. 14). An alternative name was 'F-head', because of the valve layout relative to the combustion chamber when seen in a section view from front or rear of the engine. In terms of Harley-Davidson models, all the singles from 1913 up to the early 1920s and the V-twins from 1911 up to 1929 were 'pocket valve' designs, but the name is usually particularly reserved for the V-twin board track racers of the period.

SAE
Society of Automotive Engineers.

Shovelhead
The third series of overhead-valve V-Twin engines to be fitted to Harley-Davidson's Big Twin models, replacing the 'Panhead' in 1965 and manufactured until 1984, being supplanted by the 'Evolution' engine. Made in 74- and 82-cubic inch (1207- and 1340 cc.) capacities, the nickname 'Shovelhead' – a continuation of the convention of naming the engines for the shape of the valve/rocker covers at the top of the engine – came about because of the prominent rocker boxes (not just valve/rocker covers), that were thought to resemble the shape of a shovel. Shovelheads can be further sub-divided into the Generator Shovelheads of 1965-69 and the Alternator Shovelheads of 1970-84, the different electrical equipment accompanying a major revision of the engine unit.

Shrine
Limited edition with fully-enclosed final drive chain case and other features for Shriners (US charity organization).

Suction (intake) valve
See 'Automatic intake valve'.

sv
Side Valves. Both intake and exhaust valves located in the cylinder barrel or block beside the combustion chamber and controlled from the camshaft by a cam follower and tappet. Usually, both valves were mounted beside each other in a valve chest formed on the side of the cylinder barrel or block – a configuration known as 'L-head', because of the valve layout relative to the combustion chamber when seen in a section view from front or rear of the engine. 'T-head' designs, in which the valves were on opposite sides of the combustion chamber, were another variation, less popular because of the attendant need for more complex valve trains. All such asymmetrical layouts caused problems with cylinder distortion due to localized overheating around the exhaust ports which were never fully resolved and remained a limitation on power outputs and engine speeds. Side-valve engines suffered further by comparison with overhead-valve designs due to their inherently poor breathing, thanks to the acute bends through which intake and exhaust gases had to pass to enter and leave the combustion chamber, and to the poor combustion chamber shape and low compression ratios possible with such a valve configuration. However, the cheapness of manufacture of side-valve engines and their relative ease of access for the frequently-needed valve overhauls meant that they enjoyed a period of popularity in the pre-war years, stimulated by the increases in power output and reliability produced by the work of engineers such as Sir Harry Ricardo in England and tuners such as Tom Sifton in the US. Harley-Davidson took the apparently retrograde step (in engineering terms) of replacing their established ioe engines with side-valve units as much to counter Charles Franklin's Indian 'Powerplus' models (which enjoyed a period of supremacy in US motorcycle competition in the 1920s after Harley-Davidson's 1921 decision to stop factory participation in racing) as to enjoy the benefits of cheaper manufacturing costs.

TT
In England, Tourist Trophy road races. In the USA, Tourist Trophy Steeplechase or Scrambles races take place on the dirt-track circuits that feature a jump and left- and right-hand turns.

Wheel and tyre sizes.
Between 1890 and 1929, beaded-edge tyres, also called 'veteran', 'old-timer' or 'vintage' tyres, were the most popular type of tyre for motorcycles. Beaded-edge tyres have large ridges, or beads, of hard rubber around the interior circumferences that fit into the clincher of a beaded-edge wheel rim. A minimum tyre pressure of 60 psi is needed to push the beads into the clincher of the rim and to keep the tyre in place. The most common cause of a beaded-edge tyre failure is lack of pressure allowing the tyre to detach itself.

Harley-Davidson seem to have changed to the modern wired-edge tyre between 1926 and 1928; these, with their generally-greater cross-sections and requiring much lower pressures, were sometimes referred to as 'balloon' tyres. The American motorcycle manufacturers' preference for tyres of significantly larger cross-section than those generally fitted to European motorcycles has led to the term 'balloon' tyres being applied (incorrectly) only to fatter tyres.

A beaded-edge tyre uses measurements in inches that refer to the overall wheel/tyre dimensions and not to wheel rim diameter; thus a '3x28' tyre is actually a 3-inch section tyre on a 22-inch diameter wheel rim. This convention seems to have carried on for a few years after the change to wired-edge tyres had been made. N.B. With one or two exceptions from the late-1960s onwards, all the tyres mentioned in this book are of 95-100% aspect ratio; i.e., the tyre's height is (nominally) the same as its width.

Further expressions are either explained in the text or self-explanatory from the context and photographs.

Dieter Rebmann contributed all illustrations. It can happen that in individual cases, the owners of the vehicles have made changes so that those vehicles no longer correspond in all respects to the factory original.

The story up to 1945

The history of the Harley-Davidson marque began in 1901, when Bill Harley and Arthur Davidson built a small clip-on engine for bicycles. This had a capacity of 7 cubic inches (116 cc.) and delivered perhaps ¾ horsepower. However it had one or two small problems; it ran unreliably and it was not powerful enough to do what the pair wanted.

A couple of blocks away from Harley's workplace in Milwaukee, in the State of Wisconsin, worked a man named Ole Evinrude, who would later earn fame for his outboard motors, with his partner in the Motor Car Power Equipment Company. Evinrude had been a friend of Arthur Davidson's since their days together in the nearby small town of Cambridge.

Evinrude was an internal combustion engine enthusiast and already had some experience in their construction. He was also the right man to help Messrs Harley and Davidson build a larger and more substantial engine, contributing his experience in making carburettors as well as the blueprints of his own water-cooled engine.

By 1902 the three of them had developed from these beginnings an air-cooled engine suitable for motorcycle use, which they then fitted to the first-ever Harley-Davidson motorcycle. The new engine had a capacity of 10 cubic inches (167 cc.) and a power output of about one horsepower that made a speed of at least 25 mph (40 km/h) attainable.

After the installation of the prototype engine in the first complete machine, an even bigger engine appeared in 1903 that produced about 3 horsepower from its 25 cubic inches (405 cc.); this finally satisfied the three engineers that they had a motorcycle fit for sale.

Today, Harley-Davidson count the two machines of 1902-3 and 1903-4 as prototypes, regarding the Motor Company's production history as starting in 1905. The first factory building was a small wooden shed erected by the Davidson brothers' father in his backyard.

In the beginning only a few machines were sold. The response to the first advertisements in 1905 ensured manufacture of even more; it is only from this year that manufacture in commercial quantities became more or less certain.

In addition came the first competition success (in Chicago) and the shed's floor area had to be doubled by an extension. The production of carburettors, marine engines and propellers was undertaken to provide additional sources of income.

By 1906 a full-time employee had to be taken on, while the Founders still worked at their regular employment to make ends meet.

Up to 1905 only black-painted motorcycles were made. In 1906 another colour was offered, with the illustrious name of Renault Gray (the French marque Renault being synonymous in the early 20th century

In the beginning: the first Harley factory

USA with expensive, high-quality and quiet luxury cars). This, with William Harley's insistence from the start on effective silencers to make motorcycles as acceptable as possible to the general public, later earned Harleys the nickname 'The Silent Gray Fellow'.

Demand rose, and with it the need for more space. A new, (but still wooden) factory building in Chestnut Street, later West Juneau Avenue, provided the answer. It is still the headquarters of the Harley-Davidson Motor Company today. The first newspaper adverts had already been published.

A year later William A., the eldest of the Davidson brothers, joined the motorcycle project, and in this year altogether 150 machines were manufactured, which in the future would run under the now officially-licensed brand name of Harley-Davidson.

Walter Davidson was elected as the first President and General Manager of the new Harley-Davidson Motor Company, with William A. Davidson as Vice-President and Works Manager. William S. Harley took the position of Chief Engineer and Designer, while Arthur Davidson became the first Secretary and General Sales Manager. The first shareholders' meeting took place on the 17th September, during which it was decided to expand the factory building by adding more floors, this to be financed by a share issue. In addition it was decided to print a sales brochure. By now Harley-Davidson had eighteen employees. In 1907 came the first police sale, a single motorcycle to the Pittsburgh force.

Soon after this, the foundation stone was laid for a new, brick-built, factory building on the Chestnut Street site. In the same year, 1908, the young Company numbered 35 employees. The Founders now felt it safe to give up their jobs and to throw themselves full-time into their new enterprise. Even so, Walter Davidson found time to combine business with pleasure by entering first a two-day endurance run, then an economy run the following week, that were staged by the Federation of American Motorcyclists in New York State in June 1908. He won both (see photo on page 19), earning the Motor Company valuable publicity.

Just like practically all motorcycle-manufacturers of the period, the Motor Company started to experiment with a new twin-cylinder engine of V configuration, a layout promising substantially greater power output from a compact unit not very much larger than a single. This appeared for the first time in 1909 with an engine of 54 cu. in. (880 cc.). In the same year rear stands were fitted for the first time to all models, which allowed the engine to be started while the motorcycle's rear wheel was raised clear of the ground (making life easier with a machine which still lacked any kind of device for interrupting the drive on starting-off). A Bosch magneto was also available as an option.

In 1910 the factory offered acetylene lighting for the first time. Acetylene gas was produced in a cylindrical generator in which water is allowed to drip onto calcium carbide; the gas was burned by a naked flame in the headlamp and (where fitted) the tail lamp. The V-twin model was apparently not yet fit for sale; in any case it was not offered to customers in 1910. The problems were probably finally resolved in 1910; in the following year the 50 cu. in. (811 cc.) Model 7D appeared, its ioe valve operation then becoming the basis of all Harley V-twins up to 1929. That year was

Already extended by 1905

The Founders: William A. Davidson, Walter Davidson, Arthur Davidson and William S. Harley

also the first time that one could take a companion along on a ride; provided, of course, that the proud Harley owner purchased a pillion saddle from the new in-house accessory catalogue. Provision for carrying a passenger was only provided ex-works from 1918-onwards. However, from 1914-onwards, at a cost of 85 dollars, a motorcycle could be ordered with a sidecar made by the Rogers Company of Chicago and fitted at the factory. In previous years, owners had had to resort to the relevant aftermarket accessory suppliers; after 1914 a much wider range of sidecars was offered by the factory, for the transport of people as well as of goods of all sorts. Renault Gray remained the only colour option, but customers could exercise some choice in the colour and style of pinstriping.

In 1911 Harley-Davidson now had 481 employees on the books and the factory was once more extended, substantially. The following year saw the number employed increase to well over 1,000 and the factory's floor space was again doubled. Exports to Japan (temporarily suspended in 1917) started at about this time, as did exports to Great Britain via Duncan Watson of London; some 350 Harleys, principally V-twins, being sold in the United Kingdom before the declaration of war in August 1914. The number of domestic market dealers had risen from only one in 1904 to over 200, a figure which clearly shows the unbelievable spirit of optimism prevailing in the pre-World War 1 USA. All forms of racing became ever more popular, each large city organizing regular motorcycle races and trials. It was logical under such circumstances for Harley-Davidson to create its own racing team in 1913. The first racing manager, to become a legendary figure, was called William Ottaway. This official support for racing naturally led to the factory producing racing machines in response to demand (this side of the Motor Company's history will be covered in a future volume), best known being the 11K singles and V-twins of 1915, although production of racers continued through to 1917. A spin-off of this work was the creation of the Fast Motor programme to cater for the need of some customers for more power than was provided by the standard offering. To the disappointment of the fans in the autumn of 1916, the factory ended its competition activities until further notice (as did all the motorcycle manufacturers, by mutual agreement in view of the worsening international situation), although the contracted riders were allowed to buy the works racers and ride them as private entrants. Perhaps the fans gained some consolation from the fact that they could now buy, for five cents, a publication devoted entirely to their favourite brand. The newsletter was called *The Enthusiast* and is still today very much a part of the Harley myth.

For 1912, the twin's capacity was increased to a full 60 cu. in. (989 cc.), which meant a not insignificant increase in performance. A year later the 50 cu. in. (811 cc.) version had been dropped, leaving the bigger engine as the only V-twin power unit in the range. On the singles of 1913 there was for the first time the choice between the familiar belt drive and an unenclosed roller chain drive. The former was completely supplanted by 1915; three years later the dry-cell battery-and-coil ignition system was also consigned to the scrap bin.

With the entry of the United States into the First World War in 1917, things also changed a lot at Harley-

This Brush lorry took care of Harley's transport needs from 1909

Davidson. For one thing, naturally, production for the Army increased sharply. This had less obvious consequences in that a policy of standardization of parts was instituted to help speed up production and reduce costs (to get the maximum benefit from Government contracts); the same frame, footboards, etc. began to be used on all models. On top of that, the Motor Company capitalized on the surge in patriotism following the entry of the United States into the War (and highlighted its contribution to the war effort) by painting all models, civilian and military, the patriotic 'soldier color' of olive green, completely replacing the by-now traditional Renault Gray as the standard colour for Harley-Davidson products until 1922. The Fast Motor concept was extended to the single-cylinder range (whose power output was augmented somewhat by numerous changes to the valvegear), these becoming the 500 models (singles and V-twins) of 1918 and 1919. A further consequence was the institution of the Service School; founded in the first instance to train the Army's mechanics, this Company training centre nevertheless continued until 1941 to train dealers' workshop staff.

Between 1917 and 1923 Harley-Davidson even sold bicycles, but were not especially successful with them. For a price of between 30 and 45 bucks one could become the proud owner of an engineless Harley. On the other hand, motorcycle sales were increasing rapidly. At the end of 1918, with the War over and peacetime production resuming, H-D could count on 1,000 contracted dealers. At this time the race for supremacy of the American motorcycle market was practically won. Indian, the long-time market leader had, like H-D, delivered large numbers of motorcycles to the Army, but due to chaotic business methods had made virtually no money out of them. Furthermore, in its concentration on obtaining Government orders, Indian had criminally neglected the care and development of its dealer organization, so that it was made comparatively simple for H-D to persuade existing Indian dealers to change allegiance. The Founders made good use of the opportunity. And the business was booming; in 1919 16,095 sidecars were produced. On the other hand from 1919 to 1920 no Harley singles were manufactured. In this year the number of employees – 2,400 – reached its highest-ever level. Dealers could buy Open-Loop racing frames and either begin racing under their own banner or sell the machines on to suitably-courageous private entrants. Between 1919 and approximately mid-1923 there was once again a Harley works team, which while entering in all the various branches of the sport (long track, speedway, hill-climbing, etc.) never quite managed to recapture the aura of former glories.

After a short-lived post-War boom the country went into slump and production dropped, as did the number of employees, which fell by 1923 to an all-time low of 1,000; as early as 1921 the factory had been closed completely for four weeks. As the domestic market slumped, exports assumed greater importance. Exports to Great Britain had resumed after the War's end and significant numbers of Harleys were sold by Duncan Watson's organization; principally V-twins, but the Model W also found favour in Europe and sold well there until growing economic problems and worsening labour relations provoked the Chancellor of the Exchequer into imposing a 33⅓% import duty on

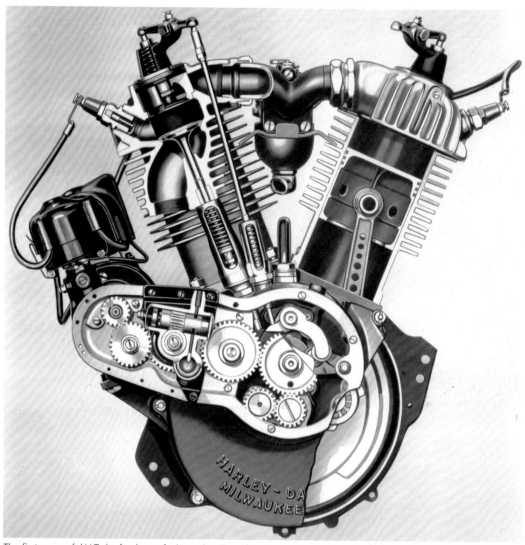

The first successful V-Twin, forebear of a legendary line of engines. Note the ioe valve layout

foreign-made motorcycles in the spring of 1925, which ended the import of American machines almost overnight. The resumption of exports to Japan in 1922 did not provide a good return either, first because it never was then a market capable of absorbing large numbers of machines and also because in 1922 Japan was shaken by a devastating earthquake that damaged the entire infrastructure of the country – and there weren't many decent roads, on which one could drive then – such that little remained unaffected.

Equally unsuccessful was the attempt by the two Davidson brothers to create a cartel with Frank Weschler, Indian's General Manager. Their competition could not have resisted the combined power of both enterprises over the market. But they could not come to an agreement, which was all the worse for them as both were threatened more and more by the waves of cheap Fords and Chevrolets.

The import of Harley-Davidson motorcycles into Germany began in 1924. In 1926, H-D commenced production of its own sidecars. There was no need to build a special facility for these as there was now plenty of spare capacity in Milwaukee. The long-standing connection with the former sidecar producer, the Rogers Company, was broken off.

At the close of the decade experiments were being conducted with V-four engines of 80 or 90 cubic inches (1300 cc. or 1475 cc.) capacity, but these however were never to reach the production stage. The reason was that things were getting much worse; on the 24th October 1929, the famous Black Thursday, the US stock market crashed. The value of stocks collapsed and many up-to-then wealthy people lost their entire fortune overnight. Unemployment reached all-time highs and the larger cities saw soup kitchens for the poor, paid for by

The Single: the engine on which Harley-Davidson's first 15 years were based. Again, this is the later ioe version

Government and organized by welfare organisations or the Salvation Army.

The motor vehicle industry suffered particularly badly. If the carmakers managed to roll out about 4.5 million passenger cars in 1929, by the end of 1932 this number had shrunk to a pitiful one million. The prices of factory-fresh vehicles halved due to the huge discounts offered and to the pressure of competition. Countless companies were forced to close for ever.

Things weren't much better for motorcycling. Apart from Indian and Harley-Davidson, none of the respected names in US motorcycling would survive 1934. Following the first, post-First World War, depression, sales had recovered to acceptable levels by 1929 (something over 21,000 machines), even though Harley-Davidson could no longer reach the record number of over 28,000 motorcycles manufactured in 1920. By 1933, production had collapsed to 3,700 units – too little to live on, too many to die.

It didn't help a great deal that in 1930 a record number of over 3,000 police forces had already ordered Harleys. Racing victories should also be recorded, although H-D officially no longer supported a works team. Every child knew the name of Joe Petrali, the Harley rider dominating the sport at that time. However in 1931 the factory was working at ten per cent of capacity, and in 1932 losses reached over 320,000 dollars.

During the worst years of the Depression, between

New but already outgrown: new construction 1909

The Harley-Davidson factory, circa 1920

1930 and 1933, Harley even catalogued stationary engines for industry; these, however, did not sell well. It didn't help the situation that just as sales were beginning to suffer from the Depression, the factory inflicted even further injuries on itself with the launch of the less-than-satisfactory Model D range and the disastrous launch of the Model Vs in 1928-30. Just as it needed good, saleable models and committed, enthusiastic dealers most, Harley found itself with major defects in its two main model ranges and a network of very disgruntled dealers (many of whom went out of business or defected to Indian as a result of H-D's woes). The only bright spots on Harley's horizon were that their major rivals, Indian, were in an even worse mess and the owner of Excelsior, the most important of Harley's lesser rivals, chose this moment to quit motorcycle manufacture.

The family business survived nevertheless, West Juneau Avenue even managing to invest in new models. But the singles fell by the wayside; from 1935 they no longer featured in the catalogue.

Overseas exports in 1932 nevertheless amounted to 1,974 units (out of a total of 6,841 motorcycles produced). In the same year a contract was agreed with the Japanese company Sankyo for the licensed production of Harleys under the name Rikuo and the first payments, to a value of 3,000 dollars, crossed the Pacific in 1933. Exports suffered from the imposition in the spring of 1932 of a 10% import tax on all foreign-built vehicles entering Great Britain, and thanks to a request to the members of the British Commonwealth of Nations to 'Buy British', this was extended to Canada, Australia and New Zealand, all of which were previously significant takers of exported Harleys. From then on Harley exports were chiefly to Japan (but only up to 1937, when the new militaristic government effectively forced the ending of foreign imports), South America, Spain and some European countries. Nevertheless, by 1935-7 exports for some years represented over 50% of H-D production, such was the state of the domestic market.

In the autumn of 1935, as though it were with the

Proud of their achievements

Last photo of the Founders together: 1936. That is (allegedly) the first of the Model Es coming off the production line

The instrument panel of the legendary 1936 Knucklehead

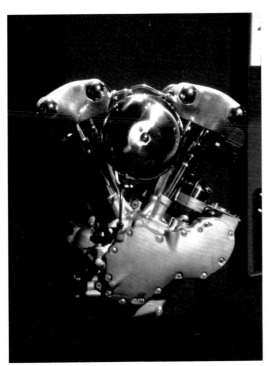

When engines still had names. The first Knucklehead of 1936

last of its strength, H-D pushed though the concept of a revolutionary (for Harley, in those days) overhead-valve-engined motorcycle with a four-speed gearbox. This found itself warmly welcomed in the motorcycling world and later by its own survival contributed to the demise of Harley's last competitor, Indian. In spite of its many teething troubles its fortunes rose with the slowly-rallying economy and sufficient numbers of machines eventually found buyers to contribute materially to the Motor Company's recovery. Joe Petrali achieved the peak of his unique racing career on it and won practically all the racing categories worthy of the name.

After further spectacular victories he retired from active racing in 1938; he then went to work for Howard Hughes. In the previous year, 1937, William A. Davidson died; the first Company Founder to do so. Walter Davidson died in 1942, on the very day on which the US War Department ordered the end of any kind of production of vehicles, cars and motorcycles for private use. In 1943 William S. Harley died, leaving just one of the four Founding Fathers to carry on.

During the War civilian production was reduced to a minimum, since all production capacity was required for Army vehicles. Production reached hitherto undreamed-of heights; both 1942 and 1943 each saw more than 29,000 motorcycles produced, topping the previous record year of 1920 by more than 1,000 machines each. Not until 1966 would similar numbers again be produced.

Ready for the off: A proud Walter Davidson poses with a 1908 single

The early singles

1903 to 1922

Model description and technical information

The origin of Messrs Harley and Davidson's first motorcycle is covered in the previous Chapter *The Story up to 1945*. This Chapter is concerned only with the technical aspects of the story.

The cast-iron engine, gradually developed in several distinct stages, was now robust enough to undergo extensive test trips. While the two machines of 1902-3 and 1903-4 were always regarded as prototypes, this did not prevent the makers from selling them when the opportunity arose. Since motorcycles at the beginning of the century were substantially cheaper than the hand-built cars of the period, prospective purchasers didn't have to have quite such deep pockets to be able to afford one.

The carburettor developed by Evinrude, Harley and Davidson together did at least serve for the first few years, but was already by 1906 replaced by a Schebler instrument. This company developed early into one of the largest carburettor suppliers in the USA.

The valve layout consisted of the overhead automatic or atmospheric intake valve opened by atmospheric pressure and closed by a spring, while the exhaust valve was located at the side of the cylinder barrel and controlled by a cam driven from the crankshaft.

The engine, even with its grey cast-iron piston (normal for the period) proved so sound that the basic design lasted, notwithstanding several capacity increases, until 1918. The bearing surfaces were lubricated by splash, the oil being introduced by a drip-feed with a hand pump being used to inject more when the circumstances demanded.

The frame was a single-loop full-cradle type looking very much like the strengthened bicycle frame that it was, except for the front downtube which curved closely around the crankcase. The frame's front downtube was bent to curve parallel to the conventional bicycle-style front mudguard and

1903: The first ever Harley-Davidson

1904: The second prototype

then followed a reverse curve down under the engine and back up to meet the straight diagonal of the saddle tube. The individual tubes were brazed into lugs in time-honoured fashion and were constantly strengthened to keep pace with the increase in weights and loads. For example, in 1911, the steering head was reinforced with additional gusseting. However, the basic design remained unchanged right up to 1924.

The oval front fork tubes curved noticeably towards the front at their lower ends. In 1907 Harley-Davidson fitted for the first time a Sager Cushion fork with coil springs enclosed in rather slim tubes (J. Harry Sager being an inventor and, as the owner of J. H. Sager Co., of Rochester, New York, a manufacturer of proprietary equipment for bicycles and motorcycles). A year later, in 1908, the Sager Cushion fork was beefed up with heavier gauge tubing and longer springs. Thus was born Harley's trademark short leading-link, or Springer, fork, which remained in use until the early 1950s and was resurrected in 1988. The fork comprised a pair of rigid (sprung) legs forming a single member on either side of the steering head with two short links projecting forwards from their pivots on the lower ends of the rigid legs and carrying the wheel spindle at their front end (the wheel spindle being in front of, or leading the suspension pivot, gives these forks their name). The smaller-diameter tubular spring legs, mounted in front of the rigid legs and also pivoted on the links, carried four one-inch (25 mm) diameter springs, two 16.5-inch (419 mm) long compression springs and two 4-inch (102 mm) long rebound springs. This permitted a few inches of suspension movement to take a lot of the shock out of the unmade road surfaces of those days.

The rear wheel was, like all standard production Harleys up to 1951, rigidly mounted. There was no front springing on any machine up to and including 1906 models. The single-speed, clutchless machine had to be simply push-started or else started like a moped by pedalling furiously. Drive was by a leather belt which ran directly from the crankshaft to the rear wheel. Naturally one had to watch constantly that the belt was correctly tensioned, or it would slip; if it rained, the belt would slip anyway, regardless of how well it was tensioned. The dust and dirt endemic on the unsurfaced roads of the day made for rapid belt wear, necessitating frequent stops for repairs.

The machines built in the first three years (up to 1905) were all finished in Piano Black. Janet Davidson, an aunt of the Davidson brothers, applied the pinstriping, in the form of double lines outlining tank and mudguards.

Today, Harley-Davidson count the few machines built in 1902-3 and 1903-4 as prototypes. Genuine series production is considered to have commenced in 1905, which is why the Motor Company began its model designations in this year with the number I.

At first only a few machines were sold, but subsequent to the first advertisements appearing in 1905, orders began to come in and production took off. It should be noted that the production figures given for the years 1903 to 1905 in the data tables of this book are the subject of much heated dispute amongst the experts.

Model development

1906 model year
The first capacity increase to 27 cu. in. (440 cc.) brought a corresponding, but minimal, boost in power output – no more than one-quarter of a horsepower. An optional extra – at extra cost – was a starting handle for the engine. Another option was the colour, Renault Gray (with carmine striping) now being offered. This subsequently became the standard finish until 1917; Piano Black being offered as an option at extra cost.

1907 and 1908 model years
The starting handle, formerly only available at extra cost, was now included as standard equipment.

Front suspension was provided for the first time, in the form of a Sager Cushion fork with enclosed coil springs. In 1908 the fork was strengthened with heavier-gauge tubing and longer springs.

1909 model year
This year saw the fitting of rear stands, which allowed the engine to be started while the motorcycle's rear wheel was raised clear of the ground (making life easier with a machine which still lacked any kind of device for interrupting the drive on starting-off). New fuel tanks (in two halves) were introduced, and throttle control was now by hand, via a handlebar-mounted twistgrip and cable.

The engine was now 30 cu. in. (494 cc.) and was fitted with the improved Schebler Model H carburettor, or the Schebler Deluxe (with an air cleaner) offered as an option by H-D, could be specified. A Bosch magneto was also available, again as an option, at extra cost. The frame was further strengthened, a second tube being added to run horizontally under the tank.

1910 model year
Drive was now by a 1¾-inch (44.5 mm) wide leather belt and a belt tensioner/idler assembly, operated by a large lever on the left-hand side of the tank, was fitted to aid starting and stopping. For the first time acetylene lighting was available ex-works. Acetylene gas was produced in a cylindrical generator in which water is allowed to drip onto calcium carbide; the gas was burned by a naked flame in the headlamp and (where fitted) the tail lamp.

1911 model year
From 1911 on the cylinder head had vertical fins instead of the original horizontal ones. The formerly curved frame front downtube was now straight. The exhaust valve was now a single-piece (steel) component.

Front brakes weren't introduced by H-D until 1928. While all H-Ds up to 1918 were single-seater machines, from about 1911 a pillion saddle could be ordered from the new in-house accessory catalogue. For those for whom that was too uncomfortable or inelegant, however, owners could resort to the relevant aftermarket accessory suppliers to find a sidecar, but that would have to be fitted either by the owner or by the friendly local Harley-Davidson dealer.

Renault Gray remained the only colour option, but customers could exercise some choice in the colour and style of pinstriping.

1905: The last year of the original model

1908 Model 4

1909 Model 5

1911 Model 7

1912 model year

On all models, the frame was reinforced again by the addition of gusseting to the tank top tubes in the area of the steering head, and the frame top tube being sloped downwards at the rear lowered the seat height. Rider comfort was further improved by the fitting of the Troxel saddle suspension system in which the (still sprung) saddle was mounted on a long adjustable spring post, with a 14-inch (356 mm) coil spring, fitted into the frame saddle tube; this was dubbed the Ful-Floteing (sic) seat by H-D. The front mudguard was treated to skirted extensions to its sides and bottom tip to keep dirt and water off the engine. The shape of the tank was altered to conform to the new frame, although its capacity remained unchanged.

For an additional 10 bucks owners could now order their machines fitted with a multi-plate free-wheel clutch enclosed in the rear hub and operated by a lever on the left-hand side of the tank. Machines with this device were identified by the addition of the letter X to their model code; i.e. Models X8 and X8A. This device now made possible the use of roller chains, a far superior form of final drive than the leather belts of those days.

1913 model year

The Model 9s featured an enlarged 34-cubic inch (565 cc.) engine which is often known as the '5-35',

Another view of a 1911 Model 7

on the basis that it was thought to produce five horsepower from its (rounded-up) thirty-five cubes. The big change, however, was the loss of the unlamented automatic, or atmospheric, intake valve and its replacement by the much more reliable ioe system of operation, clearly identifiable by the external pushrod running from the timing chest up to the rocker assembly on top of the cylinder head. This had been proved by two years of service on the V-twin engines; all Harley-Davidson models, the singles from 1913 up to the early 1920s and the V-twins from 1911 up to 1929, were now ioe, or pocket valve designs.

In this year customers had the choice for the first time between the familiar belt drive and a new-fangled (and unenclosed) roller chain drive; the cost being the same for either. What motorcyclists thought of belt drive was shown by the fact that by 1915, belt drive was no longer offered, even as an option. The dry-cell battery-and-coil ignition system was also no longer to be had.

1914 model year
The bicycle-type back-pedal rear brake was replaced by an externally-contracting band brake acting on a separate rear drum and operated by a footboard via a rod. Late in the 1913 model year a prototype two-speed transmission mounted in the rear hub had been introduced; this was now offered as standard on the Model 10C. Except for the Model 10A, still using up stocks of old belt-drive components, Harley now made its own wheel hubs and brakes.

The clutch could now be operated either by hand or by foot. The Step Starter (an early form of kickstart, but with a forward stroke) was introduced which eliminated the need to haul the machine up on to its rear stand every time one wanted to start it.

For the first time, the carburettor was fitted with a choke, for starting.

The engine was now fitted with a one-piece connecting rod and roller big-end bearings.

For this year, H-D ordered the enormous number of 2,500 sidecars from the Rogers Company of Chicago so that, at a cost of 85 dollars, a motorcycle could be ordered with a sidecar that would be fitted at the factory.

1915 model year
The First World War having interrupted supplies from Germany of Bosch magnetos, Harley had to find alternative suppliers and Remy were chosen initially, with Dixie units being found on later

1912 Model 8

1913 Model 9B

1916 Model 16

models. However, until they could expand capacity to meet the sudden demand from the US automotive and motorcycle industries, Harley had to offer battery-and-coil ignition systems again to make up the shortfall. Acetylene lighting was available with either ignition system.

The lubrication system was now supplied automatically by an engine-driven pump (although the oil tank-mounted hand pump was retained for emergency use), thus releasing the rider from the need to monitor constantly the oil supply and to pump in more by hand.

Belt drive machines being no longer wanted by buyers, only chain drive models were offered from now on (until toothed belts were rediscovered in the early 1980s).

The rear brake was improved by being fitted with two hinged brake bands and the operating mechanism was made double-acting (the Duplex brake).

1916 model year

For 1916, H-D altered their model designations to follow the calendar instead of H-D years of production from 1905. Thus machines sold in 1916, which should have been Model 12s, became Model 16s instead. This gap has caused confusion in the past, but there are no 'missing' models.

The major change was that a proper 3-speed sliding-mesh gearbox in a cast-iron housing was available as an alternative to the direct-drive standard model; unhappily for the rear hub-mounted two-speed transmission, it was redundant after only two years and was dropped from the catalogue. The Step Starter with this gearbox was operated by a rearward stroke.

Wider front forks, stronger wheel rims and front hubs and wider front mudguards were amongst many less-obvious improvements introduced. Standardization of frame and brake components (singles and twins now used the same component parts) brought technical improvements and at the same time cost savings.

A new tank was fitted, with more rounded lines and greater fuel capacity, in spite of its now combining oil and fuel compartments in the left-hand half.

Starting with this year, those who ordered sidecars got an engine with lowered compression, which naturally meant a slightly lower power output. The sidecars themselves could be ordered fitted with their own handbrake, which included a locking device.

1917 model year
In this first year of war for the United States olive green replaced the by-now traditional Renault Gray as the standard colour for Harley-Davidson motorcycles; in some cases the paint was even applied to engine and transmission parts such as the crankcases. The Fast Motor concept was introduced to the single cylinder range, becoming the 500 models of 1918 and 1919. Power outputs were boosted somewhat by numerous changes to the valvegear.

1918 model year
Minor detail changes only were made to the single-cylinder models in this, their final year of production.

No single-cylinder Harleys were made between 1919 and 1920, the smallest model in the H-D range being the 36 cu. in. (584 cc.) Model W Sport Twin.

1921 to 1922 model years
The Model CD was launched on to the market by H-D at the end of 1920 as a relatively short-lived stop-gap model intended purely for commercial work and tradespeople, hence the Commercial designation.

Its 37 cu. in. (608 cc.) engine was one-half of the new 74 cu. in./1217cc. V-twin (which, ironically, was a practice derived from those in racing circles looking for a more tuneable engine than those commercially-available on which to compete in smaller capacity classes). To keep costs to a minimum, the frame and cycle parts of the Big Twins were also used on the single.

The heavy single-cylinder Commercial model did not even appear in the Motor Company's catalogue and price lists. It is hardly surprising, therefore, that it did not sell in large numbers; not much is known of the actual details, but a figure of 39 is mentioned as being sold in its second and last year. This bears out the suspicion that the model was not sufficiently-well advertised and was probably too expensive. For only 20 bucks more, one could have a V-twin Seventy-Four of nearly twice the power. The machine was dropped at the end of the 1922 model year.

Model range

1903 to 1908
These engines have ioe valve layouts, but with automatic intake valves and side exhaust.

Model	Specification	Year
Model 0	25 cu. in./405 cc. engine, magneto ignition, direct belt drive	(1903-04)
Model 1	25 cu. in./405 cc. engine, magneto ignition, direct belt drive	(1905)
Model 2	27 cu. in./440 cc. engine, magneto ignition, direct belt drive	(1906)
Model 3	27 cu. in./440 cc. engine, magneto ignition, direct belt drive	(1907)
Model 4	27 cu. in./440 cc. engine, magneto ignition, direct belt drive	(1908)

1909 to 1912

Model	Specification	Year
Model 5	30 cu. in./494 cc. engine, battery-and-coil ignition, direct belt drive, 28-inch tyres	(1909)
Model 5A	30 cu. in./494 cc. engine, magneto ignition, direct belt drive, 28-inch tyres	(1909)
Model 5B	30 cu. in./494 cc. engine, battery-and-coil ignition, direct belt drive, 26-inch tyres	(1909)
Model 5C	30 cu. in./494 cc. engine, magneto ignition, direct belt drive, 26-inch tyres	(1909)
Model 6	30 cu. in./494 cc. engine, battery-and-coil ignition, direct belt drive, 28-inch tyres	(1910)
Model 6A	30 cu. in./494 cc. engine, magneto ignition, direct belt drive, 28-inch tyres	(1910)
Model 6B	30 cu. in./494 cc. engine, battery-and-coil ignition, direct belt drive, 26-inch tyres	(1910)
Model 6C	30 cu. in./494 cc. engine, magneto ignition, direct belt drive, 26-inch tyres	(1910)
Model 7	30 cu. in./494 cc. engine, battery-and-coil ignition, direct belt drive, 28-inch tyres	(1911)
Model 7A	30 cu. in./494 cc. engine, magneto ignition, direct belt drive, 28-inch tyres	(1911)
Model 7B	30 cu. in./494 cc. engine, battery-and-coil ignition, direct belt drive, 26-inch tyres	(1911)
Model 7C	30 cu. in./494 cc. engine, magneto ignition, direct belt drive, 26-inch tyres	(1911)
Model 8	30 cu. in./494 cc. engine, battery-and-coil ignition, direct belt drive	(1912)
Model X8	30 cu. in./494 cc. engine, battery-and-coil ignition, direct belt drive	(1912)
Model 8A	30 cu. in./494 cc. engine, magneto ignition, direct belt drive, rear hub clutch	(1912)
Model X8A	30 cu. in./494 cc. engine, magneto ignition, direct belt drive, rear hub clutch	(1912)

1913 to 1918
These engines have ioe valve layouts, with pushrod-operated overhead intake valves and side exhaust.
Model 9A   34 cu. in./565 cc. engine, magneto ignition, direct belt drive   (1913)
Model 9B   34 cu. in./565 cc. engine, magneto ignition, direct chain drive   (1913)
Model 10A   34 cu. in./565 cc. engine, magneto ignition, direct belt drive   (1914)
Model 10B   34 cu. in./565 cc. engine, magneto ignition, direct chain drive   (1914)
Model 10C   34 cu. in./565 cc. engine, magneto ignition, 2-speed gearbox, chain drive   (1914)
Model 11B   34 cu. in./565 cc. engine, magneto ignition, direct chain drive   (1915)
Model 11C   34 cu. in./565 cc. engine, magneto ignition, 2-speed gearbox, chain drive   (1915)
Model 16B   34 cu. in./565 cc. engine, magneto ignition, direct chain drive   (1916)
Model 16C   34 cu. in./565 cc. engine, magneto ignition, 3-speed gearbox, chain drive   (1916)
Model 17B   34 cu. in./565 cc. engine, magneto ignition, direct chain drive   (1917)
Model 17C   34 cu. in./565 cc. engine, magneto ignition, 3-speed gearbox, chain drive   (1917)
Model 18B   34 cu. in./565 cc. engine, magneto ignition, direct chain drive   (1918)
Model 18C   34 cu. in./565 cc. engine, magneto ignition, 3-speed gearbox, chain drive   (1918)

1921 and 1922
Model 21CD 37 cu. in./608 cc. engine, magneto ignition, 3-speed gearbox   (1921)
Model 22CD 37 cu. in./608 cc. engine, magneto ignition, 3-speed gearbox   (1922)

1903-1904 Model 0

Engine	
Engine	Air cooled ioe single
Bore x stroke	3 x 3½ (3 x 3.5) in.
Bore x stroke	76.2 x 88.9 mm
Capacity	24.74 cu. in. (405 cc)
Power output (SAE)	3 hp (2 kW)/1600 rpm
Carburettor	Harley-Davidson
Ignition	Battery-and-coil
Transmission	
Gearbox	None
Gearchange	None
Final drive	Belt
Cycle parts	
Frame construction	Curved single-loop
Front fork	Bicycle-type, unsprung
Rear suspension	None
Front brake	None
Rear brake	Back-pedal
General information	
Unladen weight	63 kg
Front wheel/tyre	2 x 28 optional 2½ x 28 or 26-inch
Rear wheel/tyre	2 x 28 optional 2½ x 28 or 26-inch
Maximum speed	56 km/h
Tank capacity	5.7 litres
Catalogue price	$200 (1904)
Production figures	1 (1903)
	1 (1904)

1905 Model 1

Engine	
Engine	Air cooled ioe single
Bore x stroke	3 x 3½ (3 x 3.5) in.
Bore x stroke	76.2 x 88.9 mm
Capacity	24.74 cu. in. (405 cc)
Power output (SAE)	3 hp (2 kW)/1600 rpm
Carburettor	Harley-Davidson
Ignition	Battery-and-coil
Transmission	
Gearbox	None
Gearchange	None
Final drive	Belt
Cycle parts	
Frame construction	Curved single-loop
Front fork	Bicycle-type, unsprung
Rear suspension	None
Front brake	None
Rear brake	Back-pedal
General information	
Unladen weight	63 kg
Front wheel/tyre	2-inch optional 2½-inch beaded-edge
Rear wheel/tyre	2-inch optional 2½-inch beaded-edge
Maximum speed	56 km/h
Tank capacity	5.7 Litres
Catalogue price	$200
Production figures	5

1906 Model 2

Engine
Engine	Air cooled ioe single
Bore x stroke	3⅛ x 3½ (3.13 x 3.5) in.
Bore x stroke	79.38 x 88.9 mm
Capacity	26.84 cu. in. (440 cc)
Power output (SAE)	3¼ hp (2 kW)/1800 rpm
Carburettor	Schebler
Ignition	Battery-and-coil

Transmission
Gearbox	None
Gearchange	None
Final drive	Belt

Cycle parts
Frame construction	Single-loop, curved downtube
Front fork	Bicycle-type, unsprung
Rear suspension	None
Front brake	None
Rear brake	Back-pedal

General information
Unladen weight	63 kg
Front wheel/tyre	2 x 28 optional 2¼ x 28 or 26-inch
Rear wheel/tyre	2 x 28 optional 2¼ x 28 or 26-inch
Maximum speed	64 km/h
Tank capacity	5.7 litres
Catalogue price	$210
Production figures	50

1907 Model 3

Engine
Engine	Air cooled ioe single
Bore x stroke	3⅛ x 3½ (3.13 x 3.5) in.
Bore x stroke	79.38 x 88.9 mm
Capacity	26.84 cu. in. (440 cc)
Power output (SAE)	3¼ hp (2 kW)/1800 rpm
Carburettor	Schebler
Ignition	Battery-and-coil

Transmission
Gearbox	None
Gearchange	None
Final drive	Belt

Cycle parts
Frame construction	Single-loop, curved downtube
Front fork	Sager Cushion fork
Rear suspension	None
Front brake	None
Rear brake	Back-pedal

General information
Unladen weight	65 kg
Front wheel/tyre	2¼ x 28 optional 2¼ x 26
Rear wheel/tyre	2¼ x 28 optional 2¼ x 26 or 26-inch
Maximum speed	64 km/h
Tank capacity	5.7 litres
Catalogue price	$210
Production figures	150

1908 Model 4

Engine
Engine	Air cooled ioe single
Bore x stroke	3⅛ x 3½ (3.13 x 3.5) in.
Bore x stroke	79.38 x 88.9 mm
Capacity	26.84 cu. in. (440 cc)
Power output (SAE)	3¼ hp (2 kW)/1800 rpm
Carburettor	Schebler
Ignition	Battery-and-coil

Transmission
Gearbox	None
Gearchange	None
Final drive	Belt

Cycle parts
Frame construction	Single-loop, curved downtube
Front fork	Sager Cushion fork
Rear suspension	None
Front brake	None
Rear brake	Back-pedal

General information
Unladen weight	65 kg
Front wheel/tyre	2½ x 28 optional 2½ x 26 beaded-edge
Rear wheel/tyre	2½ x 28 optional 2½ x 26 beaded-edge
Maximum speed	64 km/h
Tank capacity	5.7 litres
Catalogue price	$210
Production figures	450

1909 Model 5

Engine
Engine	Air cooled ioe single
Bore x stroke	3⁵⁄₁₆ x 3½ (3.31 x 3.5) in.
Bore x stroke	84.14 x 88.9 mm
Capacity	30.16 cu. in. (494 cc)
Power output (SAE)	4 hp (3 kW)/2000 rpm
Carburettor	Schebler Model H *
Ignition	See below

Transmission
Gearbox	None
Gearchange	None
Final drive	Belt

Cycle parts
Frame construction	Single-loop, curved downtube
Front fork	Springer, enclosed springs, tubular legs
Rear suspension	None
Front brake	None
Rear brake	Back-pedal

General information
Unladen weight	75 kg
Front wheel/tyre	2½ x 28 optional 2½ x 26 beaded-edge
Rear wheel/tyre	2½ x 28 optional 2½ x 26 beaded-edge
Maximum speed	72 km/h
Tank capacity	9.1 litres
Catalogue price	Model 5 $210
	Model 5A $250
	Model 5B $210
	Model 5C $250
Production figures	Model 5 864
	Model 5A 54
	Model 5B 168
	Model 5C 36

Models 5 and 5B have battery-and-coil ignition.
Models 5A and 5C have magneto ignition.
Models 5 and 5A have 28-inch wheels and tyres.
Models 5B and 5C have 26-inch wheels and tyres.

* Schebler Deluxe (with air filter) optional.

1910 Model 6

Engine
Engine	Air cooled ioe single
Bore x stroke	3⁵⁄₁₆ x 3½ (3.31 x 3.5) in.
Bore x stroke	84.14 x 88.9 mm
Capacity	30.16 cu. in. (494 cc)
Power output (SAE)	4 hp (3 kW)/2000 rpm
Carburettor	Schebler Model H *
Ignition	See below

Transmission
Gearbox	None
Gearchange	None
Final drive	Belt

Cycle parts
Frame construction	Single-loop, curved downtube
Front fork	Springer, enclosed springs, tubular legs
Rear suspension	None
Front brake	None
Rear brake	Back-pedal

General information
Unladen weight	75 kg
Front wheel/tyre	2½ x 28 optional 2½ x 26 beaded-edge
Rear wheel/tyre	2½ x 28 optional 2½ x 26 beaded-edge
Maximum speed	72 km/h
Tank capacity	9.1 litres
Catalogue price	Model 6 $210
	Model 6A $250
	Model 6B $210
	Model 6C $250
Production figures	Model 6 2302
	Model 6A 334
	Model 6B 443
	Model 6C 88

1911 Model 7

Engine
Engine	Air cooled ioe single
Bore x stroke	3⁵⁄₁₆ x 3½ (3.31 x 3.5) in.
Bore x stroke	84.14 x 88.9 mm
Capacity	30.16 cu. in. (494 cc)
Power output (SAE)	4 hp (3 kW)/2000 rpm
Carburettor	Schebler Model H *
Ignition	See below

Transmission
Gearbox	None
Gearchange	None
Final drive	Belt

Cycle parts
Frame construction	Single-loop, straight downtube
Front fork	Springer, enclosed springs, tubular legs
Rear suspension	None
Front brake	None
Rear brake	Back-pedal

General information
Unladen weight	80 kg
Front wheel/tyre	2½ x 28 optional 2½ x 26 beaded-edge
Rear wheel/tyre	2½ x 28 optional 2½ x 26 beaded-edge
Maximum speed	72 km/h
Tank capacity	9.1 litres
Catalogue price	Model 7 $225
	Model 7A $250
	Model 7B $225
	Model 7C $250
Production figures	Total Model 7 production, including the V-twin models, amounted to 5,625 motorcycles.

Models 6 (7) and 6B (7B) have battery-and-coil ignition.
Models 6A (7A) and 6C (7C) have magneto ignition.
Models 6 (7) and 6A (7A) have 28-inch wheels and tyres.
Models 6B (7B) and 6C (7C) have 26-inch wheels and tyres.

* Schebler Deluxe (with air filter) optional.

1912 Model 8

Engine
Engine	Air cooled ioe single
Bore x stroke	3⁵⁄₁₆ x 3½ (3.31 x 3.5) in.
Bore x stroke	84.14 x 88.9 mm
Capacity	30.16 cu. in. (494 cc)
Power output (SAE)	4(3)/2000 rpm
Carburettor	Schebler Model H *
Ignition	See below

Transmission
Gearbox	None
Gearchange	None
Final drive	Belt

Cycle parts
Frame construction	Single-loop, straight downtube
Front fork	Springer, enclosed springs, tubular legs
Rear suspension	None
Front brake	None
Rear brake	Back-pedal

General information
Unladen weight	80 kg
Front wheel/tyre	2½ x 28, optional 2½ x 26 beaded-edge
Rear wheel/tyre	2½ x 28, optional 2½ x 26 beaded-edge
Maximum speed	72 km/h
Tank capacity	9.1 litres
Catalogue price	Model 8 $200 Model X8 $210 Model 8A $225 Model X8A $235
Production figures	A total of 3852 Model 8s were produced, including the V-twin models.

1913 Model 9

Engine
Engine	Air cooled ioe single
Bore x stroke	3⁵⁄₁₆ x 4 (3.31 x 4) in.
Bore x stroke	84.14 x 101.6 mm
Capacity	34.47 cu. in. (565 cc)
Power output (SAE)	5(4)/2200 rpm
Carburettor	Schebler Model H *
Ignition	Magneto

Transmission
Gearbox	None
Gearchange	None
Final drive	Model 9A: Belt Model 9B: chain

Cycle parts
Frame construction	Single-loop, straight downtube
Front fork	Springer, enclosed springs, tubular legs
Rear suspension	None
Front brake	None
Rear brake	Back-pedal

General information
Unladen weight	90 kg
Front wheel/tyre	2¾ x 28, optional 2½ x 26 beaded-edge
Rear wheel/tyre	2¾ x 28, optional 2¾ x 26 beaded-edge
Maximum speed	80 km/h
Tank capacity	9.1 litres
Catalogue price	Model 9A $290 Model 9B $290
Production figures	Model 9A 1510 Model 9B 4601 **

Models 8 and X8 have battery-and-coil ignition, the others have magneto ignition.
The X model code prefix indicates that the machine is fitted with a free-wheel clutch in the rear hub (operated by a lever on the left-hand side of the tank).

* Schebler Deluxe (with air filter) optional.
** Included with the Model 9C as a form of pre-production run of 11 motorcycles was a prototype Model 10C with a two-speed transmission mounted in the rear hub.

1914 Model 10

Engine
Engine	Air cooled ioe single
Bore x stroke	3⁵⁄₁₆ x 4 (3.31 x 4) in.
Bore x stroke	84.14 x 101.6 mm
Capacity	34.47 cu. in. (565 cc)
Power output (SAE)	5(4)/2200 rpm
Carburettor	Schebler
Ignition	Magneto

Transmission
Gearbox	Models 10A and 10B: None
Gearchange	Model 10C: 2-speed (rear hub, hand change on tank)
Final drive	Model 10A: Belt Models 10B and 10C: chain

Cycle parts
Frame construction	Single-loop, straight downtube
Front fork	Springer, enclosed springs, tubular legs
Rear suspension	None
Front brake	None
Rear brake	Drum, via rod and pedal

General information
Unladen weight	90 kg	
Front wheel/tyre	3 x 28 optional, 3 x 26 beaded-edge	
Rear wheel/tyre	3 x 28 optional, 3 x 26 beaded-edge	
Maximum speed	80 km/h	
Tank capacity	9.1 litres	
Catalogue price	Model 10A	$200
	Model 10B	$210
	Model 10C	$245
	Sidecar	$85
Production figures	Model 10A	316
	Model 10B	2034
	Model 10C	877
Sidecars available	First works sidecar, no model code	
Sidecar wheel and tyre	3 x 28 only	

1915 Model 11

Engine
Engine	Air cooled ioe single
Bore x stroke	3⁵⁄₁₆ x 4 (3.31 x 4) in.
Bore x stroke	84.14 x 101.6 mm
Capacity	34.47 cu. in. (565 cc)
Power output (SAE)	5(4)/2200 rpm
Carburettor	Schebler
Ignition	Magneto

Transmission
Gearbox	Model 11B: None
Gearchange	Model 11C: 2-speed (rear hub, hand change on tank)
Final drive	Models 11B and 11C: chain

Cycle parts
Frame construction	Single-loop, straight downtube
Front fork	Springer, enclosed springs, tubular legs
Rear suspension	None
Front brake	None
Rear brake	Drum, via rod and pedal

General information
Unladen weight	95 kg	
Front wheel/tyre	3 x 28 optional, 3 x 26 beaded-edge	
Rear wheel/tyre	3 x 28 optional, 3 x 26 beaded-edge	
Maximum speed	80 km/h	
Tank capacity	9.1 litres	
Catalogue price	Model 11B	$200
	Model 11C	$230
	Sidecar (Model 11L)	$85
Production figures	Model 11B	670
	Model 11C	545
Sidecars available	Model 11L R/H single-seater Model L11L: L/H single-seater Model 11M Commercial	
Sidecar wheel and tyre	3 x 28 only	

1916 Model 16

Engine
Engine	Air cooled ioe single
Bore x stroke	3⁵⁄₁₆ x 4 (3.31 x 4) in.
Bore x stroke	84.14 x 101.6 mm
Capacity	34.47 cu. in. (565 cc)
Power output (SAE)	6(4)/2500 rpm
Carburettor	Schebler
Ignition	Magneto

Transmission
Gearbox	Model 16B: None
	Model 16C: 3-speed
Gearchange	Hand change, on tank
Final drive	Chain

Cycle parts
Frame construction	Single-loop, straight downtube
Front fork	Springer, enclosed springs, tubular legs
Rear suspension	None
Front brake	None
Rear brake	Drum, via rod and pedal

General information
Unladen weight	95 kg
Front wheel/tyre	3 x 28, optional 3 x 26 beaded-edge
Rear wheel/tyre	3 x 28, optional 3 x 26 beaded-edge
Maximum speed	80 km/h
Tank capacity	10 litres
Catalogue price	Motorcycles:
	Model 16B $200
	Model 16C $230
	Sidecar:
	Model 16L $75
	Model 16M $65

Production figures	Model 16B 292
	Model 16C 862
Sidecars available	**Chassis:**
	Model 16LC
	R/H single-seater
	Model 16LCL L/H single-seater
	Model 16SC Chassis+ Stretcher
	Model 16XT
	for luggage-carrying
	Passenger-carrying sidecars:
	Model 16L
	R/H single-seater
	Load-carrying sidecars:
	Model 16M
	Commercial standard
	Model 16RFD
	Rural Free Delivery
	Model 16P
	closed box body
	Model 16U
	open box body
	Model 16V
	double closed box body
Sidecar wheel and tyre	3 x 28 only

The engines of the sidecar-specification motorcycles of the model years 1916 to 1918 had a somewhat lower compression ratio which gave a power output of perhaps a half-horsepower less. The Fast Motors of 1917 produced approximately 8 horsepower.

1917 Model 17

Engine
Engine	Air cooled ioe single
Bore x stroke	3 5/16 x 4 (3.31 x 4) in.
Bore x stroke	84.14 x 101.6 mm
Capacity	34.47 cu. in. (565 cc)
Power output (SAE)	6(4)/2500 rpm
Carburettor	Schebler
Ignition	Magneto

Transmission
Gearbox	Model 17B: None
	Model 17C: 3-speed
Gearchange	Hand change, on tank
Final drive	Chain

Cycle parts
Frame construction	Single-loop, straight downtube
Front fork	Springer, enclosed springs, tubular legs
Rear suspension	None
Front brake	None
Rear brake	Drum, via rod and pedal

General information
Unladen weight	100 kg
Front wheel/tyre	3 x 28, optional 3 x 26 beaded-edge
Rear wheel/tyre	3 x 28, optional 3 x 26 beaded-edge
Maximum speed	80 km/h
Tank capacity	10 litres
Catalogue price	Motorcycles:
	Model 17B $215
	Model 1/C $240
	Sidecar:
	Model 17L $80
	Model 17M $70
	Model 17N $72

Production figures	Model 17B 124
	Model 17C 605
Sidecars available	**Chassis:**
	Model 17LC
	R/H single-seater
	Model 17LCL L/H single-seater
	Passenger-carrying sidecars:
	Model 17L
	R/H single-seater
	Model L17L L/H single-seater
	Load-carrying sidecars:
	Model 17M
	Commercial standard
	Model 17N
	Rural Free Delivery
	Model 17P
	closed box body
	Model 17U
	open box body
	Model 17V
	double closed box body
Sidecar wheel and tyre	3 x 28 only

1918 Model 18

Engine
Engine	Air cooled ioe single
Bore x stroke	3⁵⁄₁₆ x 4 (3.31 x 4) in.
Bore x stroke	84.14 x 101.6 mm
Capacity	34.47 cu. in. (565 cc)
Power output (SAE)	6(4)/2500 rpm
Carburettor	Schebler
Ignition	Magneto

Transmission
Gearbox	Model 18B: None
	Model 18C: 3-speed
Gearchange	Hand change, on tank
Final drive	Chain

Cycle parts
Frame construction	Single-loop, straight downtube
Front fork	Springer, enclosed springs, tubular legs
Rear suspension	None
Front brake	None
Rear brake	Drum, via rod and pedal

General information
Unladen weight	100 kg
Front wheel/tyre	3 x 28 optional 3 x 26 beaded-edge
Rear wheel/tyre	3 x 28 optional 3 x 26 beaded-edge
Maximum speed	80 km/h
Tank capacity	10 litres
Catalogue price	Model 18B $235
	Model 18C $260
	Model 18L $90 (Sidecar only)
Production figures	Model 18B 19
	Model 18C 251
Sidecars available	**Chassis:**
	Model 18LC
	R/H single-seater
	Model 18LCL L/H single-seater
	Model 18Q double-adult
	Passenger-carrying sidecars:
	Model 18L R/H single-seater
	Model L-18L L/H single-seater
	Load-carrying sidecars:
	Model 18M Commercial standard
	Model 18N Rural Free Delivery
	Model 18P closed box body
	Model 18QA Commercial closed box body
	Model 18QB Commercial open box body
	Model 18QD Parcel Car (2 box)
	Model 18QL R/H double-adult
	Model 18QC3 Two-stake platform
	Model 18QC4 Three-stake platform
Sidecar wheel and tyre	3 x 28 only

1921-1922 Model CD

Engine
Engine	Air cooled ioe single
Bore x stroke	3⁷⁄₁₆ x 4 (3.44 x 4) in.
Bore x stroke	87.31 x 101.6 mm
Capacity	37.12 cu. in. (608 cc)
Power output (SAE)	8(6)/2500 rpm
Carburettor	Schebler
Ignition	Magneto

Transmission
Gearbox	3-speed (hand change, on tank)
Final drive	Chain

Cycle parts
Frame construction	Single-loop, straight downtube
Front fork	Springer, enclosed springs, tubular legs
Rear suspension	None
Front brake	None
Rear brake	Drum, via rod and pedal

General information
Unladen weight	approx. 140 kg
Front wheel/tyre	3 x 28
Rear wheel/tyre	3 x 28
Maximum speed	90 km/h
Tank capacity	10 litres
Catalogue price	N./Av.
Production figures	Model 21CD N./Av.
	Model 22CD 39
Sidecars available	None

The first Big Twins

D series: 1909 to 1912
E series: 1912 to 1918
F series: 1913 to 1929
F series (Two Cam): 1928 to 1929
H series: 1915
J series: 1915 to 1929
J series (Two Cam): 1928 to 1929

Model description and technical information

The first prototype with a V-twin engine came into being at the end of 1906 and by February 1907 was ready to be displayed at the Chicago Motorcycle Show. Allegedly, it developed five horsepower but no further details were forthcoming. A few more V-twin-engined motorcycles were then built, however these were still for testing; none were sold. At most they were entered in motorcycle races. It wasn't until 1909 that the first V-twins were sold, although these still were officially considered as prototypes. 27 of the first Harley V-twins found buyers, at a cost of 325 dollars apiece.

The valve layout consisted of the overhead automatic or atmospheric intake valve opened by atmospheric pressure and closed by a spring, while the exhaust valve was located at the side of the cylinder barrel and controlled by a cam driven from the crankshaft. Ignition was by magneto.

As on the singles, the standard paint finish was Renault Gray with carmine striping. Piano Black was offered as an option at extra cost. All other equipment was the same as the single-cylinder models of that year, particularly the lubrication system, in which the bearing surfaces inside the cast aluminium-alloy crankcases were lubricated by splash, the oil being introduced by a drip-feed with a hand pump being used to inject more when the circumstances demanded.

1909 Model 5D: The first V-Twin

1911 Model 7D: Production version – after a year's break

1913 Model 9

The sprung front fork and the direct belt-drive transmission were the same as those used on the singles. The belt drive served as an acceptable solution until 1912, from which time the 60 cu. in. (989 cc.) engine was introduced with roller chain final drive.

At first the early Harley-Davidson carburettor was fitted. The engine initially had a capacity of just 54 cu. in. (880 cc.) and was admittedly perfectly powerful enough, but only when it was running properly. No more were built after the 27 test machines, on which the customers had been allowed to play the part of testers.

The frame was almost exactly the same as that fitted to the singles, with various brackets and accessory components strengthened and modified to cope with the heavier engine. It received in 1911 the strengthening of the steering head that was carried out on the singles in the following year, and was used in this basic form until 1924. The first prototype had had the front fork with the stronger springs with which the singles were fitted in 1908. As on the singles, a straight tube replaced the curved frame front downtube for 1911.

At the rear end, buyers searched in vain for any form of rear suspension, and, until the end of 1913, the hunt for a gearbox was similarly fruitless. Starting was either simply by pushing or by pedalling.

Model development

1909 model year
A small series of the first Harley-Davidson V-twins was built and sold. The single-cylinder models' new frame was used, with as few changes as possible.

1910 model year
Because of the technical problems encountered with the Model 5D's engine, no V-twin was officially offered for sale by H-D in 1910. The time was put to good use and in the following year Harley returned to the market with a soundly-developed design.

1911 model year
The 49 cu. in. (811 cc.) Model 7D could be identified by the (non-detachable) cylinder head having vertical fins instead of the horizontal ones of the original engine. The reduction in capacity was caused by a smaller bore size, the V-twin having reverted temporarily to the bore and stroke dimensions of the first production singles.

In contrast to the 1909 engine, the intake valve was now controlled by the camshaft, an external pushrod running from the timing chest up to the rocker assembly on top of each cylinder head; the much more reliable ioe system of valve operation.

In common with the smaller Harleys, a Schebler carburettor was fitted (the Model 5D having used the early Harley-Davidson instrument).

Drive was now by a 1¾-inch (44.5 mm) wide leather belt and a belt tensioner/idler assembly, operated by a large lever on the left-hand side of the tank, was fitted to aid starting and stopping. The formerly curved frame front downtube was now straight and the frame received other detail improvements.

1912 model year
The new (model) year brought yet another improvement to the existing frame, in which the frame top tube was sloped downwards at the rear to lower the seat height. Rider comfort was improved by the fitting of the Troxel saddle suspension system in which the (still sprung) saddle was mounted on a long adjustable spring post, with a 14-inch (356 mm) coil spring, fitted into the frame saddle tube; this was dubbed the Ful-Floteing (sic) seat by H-D. The front mudguard was treated to skirted extensions to its sides and bottom tip to keep dirt and water off the engine. The shape of the tank was altered to conform to the new frame, with the whole machine gaining a much more modern appearance as a result.

In addition, a larger-capacity V-twin engine was made available, at minimal extra cost; the 60 cu. in. (989 cc.) unit again turning out more torque and developing more power than the 49 cu. in. (811 cc.) version. As with all other Harley engines up to 1912, the bigger engine used iron pistons. Motorcycles fitted with this engine also had a decompression lever to aid starting.

For an additional 10 bucks owners could now order their machines fitted with a multi-plate free-wheel clutch enclosed in the rear hub and operated by a lever on the left-hand side of the tank. Machines with this device were identified by the addition of the letter X to their model code; i.e. Models X8D and X8E. This device now made possible the use of roller chains, a far superior form of final drive than the leather belts of those days.

1913 model year
For this year new and improved cast-iron pistons, which were made of a special alloy, were fitted for the first time, while the pistons, connecting rods and crankshaft were balanced as a single entity, rather than just the crankshaft alone, as had previously been the practice.

Since everyone knows that, in the well-known American expression 'You can't beat cubic inches', the 60 cu. in. (989 cc.) unit became the only power unit for the Harley Big Twin for nearly a decade (the special CAB engines excepted); in what was to become an equally long-standing tradition, the sales people and proud owners soon rounded the true capacity up to the full litre and the machines became known as Sixty-Ones. One consequence of the larger engine was that only chain drive models were offered from now on (until toothed belts were rediscovered in the early 1980s).

1914 Model 10

1914 Model 10

1914: Further views ...

... of the Model 10

1914 model year

The bicycle-type back-pedal rear brake was replaced for the 1914 model year by an externally-contracting band brake acting on a separate rear drum and operated by a footboard via a rod. Harley now made its own wheel hubs and brakes. The Step Starter (an early form of kickstart, but with a forward stroke) was introduced which eliminated the need to haul the machine up on to its rear stand every time one wanted to start it. Kickstarting was to remain a feature of the Harley range until 1986. Late in the 1913 model year a prototype two-speed transmission mounted in the rear hub had been introduced; this was now offered as standard on the Model 10F, for the only full model year of its existence. The clutch could now be operated either by hand or by foot.

Those who wanted sidecars could now order them with a new motorcycle, at a cost of 85 dollars, so that a sidecar (which actually came from the Rogers Company of Chicago) would be fitted at the factory. In the years to come a much wider range of sidecars was offered by the factory, for the transport of people as well as of goods of all sorts.

To cater for those who wanted even more power, there was the A Motor whose intake ports were opened out to 1 inch (25 mm).

1915 model year

As early as 1915 a proper 3-speed sliding-mesh gearbox replaced the stop-gap hub-mounted two-speed transmission. The Step Starter with this gearbox, which remained in use into the 1930s, was operated by a rearward stroke.

A steel exhaust system with foot-operated bypass made its début, as did the double-acting (Duplex) rear brake, which was also improved by being fitted with two hinged brake bands. The lubrication system was now supplied automatically by an engine-driven pump (although the oil tank-mounted hand pump was retained for emergency use), thus releasing the rider from the need to monitor constantly the oil supply and to pump in more by hand.

The First World War having interrupted supplies from Germany of Bosch magnetos, Harley had to find alternative suppliers and Remy was chosen initially, with Dixie units being found on later models. However, until they could expand capacity to meet the sudden demand from the US automotive and motorcycle industries, Harley had to offer battery-and-coil ignition systems again to make up the shortfall. This was a factor behind the launch of the Models H and J, which were fitted for the first time with a complete set of electrical equipment including lighting and battery-and-coil ignition. So far Harleys had had to put up with acetylene lighting; admittedly bright, but not always safe. Acetylene lighting remained the only lighting option with magneto ignition.

1916 model year

For 1916, H-D altered their model designations to follow the calendar instead of H-D years of production from 1905. Thus machines sold in 1916, which should have been Model 12s, became Model 16s instead. This gap has caused confusion in the past, but there are no 'missing' models.

Wider front forks, stronger wheel rims and front hubs and wider front mudguards were amongst many less-obvious improvements introduced. Standardization of frame and brake components (singles and twins now used the same component parts) brought technical improvements and at the same time cost savings.

A new tank was fitted, with more rounded lines and greater fuel capacity, in spite of its now combining oil and fuel compartments in the left-hand half.

Starting with this year, those who ordered sidecars with a V-twin as well as with a single got an engine with lowered compression, which naturally meant a slightly lower power output. The sidecars themselves could be ordered fitted with their own handbrake, which included a locking device.

1917 model year

In this first year of war for the United States olive green replaced the by-now traditional Renault Gray as the standard colour for the Big Twins as well as for all other Harley-Davidson motorcycles; in some cases the paint was even applied to engine and transmission parts such as the crankcases.

From 1917 a Schebler Deluxe carburettor with air cleaner was available as an option, the instrument fitted up to that date having to do without such a fitting; essential on all the roads of the period, but especially on the long, dusty stretches of the Mid-West. Changes to the valvegear led, amongst other things, to extended valve overlaps; this resulted in a significant power increase.

1918 model year

For those whose needs were not met by the standard motorcycles, the V-twin models from now up to 1923 could also be ordered from a complete range of Special Motors (see pages 65 to 67). Amongst other things, the use of light aluminium-alloy as a material for pistons was being introduced at Harley-Davidson (the US automobile industry having been using it successfully in this application since 1915); this opened up the prospect of more reliable and more powerful engines.

Numerous smaller improvements made the Harleys still more reliable and more robust. The Model E (with direct drive) was finding fewer and fewer buyers and so was discontinued at the end of the model year.

1919 model year

Detail changes for the 1919 model year were limited and almost entirely to do with the clutch. The

1915 Model 11F

1916 Model 16

1916 Model 16 with an early-pattern sidecar

1917 Model 17 and sidecar

1919 Model 19 and sidecar

1919 Model 19, solo, but with skis. Now that looks like fun ...

State-of-the-art at that time: detail of a an acetylene lighting set

3-speed sliding-mesh gearbox was now the only transmission choice for all Harleys.

1920 model year
For the first time a speedometer was available, at extra cost. Dealers could buy Open-Loop racing frames and either begin racing under their own banner or sell the machines on to suitably-courageous private entrants.

1921 model year
The new model year brought a larger-capacity engine, the 74 cu. in./1217 cc. Model FD and JD 'Superpowered' twins, which, as was usual for Harley-Davidson, would be sold alongside the existing 60-cube Model Fs and Js until the end of the ioe-engined models in 1929. The idea was that the Sixties were entirely adequate for solo use while the Seventy-Fours provided extra power for the sidecar pilots, but that didn't last for long.

The new engine was not just a bored-out version of the existing unit (both bore and stroke were increased), but exhibited revised crankcase halves as well as new cylinders and heads.

1922 model year
The Big Twins could be ordered with a heel-operated brake and a front stand. The standard brake, a double-acting external-contracting band device acting on a 7 5/16-inch (186 mm) diameter steel drum with a 1-inch (25 mm) wide face, could be supplemented (where demanded by local legislation), by an internal-expanding brake acting on the same drum.

The Olive Green paint was maintained as an option, but the new standard colour for all Harleys was Brewster Green 'tastily double-striped in gold'.

The V-twin engines once again got new crankcase halves, but only the larger unit got the improved, longer cylinders and the modified gearbox.

V-twin engines on machines ordered for sidecar use were fitted with compression plates of varying thicknesses under the cylinders; this naturally resulted in a somewhat lower power output.

The massively-strengthened front fork fitted this year is easily recognized by the two exposed coil springs in the centre, located between the fork tubes, to assist the coil springs enclosed in the front tubes.

1923 model year
To ease the task of rear wheel removal a two-piece hinged rear mudguard was introduced.

The larger V-twin engines got the benefit of new cylinder heads, which could be recognized by the seven cooling fins above the exhaust port.

As if anticipating the 1924 models with aluminium-alloy pistons, the Models 23FDCA and 23JDCA were launched; refer to the 1924 model year section for further details.

1924 model year
1924 brought a box-shaped silencer that would in its turn be replaced by the complete new exhaust system of the following year.

The former F and J model designations became

1924: Bill Ottaway on a Model J

FE and JE, as the E Motors, formerly (since 1918) only available at special request and produced in small numbers as part of the Specialty Models programme, were now standard production equipment. The previous year the E Motors had been fitted with cast-iron pistons; this year they were of aluminium-alloy.

Sidecar-specification machines were identified as usual by an S suffix to the model designation.

The Models FD and JD continued to be fitted with cast-iron pistons as standard equipment; if the alternative aluminium-alloy pistons were chosen, a CA suffix was added to the model designation i.e. FDCA and JDCA or, if the machine was pulling a sidecar, FDSCA and JDSCA.

The installation of Alemite grease fittings at twelve bearings on the frame and forks eased maintenance of the cycle parts, an Alemite grease gun and large can of grease being supplied free with each new motorcycle.

In addition to the Schebler Deluxe carburettor available since 1917, a Zenith carburettor was listed in the in-house accessory catalogue for the first time.

1925 model year

The management had decided in the spring of 1924

1927 Model F

1927 Model J

to update thoroughly the F and J-series models to give them a more modern appearance in line with the competition. The result, the Streamline models, appeared for the 1925 model year, at a lower price than before.

The basis was a completely new frame, wider for greater rigidity and with the seating position 3 inches (76 mm) lower than before, with drop-forged steel lugs such as the steering head replacing the previously cast items. The wheel/tyre size was now standardized at 27 x 3.50 on 20-inch wheels, further to lower the machine and provide a more comfortable ride. The highlights of this year were new handlebars that now had a more pronounced upsweep, a teardrop-shaped tank with double the fuel capacity, a cylindrical toolbox that was installed on the front forks under the horn (and therefore inevitably christened the 'St Bernard' – it was alleged that the rattling of the tools in this kept the rider awake during night-time riding!) and a completely new tubular Speedster exhaust. Improvements to the gearbox further benefited the customer.

As the aluminium-alloy pistons had not proved themselves to be sufficiently reliable, the Motor Company had to fall back on cast-iron pistons for 1925. These, however, according to the Company, were only marginally heavier than the aluminium ones and, furthermore, the customer would not

1927 Model JD

1928 Model JD

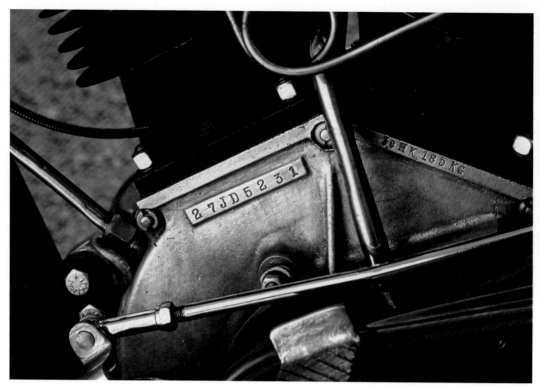

1927 Model JD

1928 Model JH

Further …

… views of …

... 1928 models ...

... and the last of the ioe V-twins ...

discern the inevitable but minimal loss of power that resulted. The model designations accordingly changed from FDCA to FDCB and JDCA to JDCB.

The flat headlamp glass was modified to give better illumination. Electric lighting was standard, but acetylene lighting could still be selected if desired.

For the first time, the sidecars were Harley-Davidson's own designs, built in-house. With the death of B. H. Rogers, the founder of the Rogers Company of Chicago who had supplied H-D with sidecars since 1914, the company ceased to manufacture sidecars and H-D were obliged to make their own, in a separate building set up for the purpose. For the first time, a windscreen was available as an option.

In 1925, as in the two preceding years, there were only Big Twins in Harley's range; the last of the old-style singles had been built in 1922, the Model W had finished in 1923 and the new, smaller, singles were still a year or so in the future.

1926 model year

For this year the range of V-twin models was strictly limited. Only the basic F, FD, J and JD models were available with or without sidecars and there were no more Specialty Models.

For the first time, the customer could choose between the new-fangled balloon tyres and the time-honoured high-pressure beaded-edge tyres.

The battery was twice as powerful as that fitted in 1925, and a generator of greater output was installed. The upswept handlebar introduced only the previous year was made somewhat flatter, and it stayed that way until the end of production of the ioe machines.

From now on the smaller F and J models began to lose their appeal in the marketplace and their sales were significantly lower than those of the Seventy-Fours.

1927 model year

The great novelty of this model year was the option of a sidestand, or jiffy stand. The customer could still choose either beaded-edge or balloon tyres. The standard colour reverted to the Olive Green used from 1917 to 1922, with maroon pinstriping which had a gold centre line edged in black. However, further colours (White and Cream) could be ordered at extra cost.

Another new feature was the simple ignition system that remained in use into the 1980s. This did without a distributor, using a contact breaker and single twin-output coil to fire both spark plugs at the same time, whether the cylinder was on the compression stroke or not. This simplified technology made some problems, such as backfiring, inevitable, but on paper the 1926 and 1927 engines developed the same power.

The range was extended by the Special models, which were identified by the suffix K (60 cu. in./989 cc. models) or L (74 cu. in./1217 cc.). They were fitted with greatly-improved Dow-metal (a 2nd-generation light magnesium alloy created by the Dow Corporation) pistons.

The cup-type Alemite grease fittings were changed for standard automotive-type fittings so that owners could also use the facilities at automobile service stations to speed up the chore of regular greasing.

1928 model year

Both Sixty and Seventy-Four models were made available in standard and sports versions, the latter identified by the suffix X.

On top of this, the Two Cam models were introduced, in two 60 cu. in./989 cc. and two 74 cu. in./1217 cc. versions, one with magneto ignition, the other with battery-and-coil. Harley engines with the valves controlled by two separate crankshaft-driven cams (each with two lobes) had been around for several years, but only on racers; this was the first time this engine was made available to the public, as the Motor Company need something to stave off the competition from the ever-faster Indian and Excelsior-Henderson fours. Fitted as standard with the domed Dow-metal pistons that had been introduced the previous year, these engines gave significantly more power than previous production models. The JDH version was reportedly capable of top speeds between 85 and 100 mph.

The frame was specific to this model, having been made lower and slimmer, so that the cutouts in the tank to clear the valvegear had to be enlarged, reducing tank capacity to 4 gallons/18 litres.

Thus the Sport and Special Sport models filled the gap between the standard models and the Two Cam machines. With the slimmer frame and the high-compression J and JD engines, the JX, JXL, JDX and JDXL models (the L suffix identifying those engines with Dow-metal pistons) provided customers with yet more choice. On all engines, the lubrication system's oil supply was made load-sensitive by linking a larger-capacity oil pump to the throttle; this gave increased oil supply as the speed increased, but at the cost of occasional over-lubrication and consequent spark-plug fouling, for example on long hills.

An equally significant new feature (if not always welcomed by the die-hards) was the fitting, at last, of an internally-expanding drum front brake, cable-operated from a handlebar lever. Tyre choice was no longer optional; Harley had settled on a standard fitment of 25 x 3.85 balloon tyres on 18-inch wheels.

An exhaust system with four-tube silencers could be ordered, at extra cost. Further new colours – Black, Grey, Blue, Green and Chestnut Brown – became available.

... of the Harley Models F and J ranges

1929 model year

The Dual Bullet double headlamps and, for this year only, an exhaust system with four-tube silencers (known at the time as the 'pipes of Pan' exhaust) identify the last of the ioe Big Twins which had come to the end of their lives and were to be replaced for 1930 by new and less complicated side-valves.

As a final innovation the Standard Model F, still available in both engine sizes, could be had with the Dow-metal pistons, such engines being identified by the L suffix, thus Models FL and FDL.

The Model FDL thus resurfaces for the first time since 1927, but what was once Special is now Standard. This repeated use of model designations is typical of H-D; for example Models F and FL return in 1941, but this time identify the 74 cu. in./1207 cc. ohv models, and so on. The aluminium-alloy piston Sport models (Models JX and JDX) had been dropped.

Only for 1929 was the special Model JDF offered, with Remy's Two-Unit electrical system, in contrast to the Single-Unit system of the Models J and JD in which the ignition coil, voltage regulator, battery, horn and connections were all mounted together in a waterproof assembly on top of the crankcases. The delicate Remy system was gradually replaced.

Model range

1909 to 1915
The Model 5D engine has an ioe valve layout, but with automatic intake valves and side exhaust.
Model 5D   54 cu. in./880 cc. engine, magneto ignition, direct drive   (1909)

All engines from the Model 7D onwards have ioe valve layout, with pushrod-operated overhead intake valves and side exhaust.
Model 7D   49 cu. in./811 cc. engine, magneto ignition, direct drive   (1911)
Model 8D and X8D   49 cu. in./811 cc. engine, magneto ignition, direct drive   (1912)
Model 8E and X8E   60 cu. in./989 cc. engine, magneto ignition, direct drive   (1912)
Model 9E   60 cu. in./989 cc. engine, magneto ignition, direct drive   (1913)
Model 9F   60 cu. in./989 cc. engine, magneto ignition, 2-speed gearbox   (1913)
Model 10E   60 cu. in./989 cc. engine, magneto ignition, direct drive   (1914)
Model 10F   60 cu. in./989 cc. engine, magneto ignition, 2-speed gearbox   (1914)
Model 11E   60 cu. in./989 cc. engine, magneto ignition, direct drive   (1915)
Model 11F   60 cu. in./989 cc. engine, magneto ignition, 3-speed gearbox   (1915)
Model 11H   60 cu. in./989 cc. engine, battery-and-coil ignition, direct drive   (1915)
Model 11J   60 cu. in./989 cc. engine, battery-and-coil ignition, 3-speed gearbox   (1915)

1916
Model 16E   60 cu. in./989 cc. engine, magneto ignition, direct drive
Model 16F   60 cu. in./989 cc. engine, magneto ignition, 3-speed gearbox
Model 16J   60 cu. in./989 cc. engine, battery-and-coil ignition, 3-speed gearbox

1917
Model 17E   60 cu. in./989 cc. engine, magneto ignition, direct drive
Model 17F   60 cu. in./989 cc. engine, magneto ignition, 3-speed gearbox
Model 17J   60 cu. in./989 cc. engine, battery-and-coil ignition, 3-speed gearbox

1918
Model 18E   60 cu. in./989 cc. engine, magneto ignition, direct drive
Model 18F   60 cu. in./989 cc. engine, magneto ignition, 3-speed gearbox
Model 18FA   60 cu. in./989 cc. engine, magneto ignition, 3-speed gearbox
Model 18FB   60 cu. in./989 cc. engine, magneto ignition, 3-speed gearbox
Model 18FCA   60 cu. in./989 cc. engine, magneto ignition, 3-speed gearbox, aluminium-alloy pistons
Model 18FCAB   69 cu. in./1130 cc. engine, magneto ignition, 3-speed gearbox, aluminium-alloy pistons
Model 18FF   60 cu. in./989 cc. engine, magneto ignition, 3-speed gearbox
Model 18FFA   60 cu. in./989 cc. engine, magneto ignition, 3-speed gearbox, aluminium-alloy pistons
Model 18J   60 cu. in./989 cc. engine, battery-and-coil ignition, 3-speed gearbox
Model 18JA   60 cu. in./989 cc. engine, battery-and-coil ignition, 3-speed gearbox
Model 18JB   60 cu. in./989 cc. engine, battery-and-coil ignition, 3-speed gearbox
Model 18JF   60 cu. in./989 cc. engine, battery-and-coil ignition, 3-speed gearbox

1919
All models have the same 3-speed gearbox and, unless otherwise specified, grey cast-iron pistons
Model 19F   60 cu. in./989 cc. engine, magneto ignition
Model 19FA   60 cu. in./989 cc. engine, magneto ignition
Model 19FB   60 cu. in./989 cc. engine, magneto ignition
Model 19FCA   60 cu. in./989 cc. engine, magneto ignition, aluminium-alloy pistons
Model 19FCAB   69 cu. in./1130 cc. engine, magneto ignition, aluminium-alloy pistons
Model 19FF   60 cu. in./989 cc. engine, magneto ignition
Model 19FFA   60 cu. in./989 cc. engine, magneto ignition, aluminium-alloy pistons
Model 19FS   60 cu. in./989 cc. engine, magneto ignition, sidecar-specification
Model 19J   60 cu. in./989 cc. engine, battery-and-coil ignition
Model 19JA   60 cu. in./989 cc. engine, battery-and-coil ignition
Model 19JB   60 cu. in./989 cc. engine, battery-and-coil ignition
Model 19JF   60 cu. in./989 cc. engine, battery-and-coil ignition
Model 19JS   60 cu. in./989 cc. engine, battery-and-coil ignition, sidecar-specification

1920
All models have the same 3-speed gearbox and, unless otherwise specified, grey cast-iron pistons
Model 20F	60 cu. in./989 cc. engine, magneto ignition
Model 20FA	60 cu. in./989 cc. engine, magneto ignition
Model 20FB	60 cu. in./989 cc. engine, magneto ignition
Model 20FCA	60 cu. in./989 cc. engine, magneto ignition, aluminium-alloy pistons
Model 20FCAB	69 cu. in./1130 cc. engine, magneto ignition, aluminium-alloy pistons
Model 20FE	60 cu. in./989 cc. engine, magneto ignition, aluminium-alloy pistons
Model 20FF	60 cu. in./989 cc. engine, magneto ignition
Model 20FFA	60 cu. in./989 cc. engine, magneto ignition, aluminium-alloy pistons
Model 20FS	60 cu. in./989 cc. engine, magneto ignition, sidecar-specification
Model 20J	60 cu. in./989 cc. engine, battery-and-coil ignition
Model 20JA	60 cu. in./989 cc. engine, battery-and-coil ignition
Model 20JB	60 cu. in./989 cc. engine, battery-and-coil ignition
Model 20JE	60 cu. in./989 cc. engine, battery-and-coil ignition, aluminium-alloy pistons
Model 20JF	60 cu. in./989 cc. engine, battery-and-coil ignition
Model 20JS	60 cu. in./989 cc. engine, battery-and-coil ignition, sidecar-specification

1921
All models have the same 3-speed gearbox and, unless otherwise specified, grey cast-iron pistons
Model 21F	60 cu. in./989 cc. engine, magneto ignition
Model 21FA	60 cu. in./989 cc. engine, magneto ignition
Model 21FCA	60 cu. in./989 cc. engine, magneto ignition, aluminium-alloy pistons
Model 21FE	60 cu. in./989 cc. engine, magneto ignition, aluminium-alloy pistons
Model 21FS	60 cu. in./989 cc. engine, magneto ignition, sidecar-specification
Model 21J	60 cu. in./989 cc. engine, battery-and-coil ignition
Model 21JA	60 cu. in./989 cc. engine, battery-and-coil ignition
Model 21JE	60 cu. in./989 cc. engine, battery-and-coil ignition, aluminium-alloy pistons
Model 21JS	60 cu. in./989 cc. engine, battery-and-coil ignition, sidecar-specification
Model 21FD	74 cu. in./1217 cc. engine, magneto ignition
Model 21FDA	74 cu. in./1217 cc. engine, magneto ignition
Model 21FDS	74 cu. in./1217 cc. engine, magneto ignition, sidecar-specification
Model 21JD	74 cu. in./1217 cc. engine, battery-and-coil ignition
Model 21JDA	74 cu. in./1217 cc. engine, battery-and-coil ignition
Model 21JDS	74 cu. in./1217 cc. engine, battery-and-coil ignition, sidecar-specification

1922
All models have the same 3-speed gearbox and, unless otherwise specified, grey cast-iron pistons
Model 22F	60 cu. in./989 cc. engine, magneto ignition
Model 22FA	60 cu. in./989 cc. engine, magneto ignition
Model 22FCA	60 cu. in./989 cc. engine, magneto ignition, aluminium-alloy pistons
Model 22FE	60 cu. in./989 cc. engine, magneto ignition, aluminium-alloy pistons
Model 22FS	60 cu. in./989 cc. engine, magneto ignition, sidecar-specification
Model 22J	60 cu. in./989 cc. engine, battery-and-coil ignition
Model 22JA	60 cu. in./989 cc. engine, battery-and-coil ignition
Model 22JE	60 cu. in./989 cc. engine, battery-and-coil ignition, aluminium-alloy pistons
Model 22JS	60 cu. in./989 cc. engine, battery-and-coil ignition, sidecar-specification
Model 22FD	74 cu. in./1217 cc. engine, magneto ignition
Model 22FDA	74 cu. in./1217 cc. engine, magneto ignition
Model 22FDS	74 cu. in./1217 cc. engine, magneto ignition, sidecar-specification
Model 22JD	74 cu. in./1217 cc. engine, battery-and-coil ignition
Model 22JDA	74 cu. in./1217 cc. engine, battery-and-coil ignition
Model 22JDS	74 cu. in./1217 cc. engine, battery-and-coil ignition, sidecar-specification

1923
All models have the same 3-speed gearbox and, unless otherwise specified, grey cast-iron pistons
Model 23F   60 cu. in./989 cc. engine, magneto ignition
Model 23FA   60 cu. in./989 cc. engine, magneto ignition

Model 23FCA   60 cu. in./989 cc. engine, magneto ignition, aluminium-alloy pistons
Model 23FE   60 cu. in./989 cc. engine, magneto ignition, aluminium-alloy pistons
Model 23FS   60 cu. in./989 cc. engine, magneto ignition, sidecar-specification
Model 23J   60 cu. in./989 cc. engine, battery-and-coil ignition
Model 23JE   60 cu. in./989 cc. engine, battery-and-coil ignition, aluminium-alloy pistons
Model 23JS   60 cu. in./989 cc. engine, battery-and-coil ignition, sidecar-specification
Model 23FD   74 cu. in./1217 cc. engine, magneto ignition
Model 23FDA   74 cu. in./1217 cc. engine, magneto ignition
Model 23FDS   74 cu. in./1217 cc. engine, magneto ignition, sidecar-specification
Model 23FDCA   74 cu. in./1217 cc. engine, magneto ignition, aluminium-alloy pistons
Model 23JD   74 cu. in./1217 cc. engine, battery-and-coil ignition
Model 23JDA   74 cu. in./1217 cc. engine, battery-and-coil ignition
Model 23JDS   74 cu. in./1217 cc. engine, battery-and-coil ignition, sidecar-specification
Model 23JDCA   74 cu. in./1217 cc. engine, battery-and-coil ignition, aluminium-alloy pistons

1924
All models have the same 3-speed gearbox and, unless otherwise specified, grey cast-iron pistons
Model 24FE   60 cu. in./989 cc. engine, magneto ignition, aluminium-alloy pistons
Model 24FES   60 cu. in./989 cc. engine, magneto ignition, sidecar-specification, aluminium-alloy pistons
Model 24JE   60 cu. in./989 cc. engine, battery-and-coil ignition, aluminium-alloy pistons
Model 24JES   60 cu. in./989 cc. engine, battery-and-coil ignition, sidecar-specification, aluminium-alloy pistons
Model 24FD   74 cu. in./1217 cc. engine, magneto ignition
Model 24FDS   74 cu. in./1217 cc. engine, magneto ignition, sidecar-specification
Model 24FDCA   74 cu. in./1217 cc. engine, magneto ignition, aluminium-alloy pistons
Model 24FDSCA   74 cu. in./1217 cc. engine, magneto ignition, sidecar-specification, aluminium-alloy pistons
Model 24JD   74 cu. in./1217 cc. engine, battery-and-coil ignition
Model 24JDS   74 cu. in./1217 cc. engine, battery-and-coil ignition, sidecar-specification
Model 24JDCA   74 cu. in./1217 cc. engine, battery-and-coil ignition, aluminium-alloy pistons
Model 24JDSCA   74 cu. in./1217 cc. engine, battery-and-coil ignition, sidecar-specification, aluminium-alloy pistons

1925
All models have the same 3-speed gearbox and grey cast-iron pistons
Model 25FE   60 cu. in./989 cc. engine, magneto ignition
Model 25FES   60 cu. in./989 cc. engine, magneto ignition, sidecar-specification
Model 25JE   60 cu. in./989 cc. engine, battery-and-coil ignition
Model 25JES   60 cu. in./989 cc. engine, battery-and-coil ignition, sidecar-specification
Model 25FDCB   74 cu. in./1217 cc. engine, magneto ignition
Model 25FDCBS   74 cu. in./1217 cc. engine, magneto ignition, sidecar-specification
Model 25JDCB   74 cu. in./1217 cc. engine, battery-and-coil ignition
Model 25JDCBS   74 cu. in./1217 cc. engine, battery-and-coil ignition, sidecar-specification

1926
All models have the same 3-speed gearbox and grey cast-iron pistons
Model 26F   60 cu. in./989 cc. engine, magneto ignition
Model 26FS   60 cu. in./989 cc. engine, magneto ignition, sidecar-specification
Model 26J   60 cu. in./989 cc. engine, battery-and-coil ignition
Model 26JS   60 cu. in./989 cc. engine, battery-and-coil ignition, sidecar-specification
Model 26FD   74 cu. in./1217 cc. engine, magneto ignition
Model 26FDS   74 cu. in./1217 cc. engine, magneto ignition, sidecar-specification
Model 26JD   74 cu. in./1217 cc. engine, battery-and-coil ignition
Model 26JDS   74 cu. in./1217 cc. engine, battery-and-coil ignition, sidecar-specification

1927
All models have the same 3-speed gearbox and grey cast-iron pistons
Model	Specification	Type
Model 27F	60 cu. in./989 cc. engine, magneto ignition	Standard
Model 27FS	60 cu. in./989 cc. engine, magneto ignition, sidecar-specification	Standard
Model 27FK	60 cu. in./989 cc. engine, magneto ignition	Special
Model 27J	60 cu. in./989 cc. engine, battery-and-coil ignition	Standard
Model 27JS	60 cu. in./989 cc. engine, battery-and-coil ignition, sidecar-specification	Standard
Model 27JK	60 cu. in./989 cc. engine, battery-and-coil ignition	Special
Model 27FD	74 cu. in./1217 cc. engine, magneto ignition	Standard
Model 27FDS	74 cu. in./1217 cc. engine, magneto ignition, sidecar-specification	Standard
Model 27FDL	74 cu. in./1217 cc. engine, magneto ignition	Special
Model 27JD	74 cu. in./1217 cc. engine, battery-and-coil ignition	Standard
Model 27JDS	74 cu. in./1217 cc. engine, battery-and-coil ignition, sidecar-specification	Standard
Model 27JDL	74 cu. in./1217 cc. engine, battery-and-coil ignition	Special

1928
All models have the same 3-speed gearbox and cast-iron pistons, except for the Special Sport models, which have pistons of Dow metal (a magnesium alloy).
Model	Specification	Type
Model 28F	60 cu. in./989 cc. engine, magneto ignition	Standard
Model 28FS	60 cu. in./989 cc. engine, magneto ignition, sidecar-specification	Standard
Model 28FH	60 cu. in./989 cc. engine, magneto ignition	Special Two Cam
Model 28J	60 cu. in./989 cc. engine, battery-and-coil ignition	Standard
Model 28JS	60 cu. in./989 cc. engine, battery-and-coil ignition, sidecar-specification	Standard
Model 28JH	60 cu. in./989 cc. engine, battery-and-coil ignition	Special Two Cam
Model 28JX	60 cu. in./989 cc. engine, battery-and-coil ignition	Sport
Model 28JXL	60 cu. in./989 cc. engine, battery-and-coil ignition, Dow metal pistons	Special Sport
Model 28FD	74 cu. in./1217 cc. engine, magneto ignition	Standard
Model 28FDS	74 cu. in./1217 cc. engine, magneto ignition, sidecar-specification	Standard
Model 28FDH	74 cu. in./1217 cc. engine, magneto ignition	Special Two Cam
Model 28JD	74 cu. in./1217 cc. engine, battery-and-coil ignition	Standard
Model 28JDS	74 cu. in./1217 cc. engine, battery-and-coil ignition, sidecar-specification	Standard
Model 28JDH	74 cu. in./1217 cc. engine, battery-and-coil ignition	Special, Two Cam
Model 28JDX	74 cu. in./1217 cc. engine, battery-and-coil ignition	Sport
Model 28JDXL	74 cu. in./1217 cc. engine, battery-and-coil ignition, Dow metal pistons	Special Sport

1929
All models have the same 3-speed gearbox and cast iron pistons, except for the models indicated, which have pistons of Dow metal (a magneisum alloy).
Model	Specification	Type
Model 29F	60 cu. in./989 cc. engine, magneto ignition	Standard
Model 29FS	60 cu. in./989 cc. engine, magneto ignition, sidecar-specification	Standardt
Model 29FL	60 cu. in./989 cc. engine, magneto ignition, Dow metal pistons	Standard
Model 29J	60 cu. in./989 cc. engine, battery-and-coil ignition	Standard
Model 29JS	60 cu. in./989 cc. engine, battery-and-coil ignition, sidecar-specification	Standard
Model 29JH	60 cu. in./989 cc. engine, battery-and-coil ignition	Special, Two Cam
Model 29JXL	60 cu. in./989 cc. engine, battery-and-coil ignition, Dow metal pistons	Special Sport
Model 29FD	74 cu. in./1217 cc. engine, magneto ignition	Standard
Model 29FDS	74 cu. in./1217 cc. engine, magneto ignition, sidecar-specification	Standard
Model 29FDH	74 cu. in./1217 cc. engine, magneto ignition	Special, Two-Cam
Model 29FDL	74 cu. in./1217 cc. engine, magneto ignition, Dow metal pistons	Standard
Model 29JD	74 cu. in./1217 cc. engine, battery-and-coil ignition	Standard
Model 29JDS	74 cu. in./1217 cc. engine, battery-and-coil ignition, sidecar-specification	Standard
Model 29JDF	74 cu. in./1217 cc. engine, battery-and-coil ignition	Two Unit
Model 29JDH	74 cu. in./1217 cc. engine, battery-and-coil ignition	Two-Cam
Model 29JDXL	74 cu. in./1217 cc. engine, battery-and-coil ignition, Dow metal pistons	Special Sport

Model 5D 1909

Engine
Engine	Air-cooled 45° ioe V-twin
Bore x stroke	3⅛ x 3½ (3.13 x 3.5) in.
Bore x stroke	79.38 x 88.9 mm
Capacity	53.68 cu. in. (880 cc)
Power output (SAE)	6½ (5)/1800 rpm
Carburettor	Harley-Davidson
Ignition	Magneto

Transmission
Gearbox	None
Gearchange	None
Final drive	Belt

Cycle parts
Frame construction	Single-loop, curved front downtube
Front fork	Springer, enclosed springs, tubular legs
Rear suspension	None
Front brake	None
Rear brake	Back-pedal

General information
Unladen weight	135 kg
Front wheel/tyre	2½ x 28 beaded-edge
Rear wheel/tyre	2½ x 28 beaded-edge
Maximum speed	80 km/h
Tank capacity	9.1 litres
Catalogue price	$325
Production figures	27

Model 7D 1911

Engine
Engine	Air-cooled 45° ioe V-twin
Bore x stroke	3 x 3½ (3 x 3.5) in.
Bore x stroke	76.2 x 88.9 mm
Capacity	49.48 cu. in. (811 cc)
Power output (SAE)	6½ (5)/2000 rpm
Carburettor	Schebler
Ignition	Magneto

Transmission
Gearbox	None
Gearchange	None
Final drive	Belt

Cycle parts
Frame construction	Single-loop, straight downtube
Front fork	Springer, enclosed springs, tubular legs
Rear suspension	None
Front brake	None
Rear brake	Back-pedal

General information
Unladen weight	135 kg
Front wheel/tyre	2½ x 28 beaded-edge
Rear wheel/tyre	2½ x 28 beaded-edge
Maximum speed	80 km/h
Tank capacity	9.1 litres
Catalogue price	$300
Production figures	Total Model 7 production, including the single-cylinder models, amounted to 5,625 motorcycles.

Models 8D and X8D 1912

Engine
Engine	Air-cooled 45° ioe V-twin
Bore x stroke	3 x 3½ (3 x 3.5) in.
Bore x stroke	76.2 x 88.9 mm
Capacity	49.48 cu. in. (811 cc)
Power output (SAE)	6½ (5)/2000 rpm
Carburettor	Schebler
Ignition	Magneto

Transmission
Gearbox	None
Gearchange	None
Final drive	Belt

Cycle parts
Frame construction	Single-loop, straight downtube
Front fork	Springer, enclosed springs, tubular legs
Rear suspension	None
Front brake	None
Rear brake	Back-pedal

General information
Unladen weight	135 kg
Front wheel/tyre	2½ x 28 beaded-edge
Rear wheel/tyre	2½ x 28 beaded-edge
Maximum speed	85 km/h
Tank capacity	9.1 litres
Catalogue price	Model 8D $275
	Model X8D $285
Production figures	A total of 3852 Model 8s were produced, including the singles.

Model X8E 1912

Engine
Engine	Air-cooled 45° ioe V-twin
Bore x stroke	3⁵⁄₁₆ x 3½ (3.31 x 3.5) in.
Bore x stroke	84.14 x 88.9 mm
Capacity	60.33 cu. in. (989 cc)
Power output (SAE)	8(6)/2200 rpm
Carburettor	Schebler 1-inch
Ignition	Magneto

Transmission
Gearbox	None
Gearchange	None
Final drive	Chain

Cycle parts
Frame construction	Single-loop, straight downtube
Front fork	Springer, enclosed springs, tubular legs
Rear suspension	None
Front brake	None
Rear brake	Back-pedal

General information
Unladen weight	140 kg
Front wheel/tyre	2½ x 28 beaded-edge
Rear wheel/tyre	2½ x 28 beaded-edge
Maximum speed	90 km/h
Tank capacity	9.1 litres
Catalogue price	Model 8E $285
Production figures	A total of 3852 Model 8s were produced, including the singles.

The X model code prefix indicates that the machine is fitted with a free-wheel clutch in the rear hub (operated by a lever on the left-hand side of the tank).

Model 9 1913

Engine
Engine	Air-cooled 45° ioe V-twin
Bore x stroke	3⁵⁄₁₆ x 3½ (3.31 x 3.5) in.
Bore x stroke	84.14 x 88.9 mm
Capacity	60.33 cu. in. (989 cc)
Power output (SAE)	9(7)/2400 rpm
Carburettor	Schebler 1-inch
Ignition	Magneto

Transmission
Gearbox	Model 9E: None
	Model 9F: 2-speed, rear hub-mounted
Gearchange	Hand change, on tank
Final drive	Chain

Cycle parts
Frame construction	Single-loop, straight downtube
Front fork	Springer, enclosed springs, tubular legs
Rear suspension	None
Front brake	None
Rear brake	Back-pedal

General information
Unladen weight	140 kg
Front wheel/tyre	2¾ x 28 beaded-edge
Rear wheel/tyre	2¾ x 28 beaded-edge
Maximum speed	90 km/h
Tank capacity	9.1 litres

Model 10 1914

Engine
Engine	Air-cooled 45° ioe V-twin
Bore x stroke	3⁵⁄₁₆ x 3½ (3.31 x 3.5) in.
Bore x stroke	84.14 x 88.9 mm
Capacity	60.33 cu. in. (989 cc)
Power output (SAE)	9(7)/2400 rpm
Carburettor	Schebler 1-inch
Ignition	Magneto

Transmission
Gearbox	Model 10E: None
	Model 10F: 2-speed, rear hub-mounted
Gearchange	Hand change, on tank
Final drive	Chain

Cycle parts
Frame construction	Single-loop, straight downtube
Front fork	Springer, enclosed springs, tubular legs
Rear suspension	None
Front brake	None
Rear brake	Drum, via rod and pedal

General information
Unladen weight	140 kg
Front wheel/tyre	3 x 28 beaded-edge
Rear wheel/tyre	3 x 28 beaded-edge
Maximum speed	90 km/h
Tank capacity	9.1 litres
Sidecars available	First works sidecar, no model code.

The Model 9F was introduced only as a prototype and so has no official list price, but was sold at approximately $385. The Model 10F was officially introduced for the 1914 model year.

Sidecar wheels and tyres were always the same as those of the solo motorcycles.

Engines with intake ports opened out to 1 inch (25 mm) for 1914 were identified by the suffix A and delivered some 11 horsepower at approximately 2800 rpm.

All ex-works racing engines of 1914 were identified by the suffix M.

Catalogue prices for all models from the 1913 model year onwards are given at the end of this Chapter.

Model 11 1915

Engine
Engine	Air-cooled 45° ioe V-twin
Bore x stroke	3⁵⁄₁₆ x 3½ (3.31 x 3.5) in.
Bore x stroke	84.14 x 88.9 mm
Capacity	60.33 cu. in. (989 cc)
Power output (SAE)	11(8)/3000 rpm
Compression ratio	3.75:1
Carburettor	Schebler 1-inch
Ignition	Models E and F: Magneto Models H and J: Battery-and-coil

Transmission
Gearbox	Models 11E and 11H: None Models 11F and 11J: 3-speed
Gearchange	Hand change, on tank
Final drive	Chain

Cycle parts
Frame construction	Single-loop, straight downtube
Front fork	Springer, enclosed springs, tubular legs
Rear suspension	None
Front brake	None
Rear brake	Drum, via rod and pedal

General information
Unladen weight	150 kg
Front wheel/tyre	3 x 28 beaded-edge
Rear wheel/tyre	3 x 28 beaded-edge
Maximum speed	95 km/h
Tank capacity	9.1 litres
Sidecars available	Model 11L R/H single-seater Model L11L: L/H single-seater Model 11M Commercial

Sidecar wheels and tyres were always the same as those of the solo motorcycles.

Model 16 1916

Engine
Engine	Air-cooled 45° ioe V-twin
Bore x stroke	3⁵⁄₁₆ x 3½ (3.31 x 3.5) in.
Bore x stroke	84.14 x 88.9 mm
Capacity	60.33 cu. in. (989 cc)
Power output (SAE) Solo motorcycle:	11(8)/3000 rpm
Sidecar:	9(7)/2700 rpm
Compression ratio: Solo motorcycle	3.75:1
Sidecar	3.45:1
Carburettor	Schebler 1-inch
Ignition	Models E and F: Magneto Model J: Battery-and-coil

Transmission
Gearbox	Model 16E: None Models 16F and 16J: 3-speed
Gearchange	Hand change, on tank
Final drive	Chain

Cycle parts
Frame construction	Single-loop, straight downtube
Front fork	Springer, enclosed springs, tubular legs
Rear suspension	None
Front brake	None
Rear brake	Drum, via rod and pedal

General information
Unladen weight	150 kg
Front wheel/tyre	3 x 28 beaded-edge
Rear wheel/tyre	3 x 28 beaded-edge
Maximum speed	95 km/h
Tank capacity	10 litres
Sidecars available	**Chassis:** Model 16LC R/H single-seater Model 16LCL L/H single-seater Model 16XT for luggage-carrying **Passenger-carrying sidecars:** Model 16L R/H single-seater **Load-carrying sidecars:** Model 16M Commercial standard Model 16RFD Rural Free Delivery Model 16P with closed box body Model 16U with open box body Model 16V with double closed box body

Model 17 1917

Engine
Engine	Air-cooled 45° ioe V-twin
Bore x stroke	3 5/16 x 3 1/2 (3.31 x 3.5) in.
Bore x stroke	84.14 x 88.9 mm
Capacity	60.33 cu. in. (989 cc)
Power output (SAE)	14(10)/3200 rpm
Compression ratio	3.75:1
Carburettor	Schebler 1-inch
Ignition	Models E and F: Magneto Model J: Battery-and-coil

Transmission
Gearbox	Model 17E: None Models 17F and 17J: 3-speed
Gearchange	Hand change, on tank
Final drive	Chain

Cycle parts
Frame construction	Single-loop, straight downtube
Front fork	Springer, enclosed springs, tubular legs
Rear suspension	None
Front brake	None
Rear brake	Drum, via rod and pedal

General information
Unladen weight	150 kg
Front wheel/tyre	3 x 28 beaded-edge
Rear wheel/tyre	3 x 28 beaded-edge
Maximum speed	100 km/h
Tank capacity	10 litres

Sidecars available

Chassis:
Model 17LC
R/H single-seater
Model 17LCL L/H single-seater

Passenger-carrying sidecars:
Model 17L
R/H single-seater
Model L17L L/H single-seater

Load-carrying sidecars:
Model 17M
Commercial standard
Model 17N
Rural Free Delivery
Model 17P with closed box body
Model 17U with open box body
Model 17V with double closed box body

Model 18 1918

Engine
Engine	Air-cooled 45° ioe V-twin
Bore x stroke	3⁵⁄₁₆ x 3½ (3.31 x 3.5) in.
Bore x stroke	84.14 x 88.9 mm
Capacity	60.33 cu. in. (989 cc)
	Power output (SAE)
	Solo motorcycle:
	15(11)/3200 rpm
	Sidecar:
	13(10)/2800 rpm
Compression ratio:	
Solo motorcycle	3.75:1
Sidecar	3.5:1
Carburettor	Schebler 1-inch
Ignition	Models E and F: Magneto
	Model J: Battery-and-coil

Transmission
Gearbox	Model 18E: None
	Models 18F and 18J: 3-speed
Gearchange	Hand change, on tank
Final drive	Chain

Cycle parts
Frame construction	Single-loop, straight downtube
Front fork	Springer, enclosed springs, tubular legs
Rear suspension	None
Front brake	None
Rear brake	Drum, via rod and pedal

General information
Unladen weight	150 kg
Front wheel/tyre	3 x 28 beaded-edge
Rear wheel/tyre	3 x 28 beaded-edge
Maximum speed	100 km/h
Tank capacity	10 litres

Sidecars available

Chassis:
Model 18LC
R/H single-seater
Model 18LCL L/H single-seater
Model 18Q double-adult

Passenger-carrying sidecars:
Model 18L
R/H single-seater
Model L-18L L/H single-seater

Load-carrying sidecars:
Model 18M
Commercial standard
Model 18N
Rural Free Delivery
Model 18P with closed box body
Model 18QA
Commercial closed
Model 18QB
Commercial open
Model 18QD
Parcel Car (2 Box)
Model 18QL
R/H double-adult
Model 18QC3
2-stake platform
Model 18QC4
3-stake platform

The 1914 to 1923 Special Motors

The following lists set out the alternatives offered by the factory from 1914 onwards for those who wanted somewhat more power than that offered by the standard models.

Even by 1914 there was a higher-performance version of the still brand-new V-twin engine, the A Motor, whose intake ports were opened out to 1 inch (25 mm).

In 1917, the first year of war for the United States, the Fast Motor concept was introduced to the single cylinder range, becoming the 500 models (single and V-twin) of 1918 and 1919. Power outputs were boosted somewhat by numerous changes to the valvegear.

From 1918 there was a complete range of V-twin Special Motors. The A Motors of 1918 to 1923 were not meant for racing, with out-and-out top speed as the object, but instead were intended primarily for the increasing numbers of police motorcycle patrolmen. The B Motors (1918 to 1920) were higher-compression units designed to compensate for the power loss encountered when working in the thinner air of the high-altitude States. The E Motors (1920 to 1923) were pure racers, giving higher power outputs for the amateur classes of roadracing; in 1924, they were put into series production as standard road models. All three Special Motors still used the then-common cast-iron pistons up to 1923.

Also starting in 1918 were a further range of engines which used aluminium-alloy pistons for the first time. This new technology was an offshoot of the aero engine industry, which had blossomed during the war. So there were from 1918 to 1920: alongside the 500 Motor (still with cast-iron pistons) was the 500A Motor, a version of the 500 Motor but with aluminium-alloy pistons (from which would come the E Motor of 1920), then the 500CA Motor (1918 to 1920) which was a forerunner of the T Motor (1921 to 1922), likewise with aluminium pistons and (probably) somewhat higher compression. From this was created the 500CAB Motor (1918 to 1920), the final stage of development of these engines, in which the extended stroke (4 in./101.6 mm) gave it substantially greater capacity (69 cu. in./1130 cc.), with a corresponding increase in power.

For 1914: Series 10 (all models)
- Model 10   A Motor with larger (1 inch/25 mm) intake ports.
- Model 10   M Motor – racer with substantially-increased power output (not for production models), offered from 1914 to 1916.

For 1917: Series 17 (all models):
- Model 17   Fast Motor with increased power output and special engine numbers (from the 1917 Fast Motor came the 1918 and 1919 500 motors).

For 1918: Series 18F (with acetylene lighting and magneto ignition):
- Model 18FA   A Motor with cast-iron pistons for the police.
- Model 18FB   B Motor with cast-iron pistons for high altitudes.
- Model 18FCA   500CA Motor with aluminium-alloy pistons.
- Model 18FCAB   500CAB Motor with aluminium-alloy pistons.
- Model 18FF   500 Motor with cast-iron pistons.
- Model 18FFA   500A Motor with aluminium-alloy pistons.

Series 18J (with electric lighting and battery-and-coil ignition):
- Model 18JA   A Motor with cast-iron pistons.
- Model 18JB   B Motor with cast-iron pistons.
- Model 18JF   500 Motor with cast-iron pistons.

For 1919: Series 19F (with acetylene lighting and magneto ignition):
- Model 19FA   A Motor with cast-iron pistons.
- Model 19FB   B Motor with cast-iron pistons.
- Model 19FCAB   500CAB Motor with aluminium-alloy pistons.
- Model 19FF   500 Motor with cast-iron pistons.
- Model 19FFA   500A Motor with aluminium-alloy pistons.

Series 19J (with electric lighting and battery-and-coil ignition):
- Model 19JA   A Motor with cast-iron pistons.
- Model 19JB   B Motor with cast-iron pistons.
- Model 19JF   500 Motor with cast-iron pistons.

For 1920: Series 20F (with acetylene lighting and magneto ignition):
    Model 20FA   A Motor with cast-iron pistons.
    Model 20FB   B Motor with cast-iron pistons.
    Model 20FCA   500CA Motor with aluminium-alloy pistons.
    Model 20FCAB   50CAB Motor with aluminium-alloy pistons.
    Model 20FE   E Motor with aluminium-alloy pistons.
    Model 20FF   500 Motor with cast-iron pistons.
    Model 20FFA   500A Motor with aluminium-alloy pistons.

    Series 20J (with electric lighting and battery-and-coil ignition):
    Model 20JA   A Motor with cast-iron pistons.
    Model 20JB   B Motor with cast-iron pistons.
    Model 20JE   E Motor with aluminium-alloy pistons.
    Model 20JF   500 Motor with cast-iron pistons.

Except for the 69 cu. in./1130 cc. CAB Motor, all engines are of the standard capacity of 60 cu. in./989 cc.

Power outputs of the Motors can be estimated as follows:

Motor	Year	Power	Piston
A Motor (High Performance)	1914	11 hp (8 kW) @ approx. 2,800 rpm	Cast-iron.
Fast Motor	1917	17 hp (13 kW) @ approx. 3,200 rpm	Cast-iron.
A Motor (Police Motor)	1918-1920	17 hp (13 kW) @ approx. 3,200 rpm	Cast-iron.
B Motor (High Altitude)	1918-1920	16 hp (12 kW) @ approx. 3,200 rpm	Cast-iron.
E Motor (Racing Motor)	1920	18 hp (14 kW) @ approx. 3,600 rpm	Aluminium-alloy.
500 Motor	1918-1920	17 hp (13 kW) @ approx. 3,400 rpm	Cast-iron.
500A Motor	1918-1920	18 hp (14 kW) @ approx. 3,600 rpm	Aluminium-alloy.
500CA Motor	1918-1920	20 hp (15 kW) @ approx. 3,600 rpm	Aluminium-alloy.
500CAB Motor	1918-1920	21 hp (15 kW) @ approx. 3,600 rpm	Aluminium-alloy.

The B and E Motors were offered with higher compression of approximately 4.0 to 4.2:1; whereas the B Motor, with its cast-iron pistons, was meant for high-altitude regions, the E Motor was conceived for racing, but intended for fast road riding, rather than matching, for example, the raw all-out power of the board track racers. Thanks to its increased stroke the even more special CAB Motor had a capacity of 69 cu. in./1130 cc. and was therefore the technical forerunner of the 74 cu. in./1217 cc. Big Twins due to appear in 1921. Its power output amounted to about 23 hp (17 kW) @ 3,400 rpm.

    With the introduction of the 74 cu. in./1217 cc. Big Twins in 1921 there were A Motors also for the new FD models (also the JD models, from 1922). From 1924 production of the Special Motors was either stopped or they carried on being built, but now as standard engines for road models (e.g., the 60 cu. in./989 cc. version of the E Motor).

For 1921: Series 21F (with acetylene lighting and magneto ignition):
    Model 21FA   A Motor with cast-iron pistons   60 cu. in./989 cc.
    Model 21FCA   T Motor with aluminium-alloy pistons   60 cu. in./989 cc.
    Model 21FDA   A Motor with cast-iron pistons   74 cu. in./1217 cc.
    Model 21FE   E Motor with aluminium-alloy pistons   60 cu. in./989 cc.

    Series 21J (with electric lighting and battery-and-coil ignition):
    Model 21JA   A Motor with cast-iron pistons   60 cu. in./989 cc.
    Model 21JDA   A Motor with cast-iron pistons   74 cu. in./1217 cc.
    Model 21JE   E Motor with aluminium-alloy pistons   60 cu. in./989 cc.

For 1922: Series 22F (with acetylene lighting and magneto ignition):
    Model 22FA   A Motor with cast-iron pistons   60 cu. in./989 cc.
    Model 22FCA   T Motor with aluminium-alloy pistons   60 cu. in./989 cc.
    Model 22FDA   A Motor with cast-iron pistons   74 cu. in./1217 cc.
    Model 22FE   E Motor with aluminium-alloy pistons   60 cu. in./989 cc.
    Series 22J (with electric lighting and battery-and-coil ignition):
    Model 22JA   A Motor with cast-iron pistons   60 cu. in./989 cc.
    Model 22JDA   A Motor with cast-iron pistons   74 cu. in./1217 cc.
    Model 22JE   E Motor with aluminium-alloy pistons   60 cu. in./989 cc.

In addition there was yet another Special Motor with the designation DCA, practically a counterpart to the 60 cu. in./989 cc. T Motor for the Series F and J. This became in the following year the T Motor of the 74 cu. in./1217 cc. models.

    Motor 22DCA   DCA Motor with aluminium-alloy pistons   74 cu. in./1217 cc.

For 1923: Series 23F (with acetylene lighting and magneto ignition):
- Model 23FA   A Motor with cast-iron pistons   60 cu. in./989 cc.
- Model 23FCA   T Motor with aluminium-alloy pistons 60 cu. in./989 cc.
- Model 23FDA   A Motor with cast-iron pistons   /4 cu. in./1217 cc.
- Model 23FDCA   T Motor with aluminium-alloy pistons 74 cu. in./1217 cc.
- Model 23FE   E Motor with aluminium-alloy pistons 60 cu. in./989 cc.

Series 23J (with electric lighting and battery-and-coil ignition):
- Model 23JDA   A Motor with cast-iron pistons   74 cu. in./1217 cc.
- Model 23JDCA   T Motor with aluminium-alloy pistons 74 cu. in./1217 cc.
- Model 23JE   E Motor with aluminium-alloy pistons 60 cu. in./989 cc.

Power outputs of the Motors can be estimated as follows:

Motor	Power	Material
A Motor (60 cu. in./989 cc.)	17 hp (13 kW) @ approx. 3,200 rpm	Cast-iron.
A Motor (74 cu. in./1217 cc.)	21 hp (16 kW) @ approx. 3,400 rpm	Cast-iron.
E Motor (60 cu. in./989 cc.)	18 hp (13 kW) @ approx. 3,600 rpm	Aluminium-alloy.
T Motor (60 cu. in./989 cc.)	20 hp (15 kW) @ approx. 3,600 rpm	Aluminium-alloy.
T Motor (74 cu. in./1217 cc.)	22 hp (16 kW) @ approx. 3,600 rpm	Aluminium-alloy.
DCA Motor (74 cu. in./1217 cc.)	22 hp (16 kW) @ approx. 3,600 rpm	Aluminium-alloy.

Models 19F and 19J 1919

Engine
Engine	Air-cooled 45° ioe V-twin
Bore x stroke	$3\frac{5}{16}$ x $3\frac{1}{2}$ (3.31 x 3.5) in.
Bore x stroke	84.14 x 88.9 mm
Capacity	60.33 cu. in. (989 cc)
	Power output (SAE)
	Solo motorcycle:
	15(11)/3000 rpm
	Sidecar:
	14(10)/3000 rpm
Compression ratio:	
Solo motorcycle	3.75:1
Sidecar	3.5:1
Carburettor	Schebler 1-inch
Ignition	Model F: Magneto
	Model J: Battery-and-coil

Transmission
Gearbox	3-speed
Gearchange	Hand change, on tank
Final drive	Chain

Cycle parts
Frame construction	Single-loop, straight downtube
Front fork	Springer, enclosed springs, tubular legs
Rear suspension	None
Front brake	None
Rear brake	Drum, via rod and pedal

General information
Unladen weight	160 kg
Front wheel/tyre	3 x 28 beaded-edge
Rear wheel/tyre	3 x 28 beaded-edge
Maximum speed	100 km/h
Tank capacity	10 litres

Sidecars available

Chassis:
Model 19LC
R/H single-seater
Model L19LC L/H single-seater
Model 19Q double-adult

Passenger-carrying sidecars:
Model 19L
R/H single-seater
Model L19L L/H single-seater

Load-carrying sidecars:
Model 19M
Commercial standard
Model 19QA
Commercial closed
Model 19QB
Commercial open
Model 19QD
Parcel Car (2 Box)
Model 19QL
R/H double-adult
Model 19QC3
2-stake platform
Model 19QC4
3-stake platform

Models 20F and 20J 1920

Engine
Engine	Air-cooled 45° ioe V-twin
Bore x stroke	3³⁄₁₆ x 3½ (3.31 x 3.5) in.
Bore x stroke	84.14 x 88.9 mm
Capacity	60.33 cu. in. (989 cc)
	Power output (SAE)
	Solo motorcycle:
	15(11)/3000 rpm
	Sidecar:
	14(10)/3000 rpm
Compression ratio:	
Solo motorcycle	3.75:1
Sidecar	3.5:1
Carburettor	Schebler 1-inch
Ignition	Model F: Magneto
	Model J: Battery-and-coil

Transmission
Gearbox	3-speed
Gearchange	Hand change, on tank
Final drive	Chain

Cycle parts
Frame construction	Single-loop, straight downtube
Front fork	Springer, enclosed springs, tubular legs
Rear suspension	None
Front brake	None
Rear brake	Drum, via rod and pedal

General information
Unladen weight	160 kg
Front wheel/tyre	3 x 28 beaded-edge
Rear wheel/tyre	3 x 28 beaded-edge
Maximum speed	100 km/h
Tank capacity	10 litres

Sidecars available

Chassis:
Model 20LC
R/H single-seater
Model L20LC L/H single-seater
Model L20LR Speedster, R/H
Model 20PC Double Bar
Model 20Q double-adult

Passenger-carrying sidecars:
Model 20L
R/H single-seater
Model L20L L/H single-seater
Model 20LR Roadster, R/H
Model 20LX Speedster, R/H

Load-carrying sidecars:
Model 20M
Commercial standard
Model 20QA
Commercial closed
Model 20QB
Commercial open
Model 20QD
Parcel Car (2 Box)
Model 20QL
R/H double adult
Model 20QC3
2-stake platform
Model 20QC4
3-stake platform

Models 21F and 21FS
Models 21J and 21JS 1921

Engine
Engine	Air-cooled 45° ioe V-twin
Bore x stroke	3 5/16 x 3 1/2 (3.31 x 3.5) in.
Bore x stroke	84.14 x 88.9 mm
Capacity	60.33 cu. in. (989 cc)
Power output (SAE)	Solo motorcycle: 16(12)/3000 rpm
	Sidecar: 15(11)/3000 rpm
Compression ratio:	
Solo motorcycle	3.8:1
Sidecar	3.7:1
Carburettor	Schebler 1-inch
Ignition	Model F: Magneto
	Model J: Battery-and-coil

Transmission
Gearbox	3-speed
Gearchange	Hand change, on tank
Final drive	Chain

Cycle parts
Frame construction	Single-loop, straight downtube
Front fork	Springer, enclosed springs, tubular legs
Rear suspension	None
Front brake	None
Rear brake	Drum, via rod and pedal

General information
Unladen weight	160 kg
Front wheel/tyre	3 x 28 beaded-edge
Rear wheel/tyre	3 x 28 beaded-edge
Maximum speed	105 km/h
Tank capacity	10.4 litres

Sidecars available

Chassis:
Model 21LC
R/H single-seater
Model 21Q double-adult
Passenger-carrying sidecars:
Model 21L
R/H single-seater
Model L21L L/H single-seater
Model 21LR
Roadster R/H
Model 21LX
Speedster R/H
Load-carrying sidecars:
Model 21M
Commercial standard
Model 21QA
Commercial closed
Model 21QB
Commercial open
Model 21QD
Parcel Car (2 box)
Model 21QL
R/H double-adult

Models 21FD and 21FDS
Models 21JD and 21JDS 1921

Engine
Engine	Air-cooled 45° ioe V-twin
Bore x stroke	3⁷⁄₁₆ x 4 (3.44 x 4) in.
Bore x stroke	87.31 x 101.6 mm
Capacity	74.24 cu. in. (1217 cc)
Power output (SAE)	Solo motorcycle: 19(15)/3200 rpm
	Sidecar: 18(13)/3000 rpm
Compression ratio:	
Solo motorcycle	3.8:1
Sidecar	3.7:1
Carburettor	Schebler 1¼-inch
Ignition	Model FD: Magneto
	Model JD: Battery-and-coil

Transmission
Gearbox	3-speed
Gearchange	Hand change, on tank
Final drive	Chain

Cycle parts
Frame construction	Single-loop, straight downtube
Front fork	Springer, enclosed springs, tubular legs
Rear suspension	None
Front brake	None
Rear brake	Drum, via rod and pedal

General information
Unladen weight	170 kg
Front wheel/tyre	3 x 28 beaded-edge
Rear wheel/tyre	3 x 28 beaded-edge
Maximum speed	113 km/h
Tank capacity	10.4 litres

Sidecars available

Chassis:
Model 21LC
R/H single-seater
Model 21Q double-adult

Passenger-carrying sidecars:
Model 21L
R/H single-seater
Model L21L L/H single-seater
Model 21LR
Roadster R/H
Model 21LX
Speedster R/H

Load-carrying sidecars:
Model 21M
Commercial standard
Model 21QA
Commercial closed
Model 21QB
Commercial open
Model 21QD
Parcel Car (2 Box)
Model 21QL
R/H double-adult

Models 22F and 22FS
Models 22J and 22JS 1922

Engine
Engine	Air-cooled 45° ioe V-twin
Bore x stroke	3 5/16 x 3 1/2 (3.31 x 3.5) in.
Bore x stroke	84.14 x 88.9 mm
Capacity	60.33 cu. in. (989 cc)
Power output (SAE)	Solo motorcycle: 16(12)/3000 rpm Sidecar: 15(11)/3000 rpm
Compression ratio:	
Solo motorcycle	3.7:1
Sidecar	3.5:1
Carburettor	Schebler 1-inch
Ignition	Model F: Magneto Model J: Battery-and-coil

Transmission
Gearbox	3-speed
Gearchange	Hand change, on tank
Final drive	Chain

Cycle parts
Frame construction	Single-loop, straight downtube
Front fork	Springer, enclosed springs, tubular legs
Rear suspension	None
Front brake	None
Rear brake	Drum, via rod and pedal

General information
Unladen weight	160 kg
Front wheel/tyre	3 x 28 beaded-edge
Rear wheel/tyre	3 x 28 beaded-edge
Maximum speed	105 km/h
Tank capacity	10.4 litres

Sidecars available

Chassis:
Model 21LC
Model 22LC
R/H single-seater
Model L22LC L/H single-seater
Model 22Q double-adult

Passenger-carrying sidecars:
Model 22L
R/H single-seater
Model L22L L/H single-seater
Model 22LR
Roadster R/H
Model L22LR
Roadster L/H
Model 22LT Tourist R/H
Model L22LT
Tourist L/H
Model 22LX
Speedster R/H

Load-carrying sidecars:
Model 22M
Commercial standard
Model 22QA
Commercial closed
Model 22QB
Commercial open
Model 22QD
Parcel Car (2 Box)
Model 22QL
R/H double-adult
Model 22QT
R/H double-adult Tourist

Models 22FD and 22FDS
Models 22JD and 22JDS 1922

Engine
Engine	Air-cooled 45° ioe V-twin
Bore x stroke	3⁷⁄₁₆ x 4 (3.44 x 4) in.
Bore x stroke	87.31 x 101.6 mm
Capacity	74.24 cu. in. (1217 cc)
Power output (SAE)	Solo motorcycle: 19(15)/3200 rpm
	Sidecar: 18(13)/3000 rpm
Compression ratio:	
Solo motorcycle	3.7:1
Sidecar	3.5:1
Carburettor	Schebler 1¼-inch
Ignition	Model FD: Magneto
	Model JD: Battery-and-coil

Transmission
Gearbox	3-speed
Gearchange	Hand change, on tank
Final drive	Chain

Cycle parts
Frame construction	Single-loop, straight downtube
Front fork	Springer, enclosed springs, tubular legs
Rear suspension	None
Front brake	None
Rear brake	Drum, via rod and pedal

General information
Unladen weight	170 kg
Front wheel/tyre	3 x 28 beaded-edge
Rear wheel/tyre	3 x 28 beaded-edge
Maximum speed	113 km/h
Tank capacity	10.4 litres

Sidecars available

Chassis:
Model 22LC
R/H single-seater
Model L22LC L/H single-seater
Model 22Q double-adult

Passenger-carrying sidecars:
Model 22L
R/H single-seater
Model L22L L/H single-seater
Model 22LR
Roadster R/H
Model L22LR
Roadster L/H
Model 22LT Tourist R/H
Model L22LT Tourist L/H
Model 22LX
Speedster R/H

Load-carrying sidecars:
Model 22M
Commercial standard
Model 22QA
Commercial closed
Model 22QB
Commercial open
Model 22QD
Parcel Car (2 Box)
Model 22QL
R/H double-adult
Model 22QT
R/H double-adult Tourist

Models 23F and 23FS
Models 23J and 23JS 1923

Engine
Engine	Air-cooled 45° ioe V-twin
Bore x stroke	3⁵⁄₁₆ x 3½ (3.31 x 3.5) in.
Bore x stroke	84.14 x 88.9 mm
Capacity	60.33 cu. in. (989 cc)
Power output (SAE)	Solo motorcycle: 16(12)/3000 rpm
	Sidecar: 15(11)/3000 rpm
Compression ratio:	
Solo motorcycle	3.7:1
Sidecar	3.5:1
Carburettor	Schebler 1-inch
Ignition	Model F: Magneto
	Model J: Battery-and-coil

Transmission
Gearbox	3-speed
Gearchange	Hand change, on tank
Final drive	Chain

Cycle parts
Frame construction	Single-loop, straight downtube
Front fork	Springer, enclosed springs, tubular legs
Rear suspension	None
Front brake	None
Rear brake	Drum, via rod and pedal

General information
Unladen weight	160 kg
Front wheel/tyre	3 x 28 beaded-edge
Rear wheel/tyre	3 x 28 beaded-edge
Maximum speed	105 km/h
Tank capacity	10.4 litres
Sidecars available	**Chassis:**
	Model 23LC
	R/H single-seater
	Model 23Q
	R/H double-adult
	Model L23LQ L/H single-seater
	Passenger-carrying sidecars:
	Model L23LC sidecar L/H
	Model 23LT Tourist R/H
	Model L23LT Tourist L/H
	Model 23LX Speedster R/H
	Load-carrying sidecars:
	Model 23M Commercial standard
	Model 23QT R/H double-adult Tourist
	Model L23LQT L/H double-adult

Models 23FD and 23FDS
Models 23JD and 23JDS 1923

Engine
Engine	Air-cooled 45° ioe V-twin
Bore x stroke	3⁷⁄₁₆ x 4 (3.44 x 4) in.
Bore x stroke	87.31 x 101.6 mm
Capacity	74.24 cu. in. (1217 cc)
Power output (SAE)	Solo motorcycle: 19(14)/3200 rpm
	Sidecar: 18(13)/3000 rpm
Compression ratio:	
Solo motorcycle	3.7:1
Sidecar	3.5:1
Carburettor	Schebler 1¼-inch
Ignition	Model FD: Magneto
	Model JD: Battery-and-coil

Transmission
Gearbox	3-speed
Gearchange	Hand change, on tank
Final drive	Chain

Cycle parts
Frame construction	Single-loop, straight downtube
Front fork	Springer, enclosed springs, tubular legs
Rear suspension	None
Front brake	None
Rear brake	Drum, via rod and pedal

General information
Unladen weight	170 kg
Front wheel/tyre	3 x 28 beaded-edge
Rear wheel/tyre	3 x 28 beaded-edge
Maximum speed	113 km/h
Tank capacity	10.4 litres
Sidecars available	**Chassis:**
	Model 23LC
	R/H single-seater
	Model 23Q
	R/H double-adult
	Model L23LQ L/H single-seater
	Passenger-carrying sidecars:
	Model L23LC sidecar L/H
	Model 23LT Tourist R/H
	Model L23LT Tourist L/H
	Model 23LX Speedster R/H
	Load-carrying sidecars:
	Model 23M Commercial standard
	Model 23QT R/H double-adult Tourist
	Model L23LQT L/H double-adult

The Models 23FDCA and 23JDCA were available in small numbers in 1923. For their technical details, refer to those given for the 24FDCA and 24JDCA.

Models 24FE and 24FES
Models 24JE and 24JES 1924

Engine
Engine	Air-cooled 45° ioe V-twin
Bore x stroke	3⁵⁄₁₆ x 3½ (3.31 x 3.5) in.
Bore x stroke	84.14 x 88.9 mm
Capacity	60.33 cu. in. (989 cc)
Power output (SAE)	FE and JE: 18(13)/3600 rpm FES and JES: 17(13)/3200 rpm
Compression ratio:	
Solo motorcycle	3.7:1
Sidecar	3.5:1
Carburettor	Schebler Std./Deluxe 1-inch optional Zenith 1-inch
Ignition	Model FE: Magneto Model JE: Battery-and-coil

Transmission
Gearbox	3-speed
Gearchange	Hand change, on tank
Final drive	Chain

Cycle parts
Frame construction	Single-loop, straight downtube
Front fork	Springer, enclosed springs, tubular legs
Rear suspension	None
Front brake	None
Rear brake	Drum, via rod and pedal

General information
Unladen weight	170 kg
Front wheel/tyre	3 x 28 beaded-edge
Rear wheel/tyre	3 x 28 beaded-edge
Maximum speed	FE and JE: 108 km/h FES and JES: 95 km/h
Tank capacity	10.4 litres
Sidecars available	**Passenger-carrying sidecars:** Model 24LT Tourist R/H Model L24LT Royal Tour L/H Model 24LX Racer R/H Model L24LX Racer L/H **Load-carrying sidecars:** Model 24M Parcel Car Model 24QT R/H double-adult Family Delight Model L24QT L/H single-seater

Models 24FD and 24FDS
Models 24JD and 24JDS
Models 24FDCA and 24FDSCA
Models 24JDCA and 24JDSCA 1924

Engine
Engine	Air-cooled 45° ioe V-twin
Bore x stroke	3⁷⁄₁₆ x 4 (3.44 x 4) in.
Bore x stroke	87.31 x 101.6 mm
Capacity	74.24 cu. in. (1217 cc)
Power output (SAE)	FD and JD: 20(15)/3400 rpm FDCA and JDCA: 22(16)/3600 rpm FDS and JDS: 19(15)/3000 rpm FDSCA and JDSCA: 21(15)/3200 rpm
Compression ratio:	
Solo motorcycle	3.7:1
Sidecar	3.5:1
Carburettor	Schebler Std./Deluxe 1¼-inch optional Zenith 1½-inch
Ignition	Model FD: Magneto Model JD: Battery-and-coil

Transmission
Gearbox	3-speed
Gearchange	Hand change, on tank
Final drive	Chain

Cycle parts
Frame construction	Single-loop, straight downtube
Front fork	Springer, enclosed springs, tubular legs
Rear suspension	None
Front brake	None
Rear brake	Drum, via rod and pedal

General information
Unladen weight	175 kg
Front wheel/tyre	3 x 28 beaded-edge
Rear wheel/tyre	3 x 28 beaded-edge
Maximum speed	113 km/h
Tank capacity	10.4 litres
Sidecars available	**Passenger-carrying sidecars:** Model 24LT Tourist R/H Model L24LT Royal Tour L/H Model 24LX Racer R/H Model L24LX Racer L/H **Load-carrying sidecars:** Model 24M Parcel Car Model 24QT R/H double-adult Family Delight Model L24QT L/H double-adult

Models 25FE and 25FES
Models 25JE and 25JES 1925
(also Commercial)

Engine
Engine	Air-cooled 45° ioe V-twin
Bore x stroke	$3\frac{5}{16}$ x $3\frac{1}{2}$ (3.31 x 3.5) in.
Bore x stroke	84.14 x 88.9 mm
Capacity	60.33 cu. in. (989 cc)
Power output (SAE)	FE and JE: 18(13)/3600 rpm FES and JES: 17(13)/3200 rpm
Compression ratio:	
Solo motorcycle	3.7:1
Sidecar	3.5:1
Carburettor	Schebler Std./Deluxe 1-inch optional Zenith 1-inch
Ignition	Model FE: Magneto Model JE: Battery-and-coil

Transmission
Gearbox	3-speed
Gearchange	Hand change, on tank
Final drive	Chain

Cycle parts
Frame construction	Single-loop, straight downtube
Front fork	Springer, enclosed springs, tubular legs
Rear suspension	None
Front brake	None
Rear brake	Drum, via rod and pedal

General information
Unladen weight	180 kg
Front wheel/tyre	27 x $3\frac{1}{2}$ beaded-edge
Rear wheel/tyre	27 x $3\frac{1}{2}$ beaded-edge
Maximum speed	FE and JE: 108 km/h FES and JES: 95 km/h
Tank capacity	20 litres
Sidecars available	**Chassis:** Model 25MA Parcel Car Chassis **Passenger-carrying sidecars:** Model 25LT Tourist R/H Model L25LT Royal Tourist L/H Model 25LX Racer R/H Model L25LX Racer L/H **Load-carrying sidecars:** Model 25QT R/H Family Delight double-adult Model L25QT L/H single-seater

Models 25FDCB and 25FDCBS
Models 25JDCB and 25JDCBS 1925
(also Commercial)

Engine
Engine	Air-cooled 45° ioe V-twin
Bore x stroke	$3\frac{7}{16}$ x 4 (3.44 x 4) in.
Bore x stroke	87.31 x 101.6 mm
Capacity	74.24 cu. in. (1217 cc)
Power output (SAE)	FDCB and JDCB: 22(16)/3600 rpm FDCBS and JDCBS: 20(15)/3200 rpm
Compression ratio	
FDCB and JDCB	3.7:1
FDCBS and JDCBS	3.5:1
Carburettor	Schebler Std./Deluxe $1\frac{1}{4}$-inch optional Zenith $1\frac{1}{2}$-inch
Ignition	Model FE: Magneto Model: JE: Battery-and-coil

Transmission
Gearbox	3-speed
Gearchange	Hand change, on tank
Final drive	Chain

Cycle parts
Frame construction	Single-loop, straight downtube
Front fork	Springer, enclosed springs, tubular legs
Rear suspension	None
Front brake	None
Rear brake	Drum, via rod and pedal

General information
Unladen weight	180 kg
Front wheel/tyre	27 x $3\frac{1}{2}$ beaded-edge
Rear wheel/tyre	27 x $3\frac{1}{2}$ beaded-edge
Maximum speed	FDCB and JDCB: 113 km/h FDCBS and JDCBS: 107 km/h
Tank capacity	20 litres
Sidecars available	Model 25MA Parcel Car Chassis **Passenger-carrying sidecars:** Model 25LT Tourist R/H Model L25LT Royal Tourist L/H Model 25LX Racer R/H Model L25LX Racer L/H **Load-carrying sidecars:** Model 25QT R/H Family Delight double-adult Model L25QT L/H single-seater

Models F and FS
Models J and JS
Model J Commercial 1926 to 1929

Engine
Engine	Air-cooled 45° ioe V-twin
Bore x stroke	3 5/16 x 3 1/2 (3.31 x 3.5) in.
Bore x stroke	84.14 x 88.9 mm
Capacity	60.33 cu. in. (989 cc)
Power output (SAE)	J and F: 18(13)/3400 rpm
	JS and FS: 17(13)/3200 rpm
	JC: 15(11)/3000 rpm
Compression ratio:	
F and J	3.7:1
FDCBS and JDCBS	3.5:1
JS, FS and J Commercial	3.5:1
Carburettor	Schebler Std./Deluxe 1-inch optional Zenith 1-inch (only 1926)
Ignition	Model F: Magneto
	Model J: Battery-and-coil

Transmission
Gearbox	3-speed
Gearchange	Hand change, on tank
Final drive	Chain

Cycle parts
Frame construction	Single-loop, straight downtube
Front fork	Springer, enclosed springs, tubular legs
Rear suspension	None
Front brake:	
Up to 1928	None
From 1928	Drum, via cable and handlebar lever
Rear brake	Drum, via rod and pedal

General information
Unladen weight	180 kg
Front and rear tyres	1926-29: 27 x 3 1/2 beaded-edge
	optional 3.85 x 27 (wired-edge)
	1927-29 also 3.85 x 25 (wired-edge)
	1928 also 3.30 x 26 (wired-edge)
	1929 also 3.00 x 26 (wired-edge)
Maximum speed	108 km/h (J and F)
	100 km/h (JS and FS)
	95 km/h (JC)
Tank capacity	20 litres

Models FD and FDS
Models JD and JDS
Model JD Commercial 1926 to 1929
Model JDF 1929

Engine
Engine	Air-cooled 45° ioe V-twin
Bore x stroke	3 7/16 x 4 (3.44 x 4) in.
Bore x stroke	87.31 x 101.6 mm
Capacity	74.24 cu. in. (1217 cc)
Power output (SAE)	JD and FD: 22(16)/3400 rpm
	JDF: 22(16)/3400 rpm
	JDS and FDS: 21(15)/3200 rpm
	JDC: 17(13)/3000 rpm
Compression ratio:	
FD, JD and JDF	3.8:1
JDC and JDS	3.7:1
Carburettor	Schebler Std./Deluxe 1 1/2-inch
	optional Zenith 1 1/4-inch (only 1926)
Ignition	Model F: Magneto
	Model J: Battery-and-coil

Transmission
Gearbox	3-speed
Gearchange	Hand change, on tank
Final drive	Chain

Cycle parts
Frame construction	Single-loop, straight downtube
Front fork	Springer, enclosed springs, tubular legs
Rear suspension	None
Front brake:	
Up to 1928	None
From 1928	Drum, via cable and handlebar lever
Rear brake	Drum, via rod and pedal

General information
Unladen weight	180 kg
Front and rear tyres	1926-29: 27 x 3 1/2 beaded-edge
	optional 3.85 x 27 (wired-edge)
	1927-29 also 3.85 x 25 (wired-edge)
	1928 also 3.30 x 26 (wired-edge)
	1929 also 3.00 x 26 (wired-edge)
Maximum speed	113 km/h (FD, JD and JDF)
	107 km/h (FDS and JDS)
	95 km/h (JDC)
Tank capacity	20 litres

Sidecar details are given at the end of the Chapter.

Model JK Special 1927
Model FK Special 1927

Engine
Engine	Air-cooled 45° ioe V-twin
Bore x stroke	3 5/16 x 3 1/2 (3.31 x 3.5) in.
Bore x stroke	84.14 x 88.9 mm
Capacity	60.33 cu. in. (989 cc)
Power output (SAE)	20(15)/3800 rpm
Compression ratio	4.2:1
Carburettor	Schebler Std./Deluxe 1-inch
Ignition	Model FK: Magneto Model JK: Battery-and-coil

Transmission
Gearbox	3-speed
Gearchange	Hand change, on tank
Final drive	Chain

Cycle parts
Frame construction	Single-loop, straight downtube
Front fork	Springer, enclosed springs, tubular legs
Rear suspension	None
Front brake	None
Rear brake	Drum, via rod and pedal

General information
Unladen weight	185 kg
Front and rear tyres	1927: 27 x 3 1/2 beaded-edge optional 3.85 x 27 (wired-edge) optional 3.85 x 25 (wired-edge) optional 3.00 x 26 (wired-edge)
Maximum speed	120 km/h
Tank capacity	20 litres

Sidecar details are given at the end of the Chapter.

Model FDL Special 1927
Model JDL Special 1927

Engine
Engine	Air-cooled 45° ioe V-twin
Bore x stroke	3 7/16 x 4 (3.44 x 4) in.
Bore x stroke	87.31 x 101.6 mm
Capacity	74.24 cu. in. (1217 cc)
Power output (SAE)	24(18)/3800 rpm
Compression ratio	4.2:1
Carburettor	Schebler Std./Deluxe 1 1/4-inch
Ignition	Model FDL: Magneto Model JDL: Battery-and-coil

Transmission
Gearbox	3-speed
Gearchange	Hand change, on tank
Final drive	Chain

Cycle parts
Frame construction	Single-loop, straight downtube
Front fork	Springer, enclosed springs, tubular legs
Rear suspension	None
Front brake	None,
Rear brake	Drum, via rod and pedal

General information
Unladen weight	185 kg
Front and rear tyres	1927/29: 27 x 3 1/2 beaded-edge optional 3.85 x 27 (wired-edge) optional 3.85 x 25 (wired-edge) optional 3.00 x 26 (wired-edge)
Maximum speed	133 km/h
Tank capacity	20 litres

Sidecar details are given at the end of the Chapter.

Model FH Two Cam 1928
Model JH Two Cam 1928 to 1929

Engine
Engine	Air-cooled 45° ioe V-twin
Bore x stroke	3⁵⁄₁₆ x 3½ (3.31 x 3.5) in.
Bore x stroke	84.14 x 88.9 mm
Capacity	60.33 cu. in. (989 cc)
Power output (SAE)	23(17)/4000 rpm
Compression ratio	4.6:1
Carburettor	Schebler Std./Deluxe 1-inch
Ignition	Model FH: Magneto Model JH: Battery-and-coil

Transmission
Gearbox	3-speed
Gearchange	Hand change, on tank
Final drive	Chain

Cycle parts
Frame construction	Single-loop, straight downtube
Front fork	Springer, enclosed springs, tubular legs
Rear suspension	None
Front brake	Drum, via cable and handlebar lever
Rear brake	Drum, via rod and pedal

General information
Unladen weight	185 kg
Front and rear tyres	3.85 x 25 (wired-edge)
Maximum speed	125 km/h
Tank capacity	18 litres

Sidecar details are given at the end of the Chapter.

Model FDH Two Cam 1928 to 1929
Model JDH Two Cam 1928 to 1929

Engine
Engine	Air-cooled 45° ioe V-twin
Bore x stroke	3⁷⁄₁₆ x 4 (3.44 x 4) in.
Bore x stroke	87.31 x 101.6 mm
Capacity	74.24 cu. in. (1217 cc)
Power output (SAE)	29(21)/4000 rpm
Compression ratio	4.6:1
Carburettor	Schebler Std./Deluxe 1½-inch
Ignition	Model FDH: Magneto Model JDH: Battery-and-coil

Transmission
Gearbox	3-speed
Gearchange	Hand change, on tank
Final drive	Chain

Cycle parts
Frame construction	Single-loop, straight downtube
Front fork	Springer, enclosed springs, tubular legs
Rear suspension	None
Front brake	Drum, via cable and handlebar lever
Rear brake	Drum, via rod and pedal

General information
Unladen weight	185 kg
Front and rear tyres	3.85 x 25 (wired-edge)
Maximum speed	137 km/h
Tank capacity	20 litres

Sidecar details are given at the end of the Chapter.

Model JX Sport 1928
Model JXL Special Sport 1928 to 1929

Engine
Engine	Air-cooled 45° ioe V-twin
Bore x stroke	3 5/16 x 3 1/2 (3.31 x 3.5) in.
Bore x stroke	84.14 x 88.9 mm
Capacity	60.33 cu. in. (989 cc)
Power output (SAE)	JX: 20(15)/3600 rpm
	JXL: 22(16)/3800 rpm
Compression ratio	4.2:1
Carburettor	Schebler Std./Deluxe 1-inch
Ignition	Battery-and-coil

Transmission
Gearbox	3-speed
Gearchange	Hand change, on tank
Final drive	Chain

Cycle parts
Frame construction	Single-loop, straight downtube
Front fork	Springer, enclosed springs, tubular legs
Rear suspension	None
Front brake	Drum, via cable and handlebar lever
Rear brake	Drum, via rod and pedal

General information
Unladen weight	185 kg
Front and rear tyres	3.85 x 25 (wired-edge)
Maximum speed	JX: 115 km/h
	JXL: 120 km/h
Tank capacity	20 litres

Sidecar details are given at the end of the Chapter.

Model JDX Sport 1928
Model JDXL Special Sport 1928 to 1929

Engine
Engine	Air-cooled 45° ioe V-twin
Bore x stroke	3 7/16 x 4 (3.44 x 4) in.
Bore x stroke	87.31 x 101.6 mm
Capacity	74.24 cu. in. (1217 cc)
Power output (SAE)	JDX: 24(18)/3600 rpm
	JDXL: 26(19)/3800 rpm
Compression ratio	4.2:1
Carburettor	Schebler Std./Deluxe 1¼-inch
Ignition	Battery-and-coil

Transmission
Gearbox	3-speed
Gearchange	Hand change, on tank
Final drive	Chain

Cycle parts
Frame construction	Single-loop, straight downtube
Front fork	Springer, enclosed springs, tubular legs
Rear suspension	None
Front brake	Drum, via cable and handlebar lever
Rear brake	Drum, via rod and pedal

General information
Unladen weight	185 kg
Front and rear tyres	3.85 x 25 (wired-edge)
Maximum speed	JDX: 128 km/h
	JDXL: 133 km/h
Tank capacity	20 litres

Sidecar details are given at the end of the Chapter.

Model FDL 1929

Engine
Engine	Air-cooled 45° ioe V-twin
Bore x stroke	3⁷⁄₁₆ x 4 (3.44 x 4) in.
Bore x stroke	87.31 x 101.6 mm
Capacity	74.24 cu. in. (1217 cc)
Power output (SAE)	24(18)/3600 rpm
Compression ratio	4.2:1
Carburettor	Schebler Std./Deluxe 1¼-inch
Ignition	Magneto

Transmission
Gearbox	3-speed
Gearchange	Hand change, on tank
Final drive	Chain

Cycle parts
Frame construction	Single-loop, straight downtube
Front fork	Springer, enclosed springs, tubular legs
Rear suspension	None
Front brake	Drum, via cable and handlebar lever
Rear brake	Drum, via rod and pedal

General information
Unladen weight	185 kg
Front and rear tyres	3.85 x 25 (wired-edge)
Maximum speed	128 km/h
Tank capacity	20 litres

Sidecar details are given at the end of the Chapter.

Motorcycle catalogue prices ($ US)

Model	1913	1914	1915	1916	1917	1918	1919	1920	1921	1922	1923
E *	350	250	240	240	255	275	-	-	-	-	-
F *	N./Av.	285	275	265	275	290	350	370	450	335	285
FS *	-	-	-	-	-	-	350	370	450	335	285
FD*	-	-	-	-	-	-	-	-	485	360	310
FDS *	-	-	-	-	-	-	-	-	485	360	310
H *	-	-	275	-	-	-	-	-	-	-	-
J *	-	-	310	295	310	320	370	395	485	365	305
JS *	-	-	-	-	-	-	370	395	485	365	305
JD *	-	-	-	-	-	-	-	-	520	390	330
JDS *	-	-	-	-	-	-	-	-	520	390	330

* Prices are for motorcycles only, less sidecars.

Motorcycle catalogue prices ($ US)

Model	1924	1925	1926	1927	1928	1929
F *	-	-	295	290	290	290
FS *	-	-	295	290	290	290
FE *	300	295	-	-	-	-
FES *	300	295	-	-	-	-
FK *	-	-	-	N./Av.	-	-
FD *	315	-	315	300	300	300
FDS *	315	-	315	300	300	300
FDCA *	325	-	-	-	-	-
FDSCA *	325	-	-	-	-	-
FDCB *	-	315	-	-	-	-
FDCBS *	-	315	-	-	-	-
FDL *	-	-	-	N./Av.	-	N./Av.
J *	-	-	315	310	310	310
JS *	-	-	315	310	310	310
JE *	320	315	-	-	-	-
JES *	320	315	-	-	-	-
JH *	-	-	-	-	360	340
JK *	-	-	-	N./Av.	-	-
JX *	-	-	-	-	320	-
JXL *	-	-	-	-	320	320
JD *	335	-	335	320	320	320
JDS *	335	-	335	320	320	320
JDCA *	345	-	-	-	-	-
JDSCA *	345	-	-	-	-	-
JDCB *	-	335	-	-	-	-
JDCBS *	-	335	-	-	-	-
JDH *	-	-	-	-	390	370
JDL *	-	-	-	-	N./Av.	-
JDF *	-	-	-	-	-	N./Av.
JDX *	-	-	-	-	320	-
JDXL *	-	-	-	-	335	320

* Prices are for motorcycles only, less sidecars.

Sidecar catalogue prices ($ US)

Model	1914	1915	1916	1917	1918	1919
L	85	85	75	80	90	110
M	-	-	65	70	-	-
N	-	-	-	72	-	-

It was not possible to discover all sidecar prices, since the literature gives hardly any details of prices actually charged between 1920 and 1933.

Sidecars available in the years 1926 to 1929
(right-hand fittings only provided)

1926	1927	1928	1929
Chassis:			
Model 26LC Chassis	Model 27LC Chassis	Model 28LC Chassis	Model 29LC Chassis
Model 26MC Chassis	Model 27MC Chassis	Model 28MC Chassis	Model 29MC Chassis
Model 26Q Chassis	Model 27Q Chassis	Model 28MWC Chassis	Model 29MWC Chassis
		Model 28Q Chassis	Model 29Q Chassis
Passenger-carrying sidecars:			
Model 26LT Tourist	Model 27LT Tourist	Model 28LT Tourist	Model 29LT Tourist
Model 26LX Speedster	Model 27LX Speedster	Model 28LX Speedster	Model 29LX Speedster
Model 26QT Double-adult	Model 27QT Double-adult	Model 28QT Double-adult	Model 29QT Double-adult
Load-carrying sidecars:			
Model 26M	Model 27M	Model 28M	Model 29M
Model 26MO	Model 27MO	Model 28MO	Model 29MO
		Model 28MW	Model 29MDC
		Model 28MWX	Model 29MNP
		Model 28MWXP	Model 29MW
			Model 29MWP
		Model 28MX	Model 29MX
		Model 28MXP	Model 29MXP

MDC	Dryer's and Cleaner's Side Van with 49-inch track
M	Commercial, closed box body
MO	Commercial, open box body
MNP	Newspaper Side Van with 49-inch track
MW	Commercial Side Van with 56-inch track
MWC	Chassis for MW
MWP	Side Van with Panels with 56-inch track
MWX	Commercial Express Van with windscreen, with 56-inch track
MWXP	Commercial Express Panel Van (without windscreen) with 56-inch track
MX	Commercial Express Van with windscreen
MXP	Commercial Express Panel Van (without windscreen)

Production figures

Model	1913	1914	1915	1916	1917	1918	1919	1920	1921	1922	1923
E	6732	5055	1275	252	68	5	-	-	-	-	-
F	49	7956	9855	9496	8527	11764	5064	7579	2413	1824	2822
FD	-	-	-	-	-	-	-	-	277	909	869
H	-	-	140	-	-	-	-	-	-	-	-
J	-	-	3719	5898	9180	6571	9941	14192	4526	3183	4802
JD	-	-	-	-	-	-	-	-	2321	3988	7458

Production figures

Model	1924	1925	1926	1927	1928	1929
F	-	-	760	246	141	191
FD	502	433	232	209	131	73
FDCA	351	-	-	-	-	-
FDCB	90	-	-	-	-	-
FE	2708	1318	-	-	-	-
J	-	-	3749	3561	4184	2886
JD	2955	9506	9544	9691	11007	10182
JDCA	3014	-	-	-	-	-
JDCB	*200	-	-	-	-	-
JE	4993	4114	-	-	-	-

* Accurate figures not available.

The Model G forecars

1913 to 1916

Model description and technical information

For the Model G, a completely new vehicle that was conceived for small business users, the normal V-twin frame was adapted in 1913, being fitted with an outrigger on each side under the steering head to permit the fitting of two front wheels. Thus the vehicle was steered with normal motorcycle-type handlebars, which probably must have made for an interesting ride. The resulting forecar was also supplied with skis on the front and a spiked tyre on the rear (driven) wheel for use in snow. The engine and transmission was that of the Model 9F, unaltered.

Although it was still equipped with the two-speed rear hub-mounted gearbox, for its second production year Harley treated the forecar to the rod-operated rear brake from which the other motorcycles also profited, and for 1915 the new three-speed sliding-mesh gearbox was fitted to the forecar, as well as to the single-track V-twins of the same year.

In the 1916 model year the forecar was allowed to die a natural death, as only a whole seven machines found buyers. It is reasonably certain that some of those were left-over models from the previous year. It would be 16 years before Harley-Davidson brought a comparable vehicle to the market.

Model range		
Model G Forecar Delivery Van		
Model 9G	60 cu .in./989 cc. ioe V-twin with magneto ignition	for the USA only (1913)
Model 10G	60 cu. in./989 cc. ioe V-twin with magneto ignition	for the USA only (1914)
Model 11G	60 cu. in./989 cc. ioe V-twin with magneto ignition	for export (1915-16)

1914: The first Harley-Davidson three-wheeler

Model 9G 1913

Engine
Engine	Air-cooled 45° ioe V-twin
Bore x stroke	3⁵⁄₁₆ x 3½ (3.31 x 3.5) in.
Bore x stroke	84.14 x 88.9 mm
Capacity	60.33 cu. in. (989 cc)
Power output (SAE)	8(6)/2200 rpm
Carburettor	Schebler 1-inch
Ignition	Magneto

Transmission
Gearbox	None, optional 2-speed (rear hub-mounted)
Gearchange	Hand change, on tank
Final drive	Chain

Cycle parts
Frame construction	Forecar front end built on to standard motorcycle frame
Front fork	Forecar with leaf springs
Rear suspension	None
Front brake	None
Rear brake	Back-pedal

General information
Unladen weight	approx. 200 kg
Front wheel/tyre	2¾ x 28
Rear wheel/tyre	2¾ x 28
Maximum speed	approx. 65 km/h
Tank capacity	9.1 litres
Catalogue price	Model 9G $425
Production figures	Model 9G 63

Model 10G 1914
Model 11G 1915 to 1916

Engine
Engine	Air-cooled 45° ioe V-twin
Bore x stroke	3⁵⁄₁₆ x 3½ (3.31 x 3.5) in.
Bore x stroke	84.14 x 88.9 mm
Capacity	60.33 cu. in. (989 cc)
Power output (SAE)	8(6)/2200 rpm
Carburettor	Schebler 1-inch
Ignition	Magneto

Transmission
Gearbox	2-speed (rear hub-mounted, 1914) 3-speed (from 1915)
Gearchange	Hand change, on tank
Final drive	Chain

Cycle parts
Frame construction	Forecar front end built on to standard motorcycle frame
Front fork	Forecar with leaf springs
Rear suspension	None
Front brake	None
Rear brake	Drum, via rod and pedal

General information
Unladen weight	approx. 200 kg
Front wheel/tyre	2¾ x 28
Rear wheel/tyre	2¾ x 28
Maximum speed	approx. 65 km/h
Tank capacity	9.1 litres
Catalogue price	Model 10G $425
	Model 11G $450
Production figures	Model 10G 171 (1914)
	Model 11G 98 (1915)
	Model 11G 7 (1916)

1914: Forecar

The VCR outfit

1931

Model description and technical information

In mid-1930 Harley-Davidson began to develop a device for road-marking. To a normal motorcycle was attached not a sidecar with a conventional body, but instead a sidecar fitted out with twin tanks and associated hoses feeding the paint sprayer. Unfortunately, the sidecar being fitted on the conventional, right-hand side, of the motorcycle, painting the dividing line in the exact centre of the carriageway required the rider always to place himself squarely in the face of the oncoming traffic.

On the other hand, demarcation markings along the edge of the road were already the norm throughout much of the USA by the 1930s.

What's more, the production VCR 'Roadmarking' machine had a Harley industrial engine mounted on its right-hand side to drive the compressor.

The VCR was coupled to a standard Model VC motorcycle. For technical information, refer to the Chapter on the Model Vs; the weight of this outfit is not known and even less is known about its price and the numbers produced.

1931: The sort of thing one will do in dire need. The VCR as road-marking outfit

Servi-Car.
The Model G range
1932 to 1973

Model description and technical information

At the same time as the 'Roadmarking' machine of 1930, H-D developed another curiosity, the 'Cycle-Tow'. This was a contraption similar to the training wheels on a child's first bicycle, in which two outriggers carrying wheels were bolted on to each side of the rear frame of a normal solo motorcycle. If the rider was an experienced motorcyclist, they could be swung up clear of the ground; if not, they could be locked down to hold the machine upright. The idea was for a garage to be able to send out one man (rather than two), not necessarily a motorcyclist, to collect a customer's car and to return with it to the garage, towing the motorcycle (outrigger wheels locked down and hitched to the car's rear bumper) behind it. When work on the car was completed, the process was reversed, the whole being theoretically much cheaper and more profitable for the garage than sending out two men in a full-sized car or light van.

Since the concept was thought to lack sales potential, the idea was shelved for a while. However Indian's success with the Dispatch Tow model (based around the same basic concept) provoked second thoughts and the scheme was resurrected, but now in the form of a distinctive three-wheeled vehicle with the front of a motorcycle and the rear, based on a conventional automotive rear axle, of a car or cargo-carrying utility vehicle, depending on bodywork. The rear axle had a track of 42 inches (1,100 mm), similar to the track of most contemporary cars, deliberately done so that an inexperienced rider could use the same tracks that had been made by cars when venturing out in heavy snow.

This novel vehicle, appropriately dubbed the Servi-Car (more formally, Model G), arrived on the market at the end of the 1932 model year and rapidly found favour, most noticeably with police forces and local

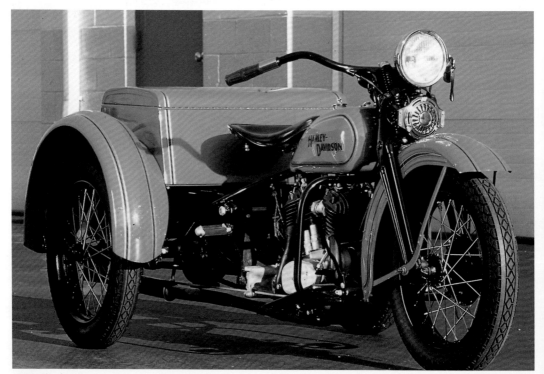

1932: The first Servi-Car, pretty in Turquoise Blue

government. Standard paint finish was Turquoise Blue, reflecting Harley's imminent adoption of brighter and more modern paint schemes rather than their traditional dull monotone greens.

Dedicated engines were not necessary; the small 45 cu. in. (742 cc.) side-valve V-twin of the Model R range was simply used, which was probably the most suitable solution since that range, acting as godfather to the new models, had also provided the front frame. In any case, the Models R and G were identical from the front, right back to the seat tube and the Servi-Cars from 1932 to 1936 even continued to use the letter R in their engine numbers.

The Servi-Cars were supplied from the outset with or without a towbar; a massive steel construction mounted on the front of the vehicle to take the shocks implicit in the original Cycle-Tow concept that would enable garages safely to provide this early form of home delivery service.

The standard equipment included a handbrake, dubbed the Parking Brake Lock. Incidentally the front and rear wheels were not interchangeable as different hubs were required. The bodywork was a hefty sheet-metal construction and the engine had its work cut out moving the vehicle when fully loaded. As a result the speeds attainable were severely limited; probably little more than 50 mph (80 km/h). A single saddle was standard equipment ex-works, but for a moderate surcharge an additional seat could be fitted.

When the Model W range arrived in the world of motorcycling in 1937, their engines were likewise also used in the Servi-Cars (except that the Model Gs from now on can be identified by the use of this letter in the engine number). This engine would remain in service for a length of time unmatched by any Harley engine before or since. The increasingly-elderly unit was pressed into service, albeit only in moderate numbers, for all of 37 years. Even the Shovelhead 'only' lasted 29 years.

Many technical alterations were made to the Servi-Cars in parallel with those made to the Model W range. Thus they were also treated to the big (in terms of air volume) tyres, and on top of that got the low-profile 5.10-16 tyres fitted to all big Harleys in 1969. Changes in carburettor equipment and jetting carried out on the Big Twins were matched on the three-wheelers.

Bottom: A 1933 Servi-Car in its originally-envisaged role (note the towbar)
Top: A 1934 Servi-Car as most were used – by the police and local government

Model range

Models GA to GE Servi-Car

Model G	45 cu. in./742 cc. sv V-twin engine, with towbar	(1932-63)
Model GA	45 cu. in./742 cc. sv V-twin engine, without towbar	(1932-63)
Model GD	45 cu. in./742 cc. sv V-twin engine, without towbar, with big body	(1933-41)
Model GDT	45 cu. in./742 cc. sv V-twin engine, with towbar and big body	(1933-41)
Model GE	45 cu. in./742 cc. sv V-twin engine, with air tank and big body, bumper and spare wheel carrier	(1932-37)
Model GE	45 cu. in./742 cc. sv V-twin engine with electric starter, without towbar	(1964-69)
Model GE (series 5A)	45 cu. in./742 cc. sv V-twin engine with electric starter, without towbar	(1970-73)
Model GE (series 8E)	Prototype 74 cu. in./1207 cc. ohv V-twin engine with electric starter, without towbar	(1973)

Model development

1933 model year
Because of their late appearance on the domestic market, there was no official price list for the 1932 Servi-Cars; in fact, it wasn't until 1934 that they even featured on the factory order form, as the same happened to the 1933 models and the modest little Servi-Car got forgotten again. Late in the model year the Schebler carburettor was changed to a similar instrument from Linkert, the new supplier. There were, as with almost every year, new tank graphics; this year's were in the new Art Deco eagle design.

1934 model year
The bumper fitted as standard equipment on the new Model GE could be ordered as an accessory for all the other models, at a bargain price of six bucks. As with the others in the range, a steering damper and lighting for the speedometer were available at extra cost. Even the standard-equipment reverse gear had to be paid for; an extra $15. A windscreen was optional, on the same basis, as were crashbars and legshields.

Technical modifications were limited to the modified Straight-bore cylinders. Apart from that, new tank graphics – the flying diamond design – and light fittings were used.

1935 model year
An additional supporting member for the rear axle mountings as well as further smaller changes identified the new models. The silencer was likewise new (designed on the reflection principle, using baffles and/or changes in exhaust/silencer diameter to reduce sound waves). The reverse gear was now included in the catalogue price.

1936 model year
Even the Servi-Cars got to enjoy the new cylinder heads with deeper cooling fins and the new-shape combustion chamber, and the Y-shaped intake

1936: Servi-Car – with towbar

1935: Servi-Car

1949: Servi-Car

manifold. In addition, the fork springs got a cover, the new tank graphics were used, the steering head angle was modified and thicker brake shoes were fitted.

1937 model year
To conform to and reinforce Harley's new Corporate Identity, the styling was revised to fit with that of the Model E, launched the previous year. On top of (literally) the new (from the new Model W range) engine, with recirculating dry-sump lubrication system instead of the previous total-loss system, was the new two-part tank (left compartment for petrol, right for oil) whose fixtures and fittings were not just a technical improvement, but also a significant aesthetic enhancement. The policy of offering accessories in options Groups was extended for the Servi-Cars, as well as for the motorcycles. It was about this time that the factory began to put pressure on its dealers only to place orders for new machines with one of the options Groups; the 'standard' machine effectively no longer existed.

1938 model year
After the Model GE (with air tank) had disappeared from the catalogue, the air tank, spare wheel holder and bumper could be ordered for the Model GDT at a cost of 40 dollars. Furthermore, the clutch and gearbox were improved, being strengthened here and there and also using stronger materials on many components. Several smaller improvements contributed to the general upgrading of these models, while the new Burgess exhaust also represented an important innovation.

1939 model year
The technical improvements made to the W Models were also carried out to the benefit of the three-wheelers, including pistons, gearbox, and so on. Apart from the optional accessory Groups, the extras available were limited to the bumper, a passenger seat and the air tank.

1940 model year
While in the previous year, as with the W Models, the front piston had had three compression rings, as opposed to the two plus oil scraper ring of the rear, for the 1940 models, the engine's lubrication system having been modified to achieve more even oiling of both cylinder bores, both pistons were treated the same with two compression rings and the oil scraper.

The front fork got improved chrome-molybdenum-steel oval-section tubular rigid (sprung) legs instead of the forged I-beam legs previously used, D-shaped footboards were introduced, and chrome-plated teardrop style metal tank badges. 5.00-16 tyres, with substantially greater air volume than the old fitment, were made available as options; by the following year they would completely replace the old 4.00-18 tyres.

1941 model year
Even in the last year of peace (for the USA), several important new features were introduced. These included the larger 7-inch (178 mm) air cleaner, the constant-mesh gearbox, the slightly-altered frame (because of the gearbox) and the Airplane-Style speedometer. The front brake and hub were now, at long last, standardized with those of the Big Twins' 7¼ inch (184 mm) diameter unit. The rear axle, formerly riveted-up, was now of welded construction; welding was slowly becoming accepted as a means of frame-building instead of brazing. However, brazing remained in use, to varying degrees, right up to 1947.

1942 model year
The sole innovation of any significance was the introduction of shock absorbers for the front fork springs.

1943 to 1945 model years
The switch to wartime production necessitated the same changes as were made to the Model W range. In 1945 at least one prototype Servi-Car was built using the flat twin engine and shaft drive from the XA instead of the WLA powerplant.

1946 model year
The various chrome trim pieces, discontinued during the war years, gradually returned to the Servi-Car as well. Latex mesh replaced the horsehair as seat padding. New bumper-mounted fittings were provided for the brake lights. The accessories were limited to Fender Skirts (panels fitting in the rear wheel arches to provide an even greater surface area for the customer's advertising signwriting), the rear bumper and the passenger seat.

1947 model year
For 1947 the gearchange was changed round so that reverse was at the back, then first, neutral and finally second and third. The box was now a one-piece component pressed from better 18-gauge steel. The new models were further enhanced by the red Speed Ball tank badges, the Tombstone taillamp and the new speedometer fitted to all 1947 Harleys.

1948 model year
The yet-again-new speedometer with edge lighting also appeared on the three-wheelers. In addition, as with the solo machines, there was a Deluxe saddle as an option. The spark plug leads were now fabric-covered.

1949 model year
The headlamp shell was made of stainless steel from now on, as was the housing for the running light on the front mudguard. For 1949 only, the timer unit (contact breaker housing) cap and generator cover were also made from this material.

Servi-Cars from 1963...

1950 model year
Only minimal alterations, apart from the fitting of the deep-skirted Air Flow front mudguard.

1951 model year
The mechanically-operated rear brake was replaced for this year by a hydraulically-operated system. The front brake now became the parking brake. In addition, steel disc wheels replaced the former wire-spoked wheels at the rear. The exhaust silencer looked like a car's, but with a horizontally-aligned fishtail.

1952 model year
The sole alteration was that the rear bumper was moved some 1½ inches (40 mm) closer to the bodywork.

1953 model year
Only the speedometer was restyled for this year.

1954 model year
All Servi-Cars were fitted with the brass Harley-Davidson Golden Anniversary 50th birthday medallion on the front mudguard. The accessory list remained practically the same as that for 1946.

1955 model year
Again, there were only a very few new features such as the oval taillamp and the big V tank badge. On the technical front, both compression rings (for each piston) were now chrome-plated.

1956 model year
In this year a new air cleaner including housing and a new speedometer were fitted. The battery case was omitted, the battery now being freely accessible. The cylinder heads were again revised and prepared for the fitting of 14-mm spark plugs.

1957 model year
Modified tank badges only.

1958 model year
After some delay, H-D finally treated the Servi-Car to the hydraulically-damped telescopic forks of the big Hydra-Glide models, which had had them since 1949. The Hydra-Glide front wheel hub and brakes were also fitted for the sake of simplicity, as was the headlamp unit, which came from now on from the F models. Besides an improved generator there were also for this year, naturally, new tank badges.

1959 model year
A new fuel tap and lines, and a carburettor with an aluminium-alloy housing were, besides the tank badges (changed again), the only changes for the '59 model year.

1960 model year
For 1960 there was an improved windscreen, with Summer and Winter variants. Further detail changes were made to the handlebars and to the nacelle around the fork tubes and headlamp.

1961 model year
Apart from the perennial changes to the tank badging, no other alterations were made.

1962 model year
No changes to the tank badging; but the Tombstone style speedometer was introduced with three round warning lights in a brushed-aluminium-alloy housing.

1963 model year
As 1961: new tank badges.

1964 model year
The former strategy of marketing two models came to an end in 1964 with the cataloguing of only a single Model G. This reverted to the GE model designation as the Servi-Car had, for the first time in its 33-year history, been given an electric starter; this meant a new gearbox and a 12-volt electrical system. The generator was fitted with a voltage regulator and the timer unit (contact breaker housing) gained an automatic ignition advance/retard mechanism.

In effect, the Servi-Car was used as a test-bed for the electric start system which would be fitted to the new Electra-Glide models due to appear the following year. The towbar continued to be listed as an option, but this configuration was no longer justified as a separate model in its own right.

1965 model year
Apart from ball-ended handlebar levers, there were no changes.

1966 model year
Only minimal changes such as tank badges, air cleaner mountings and an improved starter motor.

1967 model year
The only change for 1967, but quite a significant one, was the use of fibre-glass for the box body, which had been made of pressed-steel and wood up until then. The new bodies were introduced gradually, quite late in the model year.

1968 model year
No changes.

1969 model year
Only the stop-lamp switch and rear view mirror were changed.

1970 model year
No changes.

... to 1965 (This example seems to be missing a few vital parts ...)

The 'when I grow up I wanna be an Electra-Glide' Servi-Cars of 1966 to 1969

A 1966 to 1969 Servi-Car in police specification with front mudguard-mounted siren

1971 model year
No changes.

1972 model year
No changes. Demand for the Servi-Car was declining rapidly and from 1968 to 1972 it was not considered worthwhile to make any major changes to the range.

1973 model year
For its last year of production it was decided at H-D to treat the Servi-Car to rear disc brakes, but this didn't happen until late in the model year so there can be no more than one hundred vehicles with such equipment as standard. In this same year a prototype was even built using the Electra-Glide's 74 cu. in./1207 cc. engine, but this never made it to production; the number of likely buyers was not considered to be sufficient to warant it, and probably it would have been too expensive. So at the end of 1973 the 40-year-old Servi-Car was finally discontinued; the longest-produced Harley engine/gearbox unit was consigned to the museum, after, ironically, exactly forty-five years of production.

Production figures

Model	1932	1933	1934	1935	1936	1937	1938	1939	1940	1941	1942
G	*219	80	317	323	382	491	259	320	468	607	138
GA	-	12	40	64	55	55	83	126	156	221	261
GD	-	60	104	91	96	112	81	90	158	195	-
GDT	-	18	58	72	85	1336	102	114	126	136	-
GE	-	12	27	17	30	22	-	-	-	-	-

* This figure includes 1932 Models G and GA.

Production figures

Model	1943	1944	1945	1946	1947	1948	1949	1950	1951	1952	1953
G	22	6	26	766	1307	1050	494	520	778	515	1146*
GA	113	51	60	678	870	728	545	483	632	532	-

* This figure includes 1953 Models G and GA.

Production figures

Model	1954	1955	1956	1957	1958	1959	1960	1961	1962	1963	1964
G	*1397	394	467	518	283	288	707**	628**	703**	N./Av.	-
GA	-	647	736	674	643	524	-	-	-	N./Av.	-
GE	-	-	-	-	-	-	-	-	-	-	725

* This figure includes 1954 Models G and GA.
** This figure includes 1960 to 1962 Models G and GA.

Production figures

Model	1965	1966	1967	1968	1969	1970	1971	1972	1973
GE	625	625	600	600	475	494	500	400	425

Catalogue prices ($ US)

Model	1932	1933	1934	1935	1936	1937	1938	1939	1940	1941	1942
G	450	430	430	440	440	515	515	515	515	515	525
GA	435	415	415	425	425	500	500	500	500	500	510
GD	410	430	430	440	440	515	515	515	515	515	-
GDT	-	445	445	455	455	530	530	530	530	530	-
GE	465	485	485	495	495	570	-	-	-	-	-

Catalogue prices ($ US)

Model	1943	1944	1945	1946	1947	1948	1949	1950	1951	1952	1953
G	525	525	580	594	710	755	860	860	1095	1175	1190
GA	510	510	568	582	695	740	845	845	1080	1160	1175

Catalogue prices ($ US)

Model	1954	1955	1956	1957	1958	1959	1960	1961	1962	1963	1964
G	1240	1240	1240	1367	1465	1500	1530	1555	1555	1590	-
GA	1225	1225	1225	1352	1450	1470	1500	1525	1525	1550	-
GE	-	-	-	-	-	-	-	-	-	-	1628

Catalogue prices ($ US)

Model	1965	1966	1967	1968	1969	1970	1971	1972	1973
GE	1685	1696	1930	1930	2065	N./Av.	N./Av.	N./Av.	N./Av.

Prices and availability of optional Accessory Groups for Servi-Cars ($ US)

Group	1935	1936	1937	1938	1939	1940	1941 1942	1943 1944	1945
Group #1	27.90	-	-	-	-	-	-	-	-
Standard Group	-	29.75	21.90	18.25	12.50	-	-	-	-
Utility Group	-	-	-	-	-	13.50	18.50	-	18.50
Deluxe Group G/GA	59.90	59.90	48.00	49.00	39.50	39.75	44.00	-	44.00
Deluxe Group GD/GDT	62.90	62.90	53.00	54.00	44.50	44.75	44.00	-	44.00

Note that Standard means a specification higher than standard ex-works finish. Optional Accessory Groups for the model years 1943 and 1944, while definitely available, were not listed in factory order forms.

Prices and availability of optional Accessory Groups for Servi-Cars ($ US)

Group	1946	1947	1948	1949	1950	1951	1952	1953	1954 1955	1956 1957
Utility Group	18.50	26.00	29.50	33.60	29.75	28.50	31.00	-	-	-
Standard Group	-	-	-	-	-	-	-	31.00	35.00	35.00
Deluxe Group #G1	44.00	58.00	65.50	78.50	74.85	60.50	77.20	77.20	N./Av.	84.00
Deluxe Group #G2	-	-	-	-	-	-	-	87.65	81.00	97.00

Note that Standard means a specification higher than standard ex-works finish.

All G models 1932 to 1936

Engine
Engine	Air-cooled 45° sv V-twin
Bore x stroke	2¾ x 3¹³⁄₁₆ (2.75 x 3.81) in.
Bore x stroke	69.85 x 96.84 mm
Capacity	45.12 cu. in. (742 cc)
Power output in PS (kW)/min.[-1]	16(12)4000 rpm
Compression ratio	4.3:1
Carburettor	Schebler 1-inch (1932-33)
	Linkert 1-inch Type M-11 1934-35
	Linkert 1-inch Type M-16 1936
Exhaust system	1 small flat fishtail silencer

Transmission
Gearbox	3-speed
Overall gear ratio	5.85:1
Gearbox reduction ratios	2.46 (1)
	1.66 (2)
	1.00 (3)
	R: 2.13
Secondary reduction ratio	2.18
Gearchange	Hand change, on tank
Final drive	Chain

Cycle parts
Frame construction	Single-loop frame, curved downtube and rear axle mountings
Front fork	Springers – exposed-springs, forged I-beam legs
Rear suspension	Rigid car-type axle with differential
Front brake	Drum, via cable and handlebar lever
Rear brake	Drum, via rod and pedal

General information
Unladen weight	approx. 290 kg
Maximum weight	600 kg
Front and rear tyres	4.00-18
Wheels front and rear	2.15 x 18 wire-spoked
Maximum speed	78 km/h
Tank capacity	21.3 litres

All G models 1937 to 1948

Engine
Engine	Air-cooled 45° sv V-twin
Bore x stroke	2¾ x 3¹³⁄₁₆ (2.75 x 3.81) in.
Bore x stroke	69.85 x 96.84 mm
Capacity	45.12 cu. in. (742 cc)
Power output in PS (kW)/min.[-1]	19(14)/4400 SAE (1937-40)
	21(15)/4400 rpm 1941-48
Compression ratio	4.3:1 (1937-40)
	4.75:1 (1941-48)
Carburettor	Linkert 1-inch Type M-16
Exhaust system	Tubular car-style silencer

Transmission
Gearbox	3-speed
Overall gear ratio	5.85:1
Gearbox reduction ratios	2.46 (1)
	1.66 (2)
	1.00 (3)
	R: 2.13
Secondary reduction ratio	2.18
Gearchange	Hand change, on tank
Final drive	Chain

Cycle parts
Frame construction	Single-loop frame, curved downtube and rear axle mountings
Front fork: 1937-39	Springers – exposed-springs, forged I-beam legs
1940-41	Springers – exposed-springs, oval-tube legs
Rear suspension	Rigid car-type axle with differential
Front brake	Drum, via cable and handlebar lever
Rear brake	Drum, via rod and pedal

General information
Unladen weight	approx. 290 kg
Maximum weight	600 kg
Front and rear tyres	1937-40: 4.00-18
	1940-48: 5.00-16
Wheels front and rear	1937-40: 2.15 x 18 wire-spoked
	1940-48: 3.00 x 16 wire-spoked
Maximum speed	80 km/h
Tank capacity	12.9 litres

All G models 1949 to 1963

Engine
Engine	Air-cooled 45° sv V-twin
Bore x stroke	2¾ x 3¹³⁄₁₆ (2.75 x 3.81) in.
Bore x stroke	69.85 x 96.84 mm
Capacity	45.12 cu. in. (742 cc)
Power output in PS (kW)/min.⁻¹	21(15)/4400 SAE (M-18)
	24(18)/4600 SAE (M-54)
Compression ratio	4.75:1
Carburettor	Linkert 1-inch Type M-18 (1949-63)
	Linkert 1¼-inch Type M-54 (1951-52)
Exhaust system	2 in 1 with 1 Silencer

Transmission
Gearbox	3-speed
Overall gear ratio	5.85:1
Gearbox reduction ratios	2.46 (1)
	1.66 (2)
	1.00 (3)
	R: 2.13
Secondary reduction ratio	2.18
Gearchange	Hand change, on tank
Final drive	Chain

Cycle parts
Frame construction	Single-loop frame, curved downtube and rear axle mountings
Front fork:	
1949 to 1957	Springers – exposed-springs, oval-tube legs
1958-on	Hydraulically-damped telescopic forks
Rear suspension	Rigid car-type axle with differential
Front brake	Drum, via cable and handlebar lever
Rear brake	Drum, via rod and pedal (1949-50)
	1951-63 Hydraulically-operated drums

General information
Unladen weight	approx. 290 kg
Maximum weight	600 kg
Front and rear tyres	5.00-16
Wheels, front	1937-40: 2.15 x 18
	3.00 x 16 wire-spoked 1949-50
	3.00 x 16 wire-spoked 1951-63
Wheels, rear	3.00 x 16 steel disc
Maximum speed	85 km/h
Tank capacity	12.9 litres

All G models 1964 to 1973

Engine
Model designation	– (1964-69)
	5A (1970-73)
Engine	Air-cooled 45° sv V-twin
Bore x stroke	2¾ x 3¹³⁄₁₆ (2.75 x 3.81) in.
Bore x stroke	69.85 x 96.84 mm
Capacity	45.12 cu. in (742 cc)
Power output in PS (kW)/min.⁻¹	21(15)/4400 rpm
Compression ratio	4.75:1
Carburettor	Linkert 1-inch Type M-18 (1964-65)
	Tillotson 1-inch (1966-70)
	Bendix-Zenith 1-inch (1971-73)
Exhaust system	2 in 1 with 1 Silencer

Transmission
Gearbox	3-speed
Overall gear ratio	5.85:1
Gearbox reduction ratios	2.46 (1)
	1.66 (2)
	1.00 (3)
	R: 2.13
Secondary reduction ratio	2.18
Gearchange	Hand change, on tank
Final drive	Chain

Cycle parts
Frame construction	Single-loop frame, curved downtube and rear axle mountings
Front fork	Hydraulically-damped telescopic forks
Rear suspension	Rigid car-type axle with differential
Front brake	Drum, via cable and handlebar lever
Rear brake	Up to early 1973: Hydraulically-operated drums
	From early 1973: Hydraulically-operated discs

General information
Unladen weight	approx. 290 kg
Maximum weight	600 kg
Front and rear tyres	5.00-16 (1964-68)
	5.10-16 (1969-73)
Wheels front	3.00 x 16 wire-spoked
Wheels rear	3.00 x 16 steel disc
Maximum speed	85 km/h
Tank capacity	12.9 litres

The Model TA military three-wheeler

1940/1

Model description and technical information

The big 750 cc. BMW with its shaft final drive was the inspiration behind the (mal)development of the TA series of which, even after very poor initial test results, 15 further examples were built to fulfil the Army's order.

The first experimental machine of June 1940 was fitted with the Knucklehead engine which had been launched five years earlier in the Model E range. However, the vehicle's power output was not satisfactory and so a modified engine was installed that was based on the forthcoming (due to appear in 1941) Model F's powerplant, but with capacity reduced to 68 cu. in./1122 cc. This finally satisfied the Army's inspectors, but the enormous success and lower cost of the Jeep, developed by American Bantam but now emerging from Ford's and Willys-Overland's factories in its hundreds of thousands, sounded the death-knell of the TA project. The last three-wheelers were delivered in January 1941.

1941: Model TA – only 16 were built between June 1940 and January 1941, before the project was killed by the Jeep

Model TA 1941

Engine
Type	Knucklehead
Engine	Air-cooled 45° ohv V-twin
Bore x stroke	3⁷⁄₁₆ x 3¹¹⁄₁₆ (3.44 x 3.69) in.
Bore x stroke	87.31 x 93.66 mm
Capacity	68.44 cu. in. (1122 cc)
Power output (SAE)	approx. 40(29)/4500 rpm
Compression ratio	5.0:1
Valve layout	ohv
Carburettor	Linkert 1¼-inch Type M-97
Ignition	Battery-and-coil

Transmission
Gearbox	3-speed with reverse
Gearchange	Hand change, on tank
Final drive	Propellor shaft with two half-shafts

Cycle parts
Frame construction	Twin-loop frame with straight downtubes and mountings for rear axle
Front fork	Exposed-spring Springer forks, oval-tube legs
Rear suspension	None – rigid rear axle
Front brake	Drum, via cable and handlebar lever
Rear brake	Foot brake acting on rear wheels

General information
Unladen weight	approx. 350 kg
Front wheel/tyre	5.50-16
Rear wheel/tyre	5.50-16
Maximum speed	approx. 80 km/h
Tank capacity	18.9 litres plus 3.8 litres in reserve tank at rear
Production figures	16
Contract price	$2,260

Sport Twin. Model W
1919 to 1923

Model description and technical information

No sooner had the last singles disappeared from the catalogue than a new horizontally-opposed flat twin-cylinder machine appeared, from mid-1919 onwards, to take their place. Coming as something of a shock to the contemporary US motorcycle market, the new Sport Twin model departed so completely from existing Harley-Davidson norms that one wonders who the designer really was.

With a specially-developed new open-diamond frame which used the engine as a stressed member (known by the Americans as a keystone frame), girder front forks with trailing links acting on a single central coil spring via long tie-rods and a rocker linkage (an adventurous departure from their trademark Springer forks for Harley-Davidson and close to plagiarising the front suspension design of their bitter rivals, Indian), the machine was novel in nearly every respect. Weighing only a little more than the singles, and much less than Harley's big V-twins, and with the bulk of that weight being carried low, the machine was much easier for the novice to control; in an age of vibratory singles and V-twins, the Model W's smooth-running and quiet engine was even more of an attraction to those not searching for maximum performance. While the new model was not intended for pulling sidecars and its modest eight horsepower only gave it an equally-unexceptional 50 mph (80 km/h) top speed, in its appeal either to committed motorcyclists who could not afford or did not want a Harley Big Twin, or to small business users, the Sport Twin was some six to ten years ahead of Harley's other offerings. The non-detachable cylinder heads (with proper provision for easy valve overhauls) and the one-piece cast intake and exhaust manifolds exemplified the engine's simple construction. Side valves controlled by a single crankshaft-driven camshaft were used for the first time in a Harley and were particularly suited to the flat twin layout.

Engine and gearbox were joined together in semi-unit construction, with the three-speed sliding-mesh gearbox on top, an external flywheel and wet multi-plate clutch. Helical-cut primary drive gears were used – yet another novelty – to reduce mechanical noise. A fully-enclosed final drive chaincase was another practical and innovative fitting; unfortunately this was discontinued in 1922 on grounds of cost.

The magneto-equipped engines of the first year of production (1919) were identified by the letter M. Right in the middle of its second model year (early 1920), H-D split the Model W range into two; Models WF and WJ, the first being the successor to the Model W, with acetylene lighting, the second having battery-and-coil ignition and electric lighting. Both generator and coils were now made in-house by Harley themselves.

The standard brake, a double-acting external-contracting band device acting on a $7\frac{5}{16}$-inch (186 mm) diameter steel drum with a 1-inch (25 mm) wide face, could be supplemented (where demanded by local legislation), by an internal-expanding brake acting on the same drum. The standard colour was Brewster Green 'striped in gold'.

Between 1921 and 1923 there were only detail alterations and improvements. At the end of 1923 the entire model range disappeared. They were expensive to make and just not competitive against the 11 horsepower of the Indian rival of similar catalogue price; an important consideration for most of Harley's dealers, who were largely performance-hungry enthusiasts themselves and disappointed that the machine did not live up to its Sport Twin name. Although quite successful in export markets such as Europe, which did not have the preoccupation with the heavyweight Big Twin that characterized the US motorcycle market for most of the period covered by this book, their lack of success in the domestic market doomed them to an early end.

Not until 1926 was there another small-capacity Harley, this time once more with a single-cylinder engine. Nearly twenty years had to go by before there was another flat-twin from H-D.

Selling alongside the 36 cu. in. (584 cc.) Model W range was another 37 cu. in. (608 cc.) model, but this one a heavy single-cylinder Commercial model which did not even appear in the Motor Company's catalogue and price lists. It is hardly surprising, therefore, that the Model CD of 1921 to 1922 (covered in the Chapter on the early single-cylinder models) did not sell in large numbers.

Production figures

Model	1919	1920	1921	1922	1923
W	753	-	-	-	-
WF	-	4459	1100	388	614
WJ	-	810	823	455	481

Catalogue prices ($ US)

Model	1919	1920	1921	1922	1923
W	335	-	-	-	-
WF	-	335	415	310	275
WJ	-	365	445	340	295

Model W 1919½ to 1920*
Model WF 1920½ to 1923**

Engine
Engine	Air-cooled sv flat twin
Bore x stroke	2¾ x 3 (2.75 x 3) in.
Bore x stroke	69.85 x 76.2 mm
Capacity	35.64 cu.in. (584 cc)
Power output (SAE)	8(6)/3500 rpm
Compression ratio	3.75:1
Carburettor	Schebler
Ignition	Magneto

Transmission
Gearbox	3-speed
Gearchange	Tank-mounted hand change
Final drive	Chain

Cycle parts
Frame construction	Open diamond, straight downtube
Front fork	Girder, with trailing-link springing
Rear suspension	None
Front brake	None
Rear brake	Drum, via rod and pedal

General information
Unladen weight	115 kg
Front wheel/tyre	3 x 26 beaded-edge
Rear wheel/tyre	3 x 26 beaded-edge
Maximum speed	80 km/h
Tank capacity	10.4 litres
Sidecars available	None

Model WJ 1920½ to 1923**

Engine
Engine	Air-cooled sv flat twin
Bore x stroke	2¾ x 3 (2.75 x 3) in.
Bore x stroke	69.85 x 76.2 mm
Capacity	35.64 cu.in. (584 cc)
Power output (SAE)	8(6)/3500 rpm
Compression ratio	3.75:1
Carburettor	Schebler
Ignition	Battery-and-coil

Transmission
Gearbox	3-speed
Gearchange	Tank-mounted hand change
Final drive	Chain

Cycle parts
Frame construction	Open diamond, straight downtube
Front fork	Girder, with trailing-link springing
Rear suspension	None
Front brake	None
Rear brake	Drum, via rod and pedal

General information
Unladen weight	115 kg
Front wheel/tyre	3 x 26 beaded-edge
Rear wheel/tyre	3 x 26 beaded-edge
Maximum speed	80 km/h
Tank capacity	10.4 litres
Sidecars available	None

* Introduced mid-way through the 1919 model year, hence 1919½.
** Introduced mid-way through the 1920 model year, hence 1920½.

The Model A and B singles

1926 to 1934

Model description and technical information

With the last Harley single disappearing from the catalogue in 1922 and the small flat twin also no longer to be found, it was decided in Milwaukee at about the middle of the decade to re-enter the market with a small single. This was developed with an eye for export, but also because they didn't want to leave this field of the domestic market to the competition, least of all to their great rivals, Indian, with their 21 cubic inch (350 cc.) single-cylinder Prince of 1925.

The frame was visually similar to that of the big Harleys, but of much lighter construction. In particular the steering head lug, while a drop-forged steel component like that of the bigger-capacity models, lacked the middle strengthening rib. The front fork's thinner spring legs blended at their upper ends into the now-exposed coil springs. By making similar small weight savings throughout the entire machine, which was available only as a single-seater and not recommended for sidecars, it emerged some 130 lb (60 kg) lighter than the big V-twins. The final drive chain was on the right on the singles, whereas on the V-twins it was on the left. The rear brake was an external-contracting band device acting on a 5¾-inch (146 mm) diameter steel drum with a 1-inch (25 mm) wide face. Paint finish was Olive Green with wide maroon striping, with a centre gold stripe.

The engines were also completely new. Those with magneto ignition and acetylene lighting were designated the Model A, while those with battery-and-coil ignition and electric lighting were the Model B. Both versions were available either with the-then commonplace side-valve layout (Models A and B) or with – offered for the first time on a Harley sold to the general public – overhead valves (Models AA and AAE, BA and BAE), a layout which gave it fully half as much power again as its flathead brother and a top speed of some 65 mph (100km/h) as opposed to the sidewhacker's possible maximum of 50 mph (80 km/h). A characteristic of the singles was the exhaust system, which used a relatively large diameter downpipe leading into a perfectly-cylindrical silencer. Both were fitted with detachable cylinder heads made under licence to the famous English engine designer and researcher Sir Harry Ricardo, who had developed a combustion chamber

1926 Model B – right-hand side

Model range

Model A	21 cu. in./346 cc. sv single with magneto ignition	for the USA only (1926-30)
Model AA	21 cu. in./346 cc. ohv single with magneto ignition	for the USA only (1926-30)
Model AAE	21 cu. in./346 cc. ohv single with magneto ignition	for export (1926-28)
Model AAF	21 cu. in./346 cc. ohv single with magneto ignition	for export (1929)
Model B	21 cu. in./346 cc. sv single with battery-and-coil ignition	for the USA only (1926-34)
Model BA	21 cu. in./346 cc. ohv single with battery-and-coil ignition	for the USA only (1926-31)
Model BAE	21 cu. in./346 cc. ohv single with battery and coil ignition	for export (1926-28)
Model BAF	21 cu. in./346 cc. ohv single with battery-and-coil ignition	for export (1929-30)

configuration using an early form of squish band. It was no surprise, therefore, when racing variants (Model S) of these small, high-revving machines soon appeared. These Racing Singles appeared in the official catalogue at a cost of 300 dollars and delivered a very respectable 15 hp (11 kW) at some 4,500 rpm; to their fans, they soon became known as the 'Peashooter'. The Models AAE and BAE were created for export markets, becoming the AAF and BAF towards the ends of their lives.

Model development

1927 model year
New Ricardo cylinder heads were fitted to both Models AA and BA. All models received a strengthened frame and improved fuel tanks. Amongst further small alterations was a new exhaust with reduced back-pressure, to permit the engine to breathe more easily.

1928 model year
The B models were fitted with magnesium-alloy pistons, the flywheels and connecting rod were lightened and the number of teeth on the clutch wheel rose from 36 to 47. As with all Harleys for this year, an internally-expanding drum front brake, cable-operated from a handlebar lever, was fitted to all models.

1929 model year
The magnesium-alloy pistons were now standard equipment on Models AA and BA as well. All models apart from those for export were fitted as standard equipment with the double headlamps and four-tube silencer exhaust system fitted to the big Harleys. The clutch was strengthened with two more friction plates. The engine lubrication system's oil supply was made load-sensitive by linking the oil pump to the throttle. Handlebars and wheel rims were now painted black.

1926 Model B – left-hand side

1926 Model B engine – timing side

1927 Model B

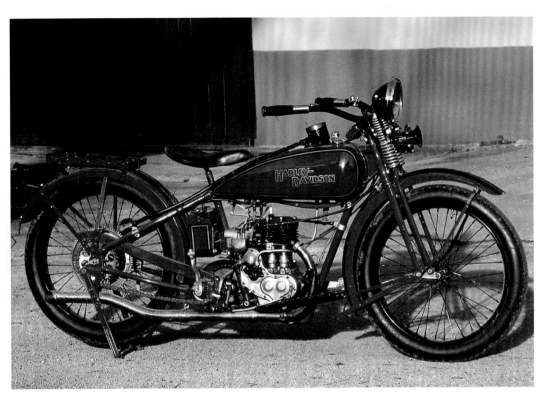

1927 Model B

1928 Model B

1930 model year
The export versions now got the four-tube exhaust, while the US-market models were fitted with a new twin-tube system. Models AA and BA reverted to a single headlamp (for this year only), while the B models could be fitted at special request with a steering damper.

1931 model year
In the pursuit of a policy of standardization of parts followed by the Motor Company at this time, the few B models exported were fitted with the larger external-contracting band rear brake, claimed to be 31% more effective, fitted to the Model C and D ranges. No B models were sold in the US in this year; even the production of A models was discontinued.

1932 model year
After a year's break in sales, the B models were now catalogued again in the US itself and sold at a loss – at $195, they were the cheapest new Harleys ever sold. Created from parts held in stock and from unsold machines brought back from the collapsed markets overseas (especially in Australia and New Zealand); it might almost have been easier to have scrapped them, but nearly 1,100 sales between 1932 and 1934 were better than nothing. The seat tube was shortened by 1 inch (25 mm), to reduce the seat height.

1933 model year
In the middle of the model year a new carburettor was fitted from Linkert, the new supplier. Frame and forks were now painted black. There were, as with almost every year, new tank graphics; this year's were in the new Art Deco eagle design.

1934 model year
No further developments for this, their final model year.

1929 Model B

1929 Model B

Production figures

Model	1926	1927	1928	1929	1930	1931	1932	1933	1934
A	1128	444	519	197	4	-	-	-	-
AA	61	32	65	21	1	-	-	-	-
AAE	146	66	4	-	-	-	-	-	-
AAF	-	-	-	5	-	-	-	-	-
B	5979	3711	3483	1592	577	2	535	123	424
BA	515	481	943	191	9	1	-	-	-
BAE	161	43	207	-	-	-	-	-	-
BAF	-	-	-	213	86	-	-	-	-

The Model BR is not listed here as it was a pure racing machine, of which approximately 95 were produced.

Model A 1926 to 1930
Model B 1926 to 1930
Model B 1931*
Model B 1932 to 1934

Engine
Engine	Air-cooled sv single
Bore x stroke	2⅞ x 3¼ (2.88 x 3.25) in.
Bore x stroke	73.03 x 82.55 mm
Capacity in cubic inch	21.10 cu. in. (346 cc)
Power output (SAE)	8(6)/4000 rpm
Compression ratio	4.0:1
Carburettor	Schebler (1926-32)
	Linkert (1933-34)
Ignition:	
Model A	Magneto
Model B	Battery-and-coil
Exhaust system:	
1926-28	Single-tube
1929-30	Four-tube
1931-34	Twin-tube

Transmission
Gearbox	3-speed
Gearchange	Hand change, on tank
Final drive	Chain

Cycle parts
Frame construction	Single-loop, straight downtube
Front fork	Springer, exposed springs, forged oval section legs
Rear suspension	None
Front brake	None
Rear brake	Drum, via rod and pedal

General information
Unladen weight	114 kg (A), 122 kg (B)
Front wheel/tyre	26 x 3.30 wired-edge (1926-29)
	20 x 3.30 wired-edge (1930-34)
Rear wheel/tyre	26 x 3.30 wired-edge (1926-29)
	20 x 3.30 wired-edge (1930-34)
Maximum speed	90 km/h
Tank capacity	11.3 litres (Teardrop)
Sidecars available	None

* Export models.

Model AA 1926 to 1930
Model AAE 1926 to 1928*
Model AAF 1929*
Model BA 1926 to 1930
Model BA 1931*
Model BAE 1926 to 1928*
Model BAF 1929 to 1930*

Engine
Engine	Air-cooled ohv single
Bore x stroke	2⅞ x 3¼ (2.88 x 3.25) in.
Bore x stroke	73.03 x 82.55 mm
Capacity in cubic inch	21.10 cu. in. (346 cc)
Power output (SAE)	12(9)/4400 rpm
Compression ratio	6.0:1
Carburettor	Schebler
Ignition:	
Model AA	Magneto
Model BA	Battery-and-coil
Exhaust system:	
1926-28	Single-tube
AA & BA from 1929-on	Four-tube
Export from 1929-on	Twin-tube
All from 1931-on	Twin-tube

Transmission
Gearbox	3-speed
Gearchange	Hand change, on tank
Final drive	Chain

Cycle parts
Frame construction	Single-loop, straight downtube
Front fork	Springer, exposed springs, forged oval section legs
Rear suspension	None
Front brake	None
Rear brake	Drum, via rod and pedal

General information
Unladen weight	111 kg (AA), 119 kg (BA)
Front wheel/tyre	26 x 3.30 wired-edge (1926-29)
	20 x 3.30 wired-edge (1930-31)
	26 x 3.30 wired-edge (1926-29)
	20 x 3.30 wired-edge (1930-31)
Maximum speed	100 km/h
Tank capacity	11.3 litres (Teardrop)
Sidecars available	None

Catalogue prices ($ US)

Model	1926	1927	1928	1929	1930	1931	1932	1933	1934
A	210	210	210	210	210	-	-	-	-
AA	250	230	230	230	230	-	-	-	-
B	235	235	235	235	235	195	195	187.50	187.50
BA	275	255	255	255	255	-	-	-	-

The Model C singles
1929 to 1934/7

Model description and technical information

Announced in the summer of 1928, a major new arrival on the market for 1929 was a big 'Thirty-Fifty' (500 cc.) single – the Model C. Its engine was a side-valve, like that of the Models A and B, and actually of 30 cubic inch (493 cc.) capacity. The 'Baby Harley' was no engineering *tour de force*, even though its 6-volt electrical system was new to H-D. With the same frame, clutch and gearbox as its smaller sisters (although this would change in the second year of production) it could manage some 56 mph (90 km/h). Conceived as a direct competitor for the 36.38 cu. in./596 cc. Indian Model 101 Scout, it was technically superior; the Indian developed less power from practically the same size engine and was discontinued at the end of the 1930 model year.

The Model C filled the gap between the 21 cubic inch (346 cc.) singles and the 45 cu. in. (742 cc.) V-twins, thus at long last giving Harley a complete range of models.

As with all the other Harley models, all Model Cs were fitted as standard equipment with the double headlamps and all, apart from those for export, used the four-tube silencer exhaust system. The engine lubrication system's oil supply was made load-sensitive by linking the oil pump to the throttle. Brakes were internal-expanding drum front and external-contracting band rear. Paint finish was Olive Green with maroon striping, with a centre gold stripe; handlebars and wheel rims were black.

Model development

1930 model year
The frame of the 45 cu. in. (742 cc.) V-twin Model Ds was designed from scratch for 1929, as there was no earlier Harley model from which components could be used or adapted. A variation of this standard frame was reserved for the 30 cu. in./493 cc. single; only a few engine mounting points and other minor details distinguish the two frames. In the second year of production both models were revised to standardize on the Model D's new heavier frame, which lowered the seat height and increased ground clearance, and improved battery access. Furthermore, the exposed-spring forks, while visually similar to those introduced on the Models A and B, were now of much heavier construction with heat-treated drop-forged I-beam-section rigid (sprung) legs. Sidecars, however, were not recommended for the Singles.

The Model C became the first Harley to be fitted with a steering lock. Engine power was now transmitted to the gearbox via an automatically-lubricated duplex primary chain, and a larger clutch was fitted to improve the service life of the transmission. The silencer was now a new twin-tube configuration. Prospective customers could choose from a range of new colours – Black, Maroon, Grey, Blue, Cream and Coach Green – all available at extra cost.

1931 model year
Just as the few B models exported this year were fitted with the Model D's larger external-contracting band rear brake, claimed to be 31% more effective, in the pursuit of a policy of standardization of parts followed by the Motor Company at this time, so also was the Model C. On the latter, the components formerly nickel-plated for decorative purposes were now chromed, a single headlamp and new handlebar grips were fitted.

1932 model year
From this year on the complete exhaust system was finished in heat-resistant black; fuel tanks could be ordered with, instead of the traditional lettering on a plain background, a new curving design – the forerunner of the following year's Art Deco eagle design – highlighting the lettering. A strengthened frame was fitted.

Model range		
Model C	30 cu. in./493 cc. sv single, battery-and-coil ignition (Standard model)	(1929-34)
Model CB	30 cu. in./493 cc. sv single, battery-and-coil ignition (Economy model)	(1933-34)
Model CC	30 cu. in./493 cc. sv single, battery-and-coil ignition (Commercial)	(1931-32)
Model CH	30 cu. in./493 cc. sv single, high-compression, battery-and-coil ignition (Standard model)	(1931)
Model CM	30 cu. in./493 cc. sv single, magneto ignition	(1929-30)
Model CMG	30 cu. in./493 cc. sv single, magneto ignition	(1931)
Model CS	30 cu. in./493 cc. sv single, (Export model for Japan)	(1932-34)

1933 model year

For this year a low-cost economy model, the Model CB, was marketed. It reverted to the old frame of the 1929 Models A and B, which was easier to manufacture than the Model C derivative of the Model D's frame, and both frame and forks were painted black. The customer thereby saved a whole 28 bucks. New Art Deco tank graphics – the eagle design – decorated the machines.

1934 model year

For their final model year there were no further developments except for the flying diamond tank graphics. Originally, it had been planned to market a further-developed single, but for financial reasons this project was now cancelled.

1935-7 model years

According to one or two sources, Model Cs carried on in production until 1935 or even 1937; for export only, and chiefly for the Japanese market where they were popular as the front ends of locally-made three-wheeler delivery vehicles of various configurations. It is, however, more likely that these were unsold domestic-market machines diverted abroad when they could not find buyers (even at discounted prices) in the USA. The racing versions might have been manufactured for a few more years, presumably to fill the gap until the appearance of the racing variants of the Model W.

Catalogue prices ($ US)

Model	1929	1930	1931	1932	1933	1934
C	255	260	260	235	225	225
CB	-	-	-	-	197	197
CH	-	-	?	-	-	-

Production figures

Model		1929	1930	1931	1932	1933	1934
C	Standard model	1570	1483	874	213	112	220
CB	Economy model	-	-	-	-	51	310
CC	Commercial	-	-	3	7	-	-
CH	High-compression	-	-	180	-	-	-
CM	Magneto ignition	12	11	-	-	-	-
CMG	Magneto ignition	-	-	2	-	-	-
CS	Export Japan	-	-	-	100	-	-

Model C 1929 to 1934
Model CB 1933 to 1934
Model CC 1931 to 1932
Model CH 1931
Model CS 1932 to 1934 (Export Japan)
Model CM 1929 to 1930
Model CMG 1931

Engine

Engine	Air cooled sv single
Bore x stroke	3³²⁄₃₂ x 4 (3.09 x 4) in.
Bore x stroke	78.58 x 101.6 mm
Capacity	30.07 cu.in. (493 cc)
Power output (SAE):	
Model CH	12 hp (9 kW)/3600 rpm
All other models	10.5 hp (8 kW)/3600 rpm
Compression ratio:	
Model CH	5.0:1 (approx.)
All other models	4.2:1
Carburettor	Schebler Deluxe 1-inch
Ignition:	
Models C, CB, CC, CH & CS	Battery-and-coil
Models CM & CMG	Magneto
Exhaust system:	
All US models 1929-1934	Four-tube
All export models 1929-1934	Twin-tube

Transmission

Gearbox	3-speed
Gearchange	Hand change, on tank
Final drive	Chain

Cycle parts

Frame construction	Models C, CC, CH, CS, CM and CMG: Single-loop, straight downtube
	Model CB: Single-loop, straight downtube (from Models A and B)
Front fork	Springer, exposed springs, forged oval section legs
Rear suspension	None
Front brake	Drum, via cable and handlebar lever
Rear brake	Drum, via rod and pedal

General information

Unladen weight:	
1929 Models & CB	340 lb (154 kg)
1930-on models except CB	430 lb (195 kg)
Front and rear tyres	25 x 3.85 wired-edge (1929)
	4.00-18 wired-edge (1930-34)
Maximum speed	97 km/h
Tank capacity	20 litres (1929)

The Forty-Fives: Model D
1929 to 1931

Model description and technical information

Harley had nearly left it too late to fill the gap in their range between the 21 cubic inch (350 cc.) singles of 1926 onwards and the forthcoming Big Twins. However, with the combined launch of the big 30 cu. in./493 cc. Model C singles and the completely new 45 cu. in. (742 cc.) Model D range in the summer of 1928, the gap could be considered to be closed.

The engines were a complete departure from the existing ioe V-twins as they were side-valve, with detachable cylinder heads and a cross-flow combustion chamber which gave a lower silhouette and greater specific power output than the old units. Ricardo-patent cylinder heads were standard equipment, as were, in the Harley jargon, Dow-metal (a 2nd-generation light magnesium alloy created by the Dow Corporation) pistons. The Model Ds were known as the 'three-cylinder' Harleys because, thanks to lack of space in the new frame to mount a generator in the conventional position, the generator was mounted vertically beside the front cylinder and driven via a very complex and failure-prone drive train from the timing gears. Unfortunately they were pitifully slow – their top speed was some 20 mph (32 km/h) slower than their immediate rival, the 45 cu. in./745 cc. Indian Model 101 Scout – which did nothing to endear them to dealers and customers when the new machines turned out to be not only slow, but unreliable as well. A publicity stunt staged in the spring of 1929 involved the younger Davidson generation, Walter, Gordon and Allan, riding brand-new Model Ds from Milwaukee to the Pacific North-West and back, being met along the way by William J. Harley and William H. Davidson in a Model VL sidecar outfit; the trip was interrupted so many times due to repairs and adjustments being needed for the machines that the publicity campaign was cancelled.

The ignition system was by battery-and-coil, naturally, for this new model range, in the form exclusive to Harley from 1927 onwards. The engine lubrication system's oil supply was made load-sensitive by linking the oil pump to the throttle. The exhaust system for the first two model years used the extremely quiet four-tube silencer; only in 1931 was this replaced by the new twin-tube silencer.

The gearbox, virtually unchanged since 1915, was also fitted to the Model Ds, with the usual kickstart; completely new, however, was the gearboxes' detachable end cover. A rear wheel stand facilitated starting. The clutch and gearbox were the lightweight versions created for the singles, and, as with the Model C, the final drive chain had been moved to the right; on

1930 Model DL, the first Harley Forty-Five. Note the 'three-cylinder Harley' vertically-mounted generator

this score alone, therefore, the existing Big Twin frame could not be used.

The tank-mounted gearchange with foot clutch would remain unchanged until the beginning of the 1950s as though this were the only possible means of selecting gears as far as H-D were concerned. The new dry clutch was, however, a genuine improvement, as was the primary drive which used a duplex chain.

The Model Ds' frame was designed from scratch for 1929, as there was no earlier Harley model from which components could be used or adapted. A variation of this standard frame was laid out for the Model C; only a few engine mounting points and other minor details distinguish the two frames. In the second year of production both models were revised to standardize on the Model D's strengthened and heavier frame. The original frame was too light for sidecar work, as were the front forks.

The single downtubes were sleeved into dropforged steering head lugs and brazed. Allegedly, welding was already being used in frame construction by 1931, for reinforcement in particular joints, but it was only from 1936 onwards that this new technology was officially acknowledged.

The double headlamps common to all Harleys from 1929-on were, naturally, fitted to the new Model Ds and during the model year the front fork was extended by one inch (25 mm). The 18-inch wheels were those of the Two Cam Big Twin model range, brakes were therefore internal-expanding drum front and external-contracting band rear. Standard paint finish was Olive Green with maroon striping, with a centre gold stripe; handlebars and wheel rims were black, but potential owners could choose from a range of new colours – Black, Maroon, Grey, Blue, Cream and Coach Green – all available at extra cost.

Model development

1930 model year

The range had only just completed its first production year when a modified frame was fitted which lowered the seat height and increased ground clearance, and improved battery access. Furthermore, the exposed-spring forks, while visually similar to those introduced on the Models A and B, were now of much heavier construction with heat-treated drop-forged I-beam-section rigid (sprung) legs, and a steering lock increased security against theft. As with the other Harleys, the primary chain was now automatically lubricated. This new frame was considered strong enough for the Model D to be considered suitable for sidecar-hauling; unfortunately, the machine was so underpowered (especially with the engine in the low-compression sidecar state of tune), that only special lightweight single-seater sidecars could be fitted.

A larger clutch was fitted to improve the service life of the transmission. The 6¾-inch-diameter, 1¹⁄₁₆-inch-wide (171 mm x 27 mm) front brake drum was now made integral with the hub.

The modifications had some effect; in the hands of a skilled rider, a good Model D was capable of amazing results, and William H. Davidson was just such a skilled rider. President of Harley-Davidson from 1942 (on the death of his uncle Walter) until 1971-3 and a lifelong motorcyclist, William had started full-time work on the shop floor in the family firm in 1928 and by 1930 was still working his way up through the ranks. In 1930 he (with William J. Harley on another factory bike) entered the Jack Pine Enduro on a brand-new and standard Model DLD and won the coveted cowbell trophy with a near-perfect score, a feat which earned him the title of American Motorcycle Association (AMA) National Enduro Champion and continued Harley's run of Jack Pine victories every year since 1924. The Jack Pine, which has been run every year (except for 1942-5) since 1923 by the Lansing Motorcycle Club (Lansing being the capital city of the State of Michigan) is the most gruelling single event in the US motorcycle competition year and victory was highly prized for its value as publicity by all manufacturers. Originally a three-day, 800-mile event (later reduced to two days and about 500 miles; in 1930 the course was 420 miles long), usually starting in Lansing on the steps of the Capitol Building, run through some of the most challenging terrain in the state of Michigan and finishing in up-state Michigan at the Lansing Motorcycle Club's grounds, riders on stripped-down Big Twins had to average 24 mph (rigidly controlled by check points) across rivers, through sand, thick brush and deep woods.

1931 model year

This was the last year in which the Schebler Deluxe carburettor was fitted as standard. The 25-year association with this supplier was about to come to an end, all Harleys from 1933 onwards being fitted with Linkert instruments. If one can believe the sources, welding was used for the first time in frame gusseting; otherwise frames were still brazed. As with all 1931-on Harleys, the over-complex double headlamps gave way to a single, larger, 7-inch unit. A larger external-contracting band rear brake, claimed to be 31% more effective, was fitted to the Model Ds; in the pursuit of a policy of standardization of parts followed by the Motor Company at this time, this was also fitted to the Model C and the few Model Bs exported. Other technical improvements were a modified clutch and a revised generator drive. On the aesthetic front, the components formerly nickel-plated for decorative purposes were now chromed and new handlebar grips were fitted, as was a completely new fishtail silencer.

Unfortunately it was all too late for the Model D; slow and unreliable, its numerous teething troubles had tried the patience of the customers too far, and the developing Depression did the rest. Its sales performance never met Harley's expectations and a replacement was hurriedly developed; the much-improved Model R.

Model range

Model D	45 cu. in./742 cc. sv V-twin engine	(1929-31)
Model DS	45 cu. in./742 cc. sv V-twin engine, sidecar-specification	(1930-31)
Model DL	45 cu. in./742 cc. sv V-twin engine, medium compression	(1929-31)
Model DLD	45 cu. in./742 cc. sv V-twin engine, high compression	(1930-31)
Model DC	45 cu. in./742 cc. sv V-twin engine, Commercial	(1931)

Production figures

Model	1929	1930	1931
D	4513	2000	715
DS	-	213	276
DL	2343	3191	1306
DLD	-	206	241
DC	-	-	11

Catalogue prices ($ US)

Model	1929	1930	1931
D	290	310	310
DS *	-	310	310
DL	290	310	310
DLD	-	310	325

* Prices are for motorcycles only, less sidecars.

Model D 1929 to 1931
Model DS 1930 to 1931
Model DC 1931

Engine
Engine	Air-cooled 45° sv V-twin
Bore x stroke	2¾ x 3¹³⁄₁₆ (2.75 x 3.81) in.
Bore x stroke	69.85 x 96.84 mm
Capacity	45.12 cu. in. (742 cc)
Power output (SAE)	15(11)/3900 rpm
Compression ratio	4.3:1
Carburettor	Schebler 1-inch
Ignition	Battery-and-coil
Exhaust system:	
1929-30	Four-tube
1931	Twin-tube, fishtail ends

Transmission
Gearbox	3-speed
Gearchange	Hand change, on tank
Final drive	Chain

Cycle parts
Frame construction	Single-loop frame
Front fork	1929: Springer, exposed springs, forged oval section legs
	1930-31: Springer, exposed springs, forged I-beam legs
Rear suspension	None
Front brake	Drum, via cable and handlebar lever
Rear brake	Drum, via rod and pedal

General information
Unladen weight	390 lb/177 kg
Front and rear tyres	25 x 4.00 beaded-edge
Wheels	18-inch wire-spoked
Maximum speed	100 km/h
Tank capacity	20 litres (1929)
	21.3 litres (from 1930)
Sidecar	1929: Type 29LS
	1930: Type 30LS
	Type 30LSC (Chassis)
	1931: Type 31LS (complete sidecar)
	Type 31LS (Body only)
	Type 31LSC (Chassis)

Model DL 1929 to 1931
Model DLD 1930 to 1931

Engine
Engine	Air-cooled 45° sv V-twin
Bore x stroke	2¾ x 3¹³⁄₁₆ (2.75 x 3.81) in.
Bore x stroke	69.85 x 96.84 mm
Capacity	45.12 cu. in. (742 cc)
Power output (SAE)	DL: 18.5(14)/4000 rpm
	DLD: 20(15)/4000 rpm
Compression ratio	DL: 5.0:1
	DLD: 6.0:1
Carburettor	DL: Schebler 1-inch
	DLD: Schebler 1¼-inch
Ignition	Battery-and-coil
Exhaust system:	
1929-30	Four-tube
1931	Twin-tube, fishtail ends

Transmission
Gearbox	3-speed
Gearchange	Hand change, on tank
Final drive	Chain

Cycle parts
Frame construction	Single-loop frame
Front fork	1929: Springer, exposed springs, forged oval section legs
	1930-31: Springer, exposed springs, forged I-beam legs
Rear suspension	None
Front brake	Drum, via cable and handlebar lever
Rear brake	Drum, via rod and pedal

General information
Unladen weight	390 lb/177 kg
Front and rear tyres	25 x 4.00 beaded-edge
Wheels	18-inch wire-spoked
Maximum speed	DL: 110 km/h
	DLD: 120 km/h
Tank capacity	20 litres (1929)
	21.3 litres (from 1930)
Sidecar	1929: Type 29LS
	1930: Type 30LS
	Type 30LSC (Chassis)
	1931: Type 31LS (complete sidecar)
	Type 31LS (Body only)
	Type 31LSC (Chassis)

The Forty-Fives: Model R

1931 to 1936

Model description and technical information

As successor to the Model Ds, there were high expectations for the Model R range, particularly in view of the several weaknesses of the first Forty-Fives. The most obvious point of identification of the new models was the location of the generator in the conventional position horizontally across the front of the engine and driven by a conventional gear drive directly from the timing gear train. This necessitated a new frame whose front downtube now curved forwards around the longer engine unit, and which was heavier, as were the front forks, to permit sidecar-hauling (even though the engine's power output was still no better-suited to the purpose). A heavier clutch and gearbox were used on the Model Rs, rather than the lightweight patterns, common to the singles, used on the Model D range.

The cylinders now featured a gap for greater cooling airflow between cylinder barrel and exhaust port. Other features to improve the reliability and performance of the new models included stronger valve springs, improved timing gear teeth patterns and other valvegear modifications, bigger flywheels, better aluminium-alloy pistons and revised crankcases with a modified breather. Breathing arrangements were improved for the gearbox as well, with the fitting of a new housing which included a breather assembly. Even the oil pump housing was strengthened and the generator was fitted with improved carbon brushes for a more even output of power. The exhaust was finished in heat-resistant black; the fishtail silencer introduced for 1931 continued for another two years before being replaced by the upswept High-Flo tailpipe; this itself only lasted for a year until being replaced for the last two model years by an even-bigger large-fishtail unit. Fuel tanks could be ordered with, instead of the traditional lettering on a plain background, a new curving design – the forerunner of the following year's Art Deco eagle design – highlighting the lettering. Colours available were Black, Maroon, Grey, Blue, Cream and Coach Green, and there was also a new range of two-tone colour schemes to choose from for this year; Teak Red/Black, Maroon/Cream, White/Gold and Blue/Grey.

The Springer fork with forged I-beam-section rigid (sprung) legs was retained almost to the end of the life of the succeeding Model W range. The upgraded engine and front frame, from front wheel right back to the seat tube, was also used for the then-brand-new Servi-Car model range.

The price of the Model Rs was slightly reduced on

Model range

Model RM	45 cu. in./742 cc. sv V-twin, low compression, magneto ignition	(1931)
Model R	45 cu. in./742 cc. sv V-twin, low compression	(1932-36)
Model RS	45 cu. in./742 cc. sv V-twin, low compression, sidecar specification	(1932-36)
Model RL	45 cu. in./742 cc. sv V-twin, medium compression	(1932-36)
Model RLD	45 cu. in./742 cc. sv V-twin, high compression	(1932-36)
Model RE	45 cu. in./742 cc. sv V-twin, magnesium-alloy pistons, low compression	(1933)
Model RESX	45 cu. in./742 cc. sv V-twin magnesium-alloy pistons, low compression, sidecar specification	(1933)
Model RLE	45 cu. in./742 cc. sv V-twin, magnesium-alloy pistons, medium compression	(1933)
Model RLDE	45 cu. in./742 cc. sv V-twin, magnesium-alloy pistons, high compression	(1933)
Model RLEX	45 cu. in./742 cc. sv V-twin magnesium-alloy pistons, medium compression	(1933)
Model RLX	45 cu. in./742 cc. sv V-twin medium compression	(1934)
Model RLDX	45 cu. in./742 cc. sv V-twin, high compression	(1934)
Model RSX	45 cu. in./742 cc. sv V-twin, low compression, sidecar specification	(1933-34)
Model RX	45 cu. in./742 cc. sv V-twin, low compression	(1934)

Special Japanese export model
Model RSR	45 cu. in./742 cc. sv V-twin, low compression, sidecar specification	(1934-36)

(All models with X suffix model codes have magnesium-alloy pistons in 1933 and aluminium-alloy pistons in 1934) that of the Model D range; an indication as much of the difficulty of selling anything at that time as well as of the Motor Company's determination for its new model range to achieve success. It should be noted that a whole half-dozen Model RMs (with magneto ignition) were sold in 1931, in advance of the introduction of the new model range proper; as with the other Harleys sold with this form of ignition at this time, it was clear that demand no longer existed and they were soon dropped from the range.

Model development

1933 model year
The headlamp glass was no longer flat but convex; curved towards the outside. In addition, the frame, forks and chainguards were painted black. There were, as with almost every year, new tank graphics; this year's were in the new Art Deco eagle design. Full chrome-plating of all decorative parts, including handlebars, chainguard, exhaust pipe and silencer, cost only 15 bucks extra. Late in the model year a Linkert instrument of similar size replaced the Schebler carburettor.

1934 model year
The Airflow taillamp was fitted and new tank graphics – the flying diamond design – were used. Improved low-expansion pistons, a new oil pump and a strengthened clutch were fitted, with the (for this year only) upswept High-Flo tailpipe exhaust. Export models could even be ordered with a front stand.

1935 model year
Although introduced late this year, the new models appearing in December rather than September, as usual, the technical highlights for 1935 were an improved three-speed constant-mesh gearbox and an internal-expanding drum rear brake of 6¾ inch (171 mm) diameter and shoe width of 1¹⁄₁₆ inch (27 mm). The large-fishtail exhaust gave a new engine note, while the carburettor air intake was enlarged, as were the oil and fuel filler necks. The toolbox was moved from under the headlamp to the rear right-hand side of the frame.

For all V-twin models, the private customer had the choice between the standard specification, the Solo Group of accessories (comprising a special speedometer and a sidestand, or jiffy stand) and the Deluxe Solo Group (comprising the Solo Group fittings plus a steering damper, the famous small running light on the mudguard, the full chrome-plating package, saddlebags and a cube-shaped gearchange knob). Prospective customers for the Model R also had a new range of two-tone colour schemes to choose from for this year; Teak Red/Black, Venetian Blue/Silver, Verdant Green/Black, Egyptian Ivory/Regent Brown and Olive Green/Black.

1936 model year
The cylinder heads and cylinder barrels were again restyled for 1936, with much deeper finning for better heat dissipation. At the same time the combustion chambers were redesigned for more efficient combustion and a Y-shaped intake manifold further improved the engine's breathing.

The front fork's coil springs were enclosed, the new tank graphics were used, the steering head angle was slightly altered to increase trail, and thicker brake shoes were fitted.

At the end of the model year the succeeding Model W range was announced, which would carry on until 1951 and which, particularly during the Second World War, would be in very high demand.

1934 Model RLX: Art Deco flying diamond tank graphics

1934 Model RL: St Bernard toolbox and High-Flo exhaust

1936 Model RLD – note enclosed front fork springs

Motorcycle catalogue prices ($ US)

Model	1932	1933	1934	1935	1936
R (E)	295	280	280	295	295
RS *	295	280	280	295	295
RL (E)	295	280	280	295	295
RLD(E)	310	290	290	305	295

* Prices are for motorcycles only, less sidecars.

Prices and availability of optional Accessory Groups for motorcycles ($ US)

Group	1935	1936
Solo Group #1	24.90	-
Standard Solo Group	-	28.00
Deluxe Solo	49.50	49.50
Police Group #1	49.90	53.00
Deluxe Police	92.50	90.50

Note that Standard means a specification higher than standard ex-works finish.

Prices and availability of optional Accessory Groups for sidecars ($ US)

Group	1935	1936
LT sidecar Group #1	16.75	-
Standard Sidecar Group	-	15.75
LT Deluxe Group	39.90	39.50
Truck Deluxe Group	26.90	26.90

Note that Standard means a specification higher than standard ex-works finish.

Production figures

Model	1931	1932	1933	1934	1935	1936
R (RE)	-	410	162	450	543	539
RS	-	111	37	302	392	437
RL (RLE)	-	628	264	743	819	355
RLD (RLDE)	-	98	68	240	177	540
RM	6	-	-	-	-	-
RESX	-	-	5	-	-	-
RLEX	-	-	8	-	-	-
RX	-	-	-	5	-	-
RLX	-	-	-	36	-	-
RLDX	-	-	-	1	-	-
RSR	-	-	-	215	50	30

Model R 1932 to 1936
Model RM 1931
Model RX 1934
Model RS 1932 to 1936
Model RSX 1933 to 1934
Model RSR 1934 to 1936

Engine
Engine	Air cooled 45° sv V-twin
Bore x stroke	2¾ x 3¹³⁄₁₆ (2.75 x 3.81) in.
Bore x stroke	69.85 x 96.84 mm
Capacity	45.12 cu. in. (742 cc)
Power output (SAE)	16(12)4000 rpm
Compression ratio	4.3:1
Carburettor	Schebler 1 (1932-33)
	Linkert 1 (1934-35), Type M-11
	Linkert 1 (1936), Type M-16
Ignition:	
Model RM	Magneto
All others	Battery-and-coil
Exhaust system:	
1932-33	1 small-fishtail silencer
1934	High-Flo exhaust
1935-36	1 large-fishtail silencer

Transmission
Gearbox	3-speed
Gearchange	Hand change, on tank
Final drive	Chain

Cycle parts
Frame construction	Single-loop frame
Front fork	Springer, exposed springs, forged I-beam legs
Rear suspension	None
Front brake	Drum, via cable and handlebar lever
Rear brake	Drum, via rod and pedal

General information
Unladen weight	408 lb/185 kg
Permissible total weight	330 kg
Front and rear tyres	4.00-18 wired-edge
Wheels	18-inch wire-spoked
Maximum speed	Solo motorcycle: 110 km/h
	Sidecar: approx. 80 km/h
Tank capacity	21.3 litres

Model RL 1932 to 1936
Model RLX 1934
Model RLD 1932 to 1936
Model RLDX 1934

Engine
Engine	Air-cooled 45° sv V-twin
Bore x stroke	2¾ x 3¹³⁄₁₆ (2.75 x 3.81) in.
Bore x stroke	69.85 x 96.84 mm
Capacity	45.12 cu. in. (742 cc)
Power output (SAE)	RL: 18.5(14)/4000 rpm
	RLD: 22(16)/4000 rpm
Compression ratio	RL: 5.0:1
	RLD: 6.0:1
Carburettor	RL: Schebler 1 (1932-33)
	RLD: Schebler 1¼ (1932-33)
	RL: Linkert 1 (1934-36) Type M-16
	RLD: Linkert 1¼ (1934-36), Type M-41
Ignition	Battery-and-coil
Exhaust system:	
1932-33	1 small-fishtail silencer
1934	High-Flo exhaust
1935-36	1 large-fishtail silencer

Transmission
Gearbox	3-speed
Gearchange	Hand change, on tank
Final drive	Chain

Cycle parts
Frame construction	Single-loop frame
Front fork	Springer, exposed springs, forged I-beam legs
Rear suspension	None
Front brake	Drum, via cable and handlebar lever
Rear brake	Drum, via rod and pedal

General information
Unladen weight	408 lb/185 kg
Permissible total weight	330 kg
Front and rear tyres	4.00-18 wired-edge
Wheels	18-inch wire-spoked
Maximum speed	RL: 115 km/h
	RLD: 120 km/h
Tank capacity	21.3 litres

Sidecars:
Model 32LS, Model 32LS (Body only), Model 32LSC (Chassis)
Model 33LS, Model 33LS (Body only), Model 33LSC (Chassis)
Model 34LS, Model 34LS (Body only), Model 34LSC (Chassis)
Model 35LS, Model 35LS (Body only), Model 35LSC (Chassis)
Model 36LS, Model 36LS (Body only), Model 36LSC (Chassis)

Model RE 1933 Solo
Model RESX sidecar-specification 1933

Engine
Engine	Air-cooled 45° sv V-twin
Bore x stroke	2¾ x 3¹³⁄₁₆ (2.75 x 3.81) in.
Bore x stroke	69.85 x 96.84 mm
Capacity	45.12 cu. in. (742 cc)
Power output (SAE)	17(13)/4000 rpm
Compression ratio	4.3:1
Carburettor	Schebler 1-inch
Ignition	Battery-and-coil
Exhaust system	1 small-fishtail silencer

Transmission
Gearbox	3-speed
Gearchange	Hand change, on tank
Final drive	Chain

Cycle parts
Frame construction	Single-loop frame
Front fork	Springer, exposed springs, forged I-beam legs
Rear suspension	None
Front brake	Drum, via cable and handlebar lever
Rear brake	Drum, via rod and pedal

General information
Unladen weight	408 lb/185 kg
Permissible total weight	330 kg
Front and rear tyres	4.00-18 wired-edge
Wheels	18-inch wire-spoked
Maximum speed	Solo motorcycle: 110 km/h
	Sidecar: approx. 80 km/h
Tank capacity	21.3 litres

Models with E suffix model codes have magnesium-alloy pistons, those without have pistons of Lynite, a special aluminium alloy.
Prices and production figures for these variations are not available. For sidecar models, see previous page.

Model RLE 1933
Model RLEX 1933
Model RLDE 1933

Engine
Engine	Air-cooled 45° sv V-twin
Bore x stroke	2¾ x 3¹³⁄₁₆ (2.75 x 3.81) in.
Bore x stroke	69.85 x 96.84 mm
Capacity	45.12 cu. in. (742 cc)
Power output (SAE)	RLE: 20(15)/4000 rpm
	RLDE: 23(17)/4000 rpm
Compression ratio	RL: 5.0:1
	RLD: 6.0:1
Carburettor	RLE: Schebler 1-inch
	RLDE: Schebler 1¼-inch
Ignition	Battery-and-coil
Exhaust system	1 small-fishtail silencer

Transmission
Gearbox	3-speed
Gearchange	Hand change, on tank
Final drive	Chain

Cycle parts
Frame construction	Single-loop frame
Front fork	Springer, exposed springs, forged I-beam legs
Rear suspension	None
Front brake	Drum, via cable and handlebar lever
Rear brake	Drum, via rod and pedal

General information
Unladen weight	408 lb/185 kg
Permissible total weight	330 kg
Front and rear tyres	4.00-18 wired-edge
Wheels	18-inch wire-spoked
Maximum speed	RLE: 115 km/h
	RLDE: 120 km/h
Tank capacity	21.3 litres

The Forty-Fives: Model W

1937 to 1951

Model description and technical information

Evidently H-D were still not happy with their 45, as the Model W was not just an incremental development of the previous model, but an intensive redesign. The all-ball-bearing engine was now lubricated by a recirculating dry-sump system instead of the previous total-loss arrangement, which reduced oil consumption to a more-bearable level. The new two-part tank used the left-hand compartment for petrol, and the right-hand for oil. The oil pump was now bolted on the outside of the timing cover.

The side-valve engine was now very robust and long-lived; mileages between overhauls of 60,000 (100,000 kilometres) were quite possible.

The Forty-Fives were delivered between 1937 and 1942 (the last pre-War model year for the USA) in one of three levels of engine compression, of which the lowest was for the sidecar-specification engines of the Models W and WS. This changed with the introduction of the Model WLS in 1941, the sidecar-specification engine of which was now the medium-compression option.

The large-fishtail exhaust silencer was similar in style to that of the Big Twins, although the silencer itself was actually shorter.

The famous instrument panel with speedometer, oil pressure gauge, ammeter, ignition switch and speedometer lighting switch lay centrally between the two tanks and was much closer to the rider's field of view than the old speedometer. The speedometer itself had a white dial and was fitted with a trip odometer.

The much heavier main frame loop also showed detail improvements. In addition, welding as a means of frame construction was introduced gradually over the entire life span of the W models up to 1952. That means that practically all Model W frames exhibit signs of brazing and of welding at various seams and joints, although in varying degrees.

The three equipment specifications (introduced in 1935 with the Model R) were extended. Thus the steering damper, since 1936, was available in the Standard Solo Group, while the Deluxe Solo Group of the same year now included a stop-lamp and the licence plate holder. It was about this time that the factory began to put pressure on its dealers only to place orders for new machines with one of the options Groups; the 'standard' machine effectively no longer existed. Standard colours available were Teak Red with Black stripe and Gold edge, Bronze Brown with Blue stripe and Gold edge, Delphine Blue with Teak Red stripe and Gold edge, Police Silver with Black stripe and Gold edge and Olive Green with Black stripe and Gold edge.

Visually, the Model Ws were very similar to the

Model range

Model W	45 cu. in./742 cc. sv V-twin engine	(1937-40)
Model WS	45 cu. in./742 cc. sv V-twin engine, sidecar-specification	(1937-40)
Model WL	45 cu. in./742 cc. sv V-twin engine	(1937-51)*
Model WLS	45 cu. in./742 cc. sv V-twin engine, sidecar-specification	(1941-51)**
Model WL-SP	45 cu. in./742 cc. sv V-twin engine, aluminium-alloy cylinder heads	(1946-51)*
Model WLD	45 cu. in./742 cc. sv V-twin engine	(1937-42)
Model WLDR	45 cu. in./742 cc. sv V-twin engine	(1937-41)***
Model WLDD	45 cu. in./742 cc. sv V-twin engine	(1939)****

Models W and WS have low compression engines.
Models WL, WLS and WL-SP have medium compression engines.
Models WLD, WLDR and WLDD have high compression engines.
*   WL models were not available for civilian use during 1943 and 1944.
**   WLS models were not available for civilian use from 1943 to 1947.
***   WLDR models were pure racing motorcycles, which were however sold in road trim for 1941.
****   One source mentions a further Forty-Five, a 1939 Model WLDD. This must, however, be a printing error; nowhere else is a hint of this model's existence to be found.

styling of the larger ohv Twins, with the obvious exception of the cylinder heads. One did not have to be too close to be able to distinguish the Flathead side-valve engine (with their characteristic upright finning) from the Knucklehead ohv engine, with the pronounced round knobs on the rocker housings.

The Model W range, particularly successful in winning Government orders for military use, represented a quantum leap forward in the development of the Harley Forty-Fives and became H-D's best-selling model range. From 1937 to 1951 (with military versions being produced until 1952), some 114,000 Model Ws of all sub-types came off the lines at Milwaukee; only the F Models, produced over a much greater period of time, ran it close in the rankings.

Model development

1938 model year
In the second year of production the clutch and gearbox were improved, being strengthened here and there and also using stronger materials on many components. While the transmission components were very similar to those fitted to the Big Twins, unlike the Model E, the Model Ws still only got three speeds.

1939 model year
Many small improvements made it into mass production, such as the restyled Cats' Eye instrument panel, the Boattail taillamp, the body of which was painted to match the mudguard, stainless-steel mudguard tips, the needle roller-bearing kickstart, Neoprene-covered ignition HT leads and the self-aligning steering head bearing cones. The carburettor got a float chamber drain fitting. As a result of problems with the engine's lubrication system causing uneven lubrication of the cylinder bores, the front cylinder piston had three compression rings as opposed to the two plus oil scraper ring of the rear cylinder piston. Ride Control, the form of friction damping for the front fork springs, was introduced as an extra-cost option; models so equipped are clearly identified by the two slotted steel plates mounted on each side of the fork springs.

1940 model year
Deeply-finned aluminium-alloy cylinder heads were introduced for the first time, but only on the high-compression Model WLD and the Model WLDR racer. The difference between front and rear cylinders' pistons no longer applied, the engine's lubrication system having been modified to achieve more even oiling of both cylinder bores; both were fitted with two compression rings and an oil scraper.

The forged I-beam-section rigid (sprung) legs introduced on the front forks of the Forty-Fives in 1930 gave way to improved chrome-molybdenum-steel oval-section tubular legs, but the four massive volute springs remained as a clear point of identification for all Harley models. D-shaped footboards were introduced, chrome-plated teardrop style metal tank badges and 5.00-16 tyres, with substantially greater air volume than the old fitment, were made available as options; by the following year they would completely replace the old 4.00-18 tyres.

1941 model year
For the first time since 1937 the frame was modified, in this case to accept the new constant-mesh gearbox. The last full year of peace brought a larger 7-inch (178 mm) air cleaner, the Airplane-Style speedometer (black face with large silver numbers), a strengthened clutch with correspondingly-modified release

1937 Model WLDR – a real racing motorcycle

1940 Model WL

mechanism and 16-inch wheels. The rear brake drum was enlarged and the front brake drum was now made integral with the formerly-separate hub.

The Model WLDR racers were sold as regular road-going machines in their last year. The aluminium-alloy cylinder heads, no doubt as a consequence of the Army insisting on them for their machines, became more widespread on WL models from this year onwards. The L suffix normally identifies a high-compression model in Harley terminology, but for the Forty-Fives it became the identifier for the medium-compression option, the high-compression engines being DLD, RLD, WLD and WLDR. Model WLs from 1937 to 1940 had a 5:1 compression ratio; from 1941-onwards those with aluminium-alloy cylinder heads kept this specification, while those with cast-iron heads were reduced to 4.75:1, thus replacing the low-compression Model W in the range.

1942 model year
The transition to production for military use forced by the entry of the US into the War caused a sharp drop in the numbers of machines released for civilian use in this equally sharply-curtailed model year. The almost 9,000 Model Ws sold in 1941 contrast with the 275, virtually unchanged, sold for 1942. Full details of the Army models are given in the chapter on Military machines.

1943 to 1945 model years
During the War, civilian production came to a standstill. One was only allowed to buy a new Harley once one had obtained special permission from the authorities. Model Ws were therefore not delivered to private customers between February 1942 and October 1945 and only a few big Harleys found their way into civilian use; colours available were restricted to Grey and Silver. Only 1,357 Model Ws were bought by motorcycle enthusiasts in the entire 1945 model year. Beyond that, however, not far off 8,000 WLAs were supplied to the Army, in order to fulfil the contract.

The aluminium-alloy cylinder heads, only available on the Models WLD and WLDR before the War, were now made available only for newly-manufactured WL models, at a surcharge of seven bucks.

1946 and 1947 model years
After the end of hostilities, the Government offered war-surplus WLAs to private customers at the discounted price of $450, with the result that sales of new machines to civilians began to suffer. This practice was continued right up to 1947, during which about 15,000 Army machines found private buyers. Nevertheless, Milwaukee could still sell a total of over 4,400 WL models in the year 1946, while for 1947 3,388 were sold.

Where the saddles of the wartime machines were

padded with horsehair, latex mesh was now permitted to be used. Gradually, the various chrome trim pieces, discontinued during the war years, were made available again. At first only Grey and Flight Red paint finishes were available, but for 1947 Brilliant Black, Flight Red, Skyway Blue and Police Silver were available for the US, with Olive Green for export models. The Utility Solo Group of accessories could be added to standard specification; now the Special Solo Group was introduced as a top-of-the-range luxury specification.

1947 model year
For 1947 the gearchange was changed round so that first was at the back of the gate and closest to the rider, then neutral and finally second and third. Amongst other minor modifications were the red 'speed ball' tank badges, the Tombstone taillamp (so beloved in later years) and the new speedometer fitted to all 1947 Harleys; the speedometer needle was now red and the numbers on the speedometer dial were enlarged to improve its readability.

1948 model year
The Deluxe Solo saddle was now available for WL models. There were also slightly altered tank badges (as with every year) and modifications to the electrics.

1949 model year
The air cleaner cover was made of stainless steel from now on, as was the timer unit (contact breaker housing) cap and generator cover housing. The exhaust system was now finished in a silicon-based heat-resistant black. The Springer front forks were fitted to the end of the Model W's life, unlike on the bigger Harleys which were now changing over to hydraulically-damped telescopic front forks.

The equipment specification levels were extended; Standard, Utility, Sport and Deluxe specifications now being made available. The last cost some 122 bucks, which raised the price of a Model WL (with aluminium-alloy cylinder heads) to over 700 dollars; only a little bit more would get a new Hydra-Glide Seventy-Four.

1950 model year
The fishtail silencer evolved into the Mellow-Tone exhaust – a cigar-shaped silencer with a rounded tailpipe, which could be ordered in silicon-based heat-resistant black or fully chrome-plated. On top of that there were new deep-skirted and seamless Air Flow mudguards.

1951 model year
The final model year brought a new Linkert Type M-54 carburettor and a few aesthetic embellishments such as the chromed tube linking the two exhaust downpipes and their respective exhaust ports. In the following year, a last contingent of 549 Army Model WLAs arrived on the 29th February at different US units. Even if that was the last of the WL range proper, the three-wheeled Servi-Cars carried on using major components of it up to 1973.

1940 Model WL with Ride Control friction damping for front fork springs

1941 Model WL – the first year aluminium-alloy cylinder heads were available for road models

1941 Model WL – it'll be pretty when it's finished!

1947 Model WL – post-war material restrictions beginning to ease off, and accessories available again

1948 Model WL

1948 Model WL

Production figures

Model	1937	1938	1939	1940	1941	1942	1945	1946	1947	1948	1949	1950	195
W	509	302	260	439	4095	-	-	-	-	-	-	-	-
WS	232	247	170	202	-	-	-	-	-	-	-	-	-
WL ***	560	309	2212	569	4277*	142**	1357	4410	3388	3388	2289	1108	1044
WLD	581	402	326	567	455	133	-	-	-	-	-	-	-
WLDR	145	139	173	87	171	-	-	-	-	-	-	-	-

Civilian production suspended in favour of war production from the beginning of 1942 to the middle of 1945.
*   includes Model 41WLS.
**   includes Model 42WLS.
***   from 1946, includes WL-SP.

Motorcycle catalogue prices ($ US)

Model	1937	1938	1939	1940	1941	1942	1945	1946	1947	1948	1949	1950	195
W	355	355	355	350	-	-	-	-	-	-	-	-	-
WS *	355	355	355	350	-	-	-	-	-	-	-	-	-
WL	355	355	355	350	350	350	396	396	490	535	590	590	548
WL-SP	-	-	-	-	-	-	-	403	498	543	598	600	555
WLD	355	355	355	395	365	365	-	-	-	-	-	-	-
WLDR	(380)	(380)	(380)	(395)	385	-	-	-	-	-	-	-	-

* Prices are for motorcycles only, less sidecars.
From 1945-on, prices for sidecar-specification motorcycles are no longer given separately, but are the same as those of the standard solo machines. From the 1948 model year, sidecar-specification Model Ws were no longer available ex-works, although one could still order the necessary items, with a sidecar to be fitted later.
For its final year of production (1941), the Model WLDR was sold in road trim.

Sidecar catalogue prices ($ US)

Model	1937	1938	1939	1940	1948	1949	1950	1951
LS	105	105	105	105	N./Av.	N./Av.	N./Av.	N./Av.
LSC Chassis	63	63	63	63	N./Av.	N./Av.	N./Av.	N./Av.
LLSC	-	-	-	105	-	-	-	-
LLS Chassis	-	-	-	63	-	-	-	-
M	-	-	-	-	197	215	215	245
MC Chassis	-	-	-	-	118	130	130	155

Chassis = Frame, wheel and fittings only (for aftermarket coach-built bodies)
All sidecars from 1940 to 1947 were reserved for military and Government production; therefore no information is given here.
LS is for fitting on the right-hand side of the motorcycle, LLS for fitting on the left.
M is the Package Truck, for commercial load-carrying use only.

Prices and availability of optional Accessory Groups for motorcycles ($ US)

Group	1937	1938	1939 1942	1940 1943	1941	1944	1945	1946	1947	1948	1949	1950	1951
Standard Solo	21.75	16.70	15.50	-	-	-	-	-	-	-	-	-	-
Standard Commercial	20.00	-	-	-	-	-	-	-	-	-	-	-	-
Utility Solo	-	-	-	11.00	14.50	14.50	14.50	14.50	34.00	24.00	24.75	23.50	30.50
Sport Solo	-	-	-	22.50	27.00	-	-	-	-	59.25	75.50	46.40	58.25
Special Solo	-	-	-	-	-	-	44.50	55.00	100.00	92.00	-	-	-
Deluxe Solo	49.00	49.75	47.00	46.00	60.00	-	-	-	-	-	122.00	81.25	97.00
Standard Police	49.50	52.25	41.00	39.50	42.50	42.50	42.50	57.00	75.00	73.00	87.00	-	-
Deluxe Police	83.50	-	-	-	-	-	-	-	-	-	-	-	-

Note that Standard means a specification higher than standard ex-works finish.

Prices and availability of optional Accessory Groups for sidecars ($ US)

Group	1937	1938	1939	1940	1948	1949	1950	1951
Standard Commercial	20.00	14.25	14.00	-	-	-	-	-
Standard Sport	17.50	17.50	17.75	-	-	-	-	-
Standard Truck	-	14.25	14.00	-	-	-	-	-
Utility	-	-	-	8.50	33.50	38.50	-	-
Utility Truck	-	-	-	8.50	33.50	38.50	-	-
Deluxe Sport	44.00	44.00	47.00	50.00	90.50	102.00	102.45	122.50
Deluxe Truck	29.00	29.00	31.75	-	-	-	-	-

Note that Standard means a specification higher than standard ex-works finish.
All sidecars from 1940 to 1947 were reserved for military and Government production; therefore no information is given here.

Model W 1937 to 1940

Engine
Engine	Air-cooled 45° sv V-twin
Bore x stroke	2¾ x 3¹³⁄₁₆ (2.75 x 3.81) in.
Bore x stroke	69.85 x 96.84 mm
Capacity	45.12 cu. in. (742 cc)
Power output (SAE)	19(14)/4400 rpm
Compression ratio	4.3:1
Carburettor	Linkert Type M-21
Ignition	Battery-and-coil
Exhaust system	1 Fishtail silencer large

Transmission
Gearbox	3-speed
Gearchange	Hand change, on tank
Final drive	Chain

Cycle parts
Frame construction	Single-loop frame
Front fork	1937-39: Springer, exposed springs, forged I-beam legs. 1940: Springer, exposed springs, oval-tube legs
Rear suspension	None
Front brake	Drum, via cable and handlebar lever
Rear brake	Drum, via rod and pedal

General information
Unladen weight	230 kg
Permissible total weight	410 kg
Front and rear tyres	1937-1940: Standard 4.00-18. 1940 optional: 5.00-16
Wheels	1937-40: 2.15 x 18 wire-spoked. 1940: optional 3.00 x 16 wire-spoked
Maximum speed	110 km/h
Tank capacity	12.9 litres

Model WS and sidecar 1937 to 1940

Engine
Engine	Air-cooled 45° sv V-twin
Bore x stroke	2¾ x 3¹³⁄₁₆ (2.75 x 3.81) in.
Bore x stroke	69.85 x 96.84 mm
Capacity	45.12 cu. in. (742 cc)
Power output (SAE)	19(14)/4400 rpm
Compression ratio	4.3:1
Carburettor	Linkert Type M-21
Ignition	Battery-and-coil
Exhaust system	1 Fishtail silencer large

Transmission
Gearbox	3-speed
Gearchange	Hand change, on tank
Final drive	Chain

Cycle parts
Frame construction	Single-loop frame
Front fork	1937-39: Springer, exposed springs, forged I-beam legs. 1940: Springer, exposed springs, oval-tube legs
Rear suspension	None
Front brake	Drum, via cable and handlebar lever
Rear brake	Drum, via rod and pedal

General information
Unladen weight	approx. 330 kg
Permissible total weight	approx. 570 kg
Front and rear tyres	1937-1940: Standard 4.00-18. 1940 optional: 5.00-16
Wheels	1937-40: 2.15 x 18 wire-spoked. 1940: optional 3.00 x 16 wire-spoked
Maximum speed	90 km/h
Tank capacity	12.9 litres

The basic Model W was used practically without exception for sidecar-hauling; as a result, it did not feature in the official catalogue after 1938. Most of these Model Ws also went to the US Army; in spite of that, it is easiest to consider it here. For sidecar details, see previous page.

Model WL 1937 to 1942

Engine
Engine	Air-cooled 45° sv V-twin
Bore x stroke	2¾ x 3¹³⁄₁₆ (2.75 x 3.81) in.
Bore x stroke	69.85 x 96.84 mm
Capacity	45.12 cu. in. (742 cc)
Power output (SAE)	22(16)/4600 SAE (1937-42)
	21(15)/4600 SAE (1941-42)
Compression ratio	1937-42: 5.0:1 *
	1941-42: 4,75:1*
Carburettor	1937-41: Linkert Type M-21
	1942: Linkert Type M-51
Ignition	Battery-and-coil
Exhaust system	1 Fishtail silencer large

Transmission
Gearbox	3-speed
Gearchange	Hand change, on tank
Final drive	Chain

Cycle parts
Frame construction	Single-loop frame
Front fork	1937-39: Springer, exposed springs, forged I-beam legs
	1940-42: Springer, exposed springs, oval-tube legs
Rear suspension	None
Front brake	Drum, via cable and handlebar lever
Rear brake	Drum, via rod and pedal

General information
Unladen weight	230 kg
Permissible total weight	410 kg
Front and rear tyres	1937-1940: Standard 4.00-18
	1940 optional: 5.00-16
	1941-42: Standard 5.00-16
Wheels	1937-40: 2.15 x 18 wire-spoked
	1940: optional 3.00 x 16 wire-spoked
	1941-42: 3.00 x 16 wire-spoked
Maximum speed	115 km/h
Tank capacity	12.9 litres

Model WLS 1941 to 1942

Engine
Engine	Air-cooled 45° sv V-twin
Bore x stroke	2¾ x 3¹³⁄₁₆ (2.75 x 3.81) in.
Bore x stroke	69.85 x 96.84 mm
Capacity	45.12 cu. in. (742 cc)
Power output (SAE)	22(16)/4600 SAE (5.0)
	21(15)/4600 SAE (4,75)
Compression ratio	1937-42: 5.0:1 *
	1941-42: 4,75:1*
Carburettor	1941: Linkert Type M-21
	1942: Linkert Type M-51
Ignition	Battery-and-coil
Exhaust system	1 Fishtail silencer large

Transmission
Gearbox	3-speed
Gearchange	Hand change, on tank
Final drive	Chain

Cycle parts
Frame construction	Single-loop frame
Front fork	Springer, exposed springs, oval-tube legs
Rear suspension	None
Front brake	Drum, via cable and handlebar lever
Rear brake	Drum, via rod and pedal

General information
Unladen weight	approx. 330 kg
Permissible total weight	approx. 570 kg
Front and rear tyres	5.00-16
Wheels	3.00 x 16 wire-spoked
Maximum speed	95 km/h
Tank capacity	12.9 litres

*Compression ratios of 5.0:1 were used from 1941 only on those engines with aluminium-alloy cylinder heads. From 1941, the compression ratio on engines with cast-iron cylinder heads was reduced to 4.75:1, which resulted in a barely-noticeable loss of power.
For sidecar details, see page 130.

Model WLD 1937 to 1942

Engine
Engine	Air-cooled 45° sv V-twin
Bore x stroke	2¾ x 3¹³⁄₁₆ (2.75 x 3.81) in.
Bore x stroke	69.85 x 96.84 mm
Capacity	45.12 cu. in. (742 cc)
Power output (SAE)	24.5(18)/4700 rpm
Compression ratio	6.0:1
Carburettor	Linkert Type M-21
Ignition	Battery-and-coil
Exhaust system	1 Fishtail silencer large

Transmission
Gearbox	3-speed
Gearchange	Hand change, on tank
Final drive	Chain

Cycle parts
Frame construction	Single-loop frame
Front fork	1937-39: Exposed-spring Springer forks, forged I-beam legs 1940-42: Exposed-spring Springer forks, oval-tube legs
Rear suspension	None
Front brake	Drum, via cable and handlebar lever
Rear brake	Drum, via rod and pedal

General information
Unladen weight	230 kg
Permissible total weight	410 kg
Front and rear tyres	1937-1940: Standard 4.00-18 1940 optional: 5.00-16 1941-42: Standard 5.00-16
Wheels	1937-40: 2.15 x 18 wire-spoked 1940: optional 3.00 x 16 wire-spoked 1941-42: 3.00 x 16 wire-spoked
Maximum speed	120 km/h
Tank capacity	12.9 litres

Model WLDR 1937 to 1941

Engine
Engine	Air-cooled 45° sv V-twin
Bore x stroke	2¾ x 3¹³⁄₁₆ (2.75 x 3.81) in.
Bore x stroke	69.85 x 96.84 mm
Capacity	45.12 cu. in. (742 cc)
Power output (SAE)	29(21)/5000 rpm
Compression ratio	6.0:1
Carburettor	Linkert Type M-21
Ignition	Battery-and-coil
Exhaust system	1 Fishtail silencer large

Transmission
Gearbox	3-speed
Gearchange	Hand change, on tank
Final drive	Chain

Cycle parts
Frame construction	Single-loop frame
Front fork	1937-39: Springers, exposed-springs, forged I-beam legs 1940-41: Exposed-spring Springer forks, oval-tube legs
Rear suspension	None
Front brake	Drum, via cable and handlebar lever
Rear brake	Drum, via rod and pedal

General information
Unladen weight	approx. 180 kg
Permissible total weight	320 kg
Front and rear tyres	1937-1940: Standard 4.00-18 1940 optional: 5.00-16 1941: Standard 5.00-16
Wheels	1937-40: 2.15 x 18 wire-spoked 1940: optional 3.00 x 16 wire-spoked 1941: 3.00 x 16 wire-spoked
Maximum speed	from 130 km/h
Tank capacity	12.9 litres

For sidecar details, see page 130.

Model WL 1937 to 1940
Model WLS and sidecar 1948 to 1951

Engine
Engine	Air-cooled 45° sv V-twin
Bore x stroke	2¾ x 3¹³⁄₁₆ (2.75 x 3.81) in.
Bore x stroke	69.85 x 96.84 mm
Capacity	45.12 cu. in. (742 cc)
Power output (SAE)	22(16)/4600 rpm (M-51/52)
	25(18)/4600 SAE (M-54)
Compression ratio	5.0:1
Carburettor	1945-48: Linkert 1-inch Type M-51
	1949-51: Linkert 1-inch Type M-52
	1951: Linkert 1¼-inch Type M-54
Ignition	Battery-and-coil
Exhaust system	1 Fishtail silencer large

Transmission
Gearbox	3-speed
Gearchange	Hand change, on tank
Final drive	Chain

Cycle parts
Frame construction	Single-loop frame
Front fork	Springer, exposed springs, oval-tube legs
Rear suspension	None
Front brake	Drum, via cable and handlebar lever
Rear brake	Drum, via rod and pedal

General information
Unladen weight	235 kg (approx. 335 kg with sidecar)
Permissible total weight	410 kg (approx. 570 kg with sidecar)
Front and rear tyres	5.00-16
Wheels	3.00 x 16 wire-spoked
Maximum speed	115 km/h (95 km/h with sidecar)
Tank capacity	12.9 litres

Model WL-SP 1946 to 1951

Engine
Engine	Air-cooled 45° sv V-twin
Bore x stroke	2¾ x 3¹³⁄₁₆ (2.75 x 3.81) in.
Bore x stroke	69.85 x 96.84 mm
Capacity	45.12 cu. in. (742 cc)
Power output (SAE)	22(16)/4600 rpm (M-51/52)
	25(18)/4600 SAE (M-54)
Compression ratio	5.0:1
Carburettor	1946-48: Linkert 1-inch Type M-51
	1949-51: Linkert 1-inch Type M-52
	1951: Linkert 1¼-inch Type M-54
Ignition	Battery-and-coil
Exhaust system	1 Fishtail silencer large

Transmission
Gearbox	3-speed
Gearchange	Hand change, on tank
Final drive	Chain

Cycle parts
Frame construction	Single-loop frame
Front fork	Springer, exposed springs, oval-tube legs
Rear suspension	None
Front brake	Drum, via cable and handlebar lever
Rear brake	Drum, via rod and pedal

General information
Unladen weight	235 kg
Permissible total weight	410 kg
Front and rear tyres	5.00-16
Wheels	3.00 x 16 wire-spoked
Maximum speed	115 km/h
Tank capacity	12.9 litres

For sidecar details, see page 130.

The Forty-Fives and Fifty-Fours: Models K and KH

1952 to 1956

Model description and technical information

Exposed to the increasing competition from the British motorcycle factories after 1945, H-D had to respond to the modern BSAs, Triumphs and Matchlesses. Although the side-valve Model W range had enjoyed an outstanding reputation for reliability during the Second World War, it was by 1952 an outmoded 15-year-old design. Its replacement, the significantly more expensive Model K range, although of thoroughly modern conception, with its unit construction, telescopic forks and swinging-arm frame, still had to get by with the old side-valve engine. While in terms of power output and handling qualities the mild-mannered K was actually a failure, it did at least hold the fort against the foreign invaders until the arrival of the first really sporty Harley, the XL-series Sportsters, in 1957.

Like its illustrious predecessor, the 1928-29 Model JH/JDH Two Cam, and all the Forty-Fives from 1932-on, the engine had four cam lobes for its four valves. It also had aluminium-alloy cylinder heads (styled after the fashion of the by then-famous WRTT racing models to reinforce the new engine's sporting pretensions) and new light alloy pistons. For the second generation, the Models KH and KHK of 1954 to 1956, the cylinders themselves were longer than those on the old Model Ws, to accommodate the much-longer stroke; this extremely undersquare bore-stroke ratio made for a flexible engine, but didn't do much for its breathing at high revs. In contrast to the design of the previous Forty-Fives and the Big Twins, the four-speed gearbox was made in unit with the engine, which made for a more compact and lighter powerplant. The handlebar lever-operated clutch, right foot gearchange and left foot rear brake were laid out with flat track racing in mind, not just to copy British conventions.

The spokes of the wheels were cadmium-plated. The front brake drum was of steel, while the rear was of cast-iron; both were of 8 inches (203 mm) diameter, with 1 inch (25 mm)-wide shoes. The tank held a full 3.74 Imp. gallons (17 litres) of petrol. Standard colours available were Persian Red, Rio Blue, Tropical Green and Brilliant Black, with Metallic Bronco Bronze, White and Metallic Marine Blue available at extra cost.

Above all, the Model K was the first Harley with a modern twin-downtube swinging-arm frame in which twin spring/damper suspension units contributed just as much to a comfortable ride as the hydraulically-damped telescopic front forks. Although this gave it a (very) temporary advantage over the British opposition – Triumph were still using their sprung hub in 1952 and BSAs then had a plunger-sprung rear end – both firms had proper swinging-arm frames by 1954. Anyway, in an age when many riders were yet to be convinced of the utility of rear suspension, the K's 30 bhp and 450 lb (204 kg) all-up weight did not stack up well against the average 34 bhp and 392 lb (178 kg) of the British 40 cu. in./650 cc. parallel twins.

There were innumerable racing versions of the Model K that were actually available for many years longer than the standard road models. For example the KRTT (KRTT meaning Model K, Racing, TT scrambles) was available right up to 1968-69.

Model range

Model K	45 cu. in./742 cc. sv V-twin engine	(1952-53)
Model KH	54 cu. in./888 cc. sv V-twin engine	(1954-56)
Model KHK	54 cu. in./888 cc. sv V-twin engine	(1955-56)

These motorcycles were never designed for sidecar use.

1952 Model K

1953 Model K: Persian Red

Model development

1953 model year
The only new features worthy of note were a quicker-action throttle twistgrip and the arrival, as an optional extra, of the famous Buddy Seat. Standard colours available were Pepper Red, Glacier Blue, Forest Green and Brilliant Black, with Cavalier Brown, White and Glamour Green available at extra cost. The factory was evidently already aware of the desire of potential K owners for better performance, as a special speed kit was made available at extra cost. Fitted at the factory (and therefore identified in some sources as the Model KK), this involved polished combustion chambers and ports, modified valves and guides, hotter cams and lightened roller tappets, modified carburettor, less chrome, flatter handlebars and a top speed of some 95 mph (153 km/h).

1954 model year
The Forty-Five became the Fifty-Four for '54 (although many round it up to class it as a 55). Thanks to a massive exercise in stroking and higher compression, the 30 bhp Model K became the 38 bhp Model KH of 54 cu. in. (888 cc.) capacity. This gave it much-improved performance and raised its game almost to the level of the imported machinery.

On top of this, intake and exhaust porting was revised, larger intake valves were fitted and stronger valve springs. The crankcase and gearbox housing castings were modified so that a removable door was fitted in the crankcase side, enabling the gearbox to be dismantled separately without having to remove the engine/gearbox unit from the frame and dismantle it completely, as had previously been the case. The frame received detail modifications and the brakes were improved.

For those who wanted even more power, for 68 dollars a new Model KH could be ordered with the Speed Kit. Again, a factory-fit option involving principally a roller-bearing bottom end, higher compression, polished ports and hotter cams, the resulting Model KHK could, depending on the skill of the engine-builder, deliver up to 52 bhp.

Standard colours available were Pepper Red, Glacier Blue, Forest Green, Anniversary Yellow and Daytona Ivory; for no extra charge fuel tanks could be ordered in one colour and mudguards in another. All models were fitted with the brass Harley-Davidson Golden Anniversary 50th birthday medallion on the front mudguard.

1955 model year
For this year, amongst other changes, heavier-gauge wheel spokes were fitted, the outsides of the cylinder heads were polished as well as the insides, and many small technical improvements and aesthetic embellishments were introduced. While all models got the new big V tank badge, KH models were also fitted with the brass V medallion on the front mudguard. Standard colours available were Pepper Red, Atomic Blue, Anniversary Yellow, Aztec Brown, Brilliant Black and Glamorous Hollywood Green; Silver and White were available for police machines.

1956 model year
The frame was modified to lower the seat height and altered in readiness for the taller ohv engine of the forthcoming Sportster models. Further model developments were a new air cleaner, shorter rear suspension units with greater damping oil capacity and strengthened gears and shafts in the gearbox. Standard colour schemes for this last year of production were Pepper Red with White Slash, Atomic Blue with Champion Yellow Slash, Black with Champion Yellow Slash and Champion Yellow with Black Slash with Flamboyant Metallic Green with White Slash at extra cost. KHK models also had a shield emblem with crossed chequered flags on the oil tank and toolbox. The K models were deleted from the catalogue at the end of the model year. Owners could only buy the racing versions from now on, but these would carry on winning races for many years to come.

Motorcycle catalogue prices ($ US)

Model	1952	1953	1954	1955	1956
K	865	875	-	-	-
KH	-	-	925	925	935
KHK	-	-	-	993	1003

Prices and availability of optional Accessory Groups for motorcycles ($ US)

Group	1952	1953	1954	1955	1956
Deluxe	67.50	67.50	75.75	88.00	94.00
Standard Solo	-	22.75	-	-	-

Note that one even had to pay extra for the Standard Group.

1956 Model KH: Pepper Red with White Slash – just like the one Elvis Presley bought!

1956 Model KHK: Champion Yellow with Black Slash and KHK shield on the oil tank

Production figures

Model	1952	1953	1954	1955	1956
K	1970	1723	-	-	-
KH	-	-	1579	616	539
KHK	-	-	-	449	714

Model K 1952 to 1953

Engine
Engine	Air-cooled 45° sv V-twin
Bore x stroke	2¾ x 3¹³⁄₁₆ (2.75 x 3.81) in.
Bore x stroke	69.85 x 96.84 mm
Capacity	45.12 cu. in. (742 cc)
Power output (SAE)	30(22)/5000 rpm
Compression ratio	6.0:1
Carburettor	Linkert 1½-inch Type M-53 (1952)
	Linkert 1½-inch Type M-53A (1952-53)
Ignition	Battery-and-coil
Exhaust system	1 tubular silencer

Transmission
Gearbox	4-speed
Gearchange	Foot change
Final drive	Chain

Cycle parts
Frame construction	Twin-loop full cradle
Front fork	Hydraulically-damped telescopic
Rear suspension	Swinging arm
Front brake	Drum, via cable and handlebar lever
Rear brake	Drum, via rod and pedal

General information
Unladen weight	230 kg
Permissible total weight	400 kg
Front and rear tyres	3.25-19 (3.50-19 rear, 1953)
Wheels	2.15 x 19 wire-spoked
Maximum speed	130 km/h
Tank capacity	17 litres

Model KH 1954 to 1956
Model KHK 1955 to 1956

Engine
Engine	Air-cooled 45° sv V-twin
Bore x stroke	2¾ x 4⁹⁄₁₆ (2.75 x 4.56) in.
Bore x stroke	69.85 x 115.89 mm
Capacity	54.20 cu. in. (888 cc)
Power output (SAE)	38(28)/5000 SAE (KH)
	52(39)/5500 SAE (KHK)
Compression ratio	6.8:1 (KH)
	approx. 8.0:1 (KHK)
Carburettor	Linkert 1½-inch Type M-53A (1954)
	Linkert 1½-inch Type M-53A1 (1955-56)
Ignition	Battery-and-coil
Exhaust system	1 tubular silencer

Transmission
Gearbox	4-speed
Gearchange	Foot change
Final drive	Chain

Cycle parts
Frame construction	Twin-loop full cradle
Front fork	Hydraulically-damped telescopic
Rear suspension	Swinging arm
Front brake	Drum, via cable and handlebar lever
Rear brake	Drum, via rod and pedal

General information
Unladen weight	235 kg
Permissible total weight	400 kg
Front and rear tyres	3.50S18
Wheels	2.50 x 18 wire-spoked
Maximum speed	155 km/h (KH)
	approx. 170 km/h (KHK)
Tank capacity	17 litres

The Model V Big Twins
1930 to 1936

Model description and technical information

Harley-Davidson presented the new big side-valve-engined Model V range in July 1929, just before the Wall Street Crash. Intended as successors to the former Models F and J, and conceived as big sisters of the 45 cu. in. (742 cc.) V-twins of the Model D range, these represented the Motor Company's final departure from their well-known ioe Big Twins and from their 27-year-old roots in the de Dion-Bouton design, as well as the completion of the four-year switch to side-valve units that had begun with the introduction in 1926 of the Models A and B.

As was usual for Harley-Davidson, the 74 cu. in. (1209 cc.) engines of the Big Twins were available in various levels of compression. Almost everything was new in this motorcycle, only the air cleaner, the spark plugs and components such as ball bearings being carried over from the previous models. Above all, the engine was obviously completely new, the side-valve layout necessitating a complete redesign.

The inefficiency of the ioe valve layout is shown by the fact that humble side-valve engines could match the power outputs of all but the best pocket-valve units, only the Two Cam models being able to beat the new arrivals. However, an increase in all-up weight of some 120 lb (55 kg) naturally made itself felt in terms of on-the-road performance.

The detachable cylinder heads were laid out following the principles of Ricardo's patents and featured an oil baffle between spark plug and valves in an attempt to cure one of the inbuilt failings of the ioe models, on which over-enthusiastic valve stem lubrication had often caused misfiring and plug fouling. Unfortunately the throttle-controlled total-loss lubrication system was one of the features which H-D did carry over from the earlier models. On the other hand, engine power was now transmitted to the gearbox via an automatically-lubricated duplex primary chain. The Thirties saw the high point of experimentation with different piston materials. Besides nickel-iron pistons, which were still to be had and which were used primarily for the low-compression Commercial models, Harley-Davidson also used pistons of aluminium-alloy (Lynite) and magnesium-alloy (Dow metal). But by the end of the life of the Model Vs, aluminium-alloy was the clear winner.

The cause of most of the Model V's weight increase was the new frame, which was some 25% heavier than its predecessor. Furthermore, the exposed-spring forks, while visually similar to those introduced on the Models A and B, were now of much heavier construction with heat-treated drop-forged I-beam-section rigid (sprung) legs, tubular

1930 Model V in original oily-rag condition! Note the pillion saddle and fittings

construction being considered for the time being as no longer strong enough, especially for sidecar work. The four massive volute springs remained as a clear point of identification for all Harley models of the early Thirties, these forks also being fitted to the Forty-Fives and the Model C. Another cause of increased weight were the new large internal-expanding and all-enclosed brakes fitted; a 7¼ inch (184 mm) diameter drum at the front and an 8-inch (203 mm) diameter drum at the rear. A practical new feature was that the wheels were quickly detachable (they could be removed just by unscrewing the wheel spindle nut, leaving brake components and final drive chain undisturbed) and interchangeable; even the sidecar wheel, when the machine was so equipped, was included. The Dual Bullet double headlamps of the F and J series were standard equipment initially; soon to be replaced, however, by larger single units. The twin-tube exhaust was also carried over for the first year; it was then replaced by the first incarnation of the subsequently-famous Harley fishtail-silencer exhaust.

Standard paint finish was Olive Green with maroon (vermilion, in some sources) striping, with a centre gold stripe; handlebars and wheel rims were black. Prospective customers could, however, choose from a range of new colours – Black, Maroon, Grey, Blue, Cream and Coach Green – all available at extra cost, while two-tone paint schemes – Black/Red, Maroon/Cream, White/Gold and Blue/Grey could also be specified. Other options available, for example, were a steering damper and a speedometer, the latter available in 80 mph or (for the truly optimistic) 100 mph versions. The standard-fit rear stand could be supplemented by a front stand or a sidestand (jiffy stand). Even passengers were catered for; a pillion saddle appeared in the options list, at extra cost, as did the air cleaner. A windscreen had already been available for several years. Magneto ignition still had its devotees, even if they were few in number. A Bosch magneto made its appearance again, but soon sank out of sight for lack of demand until the final year.

Unfortunately the new Seventy-Fours did not live up to Harley's promises. Whether it was the strain of launching three major new model ranges in such a short period of time or penny-pinching in view of the prevailing economic circumstances, the new machines did not live up to expectations and produced a storm of complaints from aggrieved customers. Already significantly more expensive than the ordinary F and J models they replaced (only the exotic Two Cam models being as expensive as the most basic Model Vs), their superior (on paper) power outputs did not translate into

Model range
Seventy-Fours:
Model V	Standard, medium compression, aluminium-alloy pistons	(1930-33)
Model VC	Low-compression Commercial, nickel-iron pistons	(1930-33)
Model VCE	Low-compression Commercial, magnesium-alloy pistons, prototype **	(1933)
Model VCM	Low-compression Commercial, nickel-iron pistons, magneto ignition	(1930)
Model VCR	Low-compression Commercial, nickel-iron pistons, road-marking outfit	(1931) *
Model VD	Standard, medium compression, aluminium-alloy pistons	(1934-36)
Model VDS	Standard, medium compression, aluminium-alloy pistons, sidecar-specification	(1934-36)
Model VE	Standard, medium compression, magnesium-alloy pistons	(1933)
Model VF	Standard, medium compression, nickel-iron pistons	(1933)
Model VFS	Standard, medium compression, nickel-iron pistons, sidecar-specification	(1933)
Model VFD	Standard, medium compression, nickel-iron pistons	(1934-36)
Model VFDS	Standard, medium compression, nickel-iron pistons, sidecar-specification	(1934-36)
Model VL	Standard, medium compression, Dow metal pistons	(1930-33)
Model VLD	Standard, high compression, Dow metal pistons	(1933-36)
Model VLE	Standard, medium compression, magnesium-alloy pistons	(1933)
Model VLM	Standard, medium compression, Dow metal pistons, magneto ignition	(1930-31)
Model VM	Standard, medium compression, nickel-iron pistons, magneto ignition	(1930)
Model VMS	Standard, medium compression, nickel-iron pistons, magneto ignition, sidecar-specification	(1930-31)
Model VMG	Standard, medium compression, Dow metal pistons, Bosch magneto ignition	(1930-31)
Model VMG	Standard, medium compression, Dow metal pistons, Bosch magneto ignition	(1936)
Model VS	Standard, medium compression, aluminium-alloy pistons, sidecar-specification	(1930-33)
Model VSE	Standard, medium compression, magnesium-alloy pistons, sidecar-specification	(1933)
Model VSR	Standard, medium compression, aluminium-alloy pistons, sidecar-specification, Japanese export	(1931)

Eighties:
Model	Description	Years
Model VDDS	Low compression, sidecar-specification	(1935)
Model VFH	Low compression, nickel-iron pistons	(1936)
Model VFHS	Low compression, nickel-iron pistons, sidecar-specification	(1936)
Model VHS	Low compression, sidecar-specification	(1936)
Model VLDD	High compression	(1935)
Model VLH	High compression	(1936)

* See Chapter on Model VCR.
** Prototypes aren't covered in the data tables.
The third suffix letter D indicates 'Dynawithe' (tuned-up) engines, because of their explosive performance! Also referred to as 'TNT' motors, for the same reason.

better road performance; in fact they were worse. Their over-light flywheels gave them good acceleration up to 50 mph (80 km/h), but after that, power fell off drastically (seriously limiting the machine's load-carrying, hill-climbing and sidecar-hauling abilities) and vibration, always a problem in a 45° V-twin, became unbearable; furthermore, clutches proved too weak, the valve springs gave trouble and heavy oil consumption rapidly clogged up the silencers with carbon deposits. Complaints from customers joined those of dealers already annoyed by the problems that were emerging in the Model D range, such that urgent action was undertaken immediately at the factory to rectify the problems. Unfortunately the answer was new, larger-diameter flywheels to give the torque and spread of power that the engines lacked so much, which necessitated new crankcases, which themselves were so much bigger that a new frame was required. After several weeks of work the new components were ready and were sent out to dealers free of charge (some 1,326 machines, at least, were involved), but the dealers had to strip each affected machine right down and rebuild it around the new components at their own expense. Any dealer who complained and asked for financial compensation for the workshop time involved was told to get on with it or to hand in their franchise; several did. So, right at the beginning of the worst Depression in history, which was already having world-wide implications, Harley found themselves not only saddled with the very heavy cost of rectifying the faults, but also with disgruntled dealers and even more unhappy customers; sales dropped like a stone.

By autumn 1934 the Model Vs, thanks to constant updating and technical improvements, had evolved into reliable and dependable machines and sales generally were increasing again. Delays in the development of the new ohv model led the factory to introduce an ultra-Big Twin, the Eighty, whose extra power was also aimed at the sidecar-driver and to head off opposition from Indian. Actually of 79 cubic inch (1293 cc.) capacity, the new model became available from the end of the 1935 model year onwards and sold in substantial numbers. The models with high-compression engines could muster nearly forty horsepower (the same as the new ohv model) and could manage some 90 mph (140 km/h) as a result.

Model development

1931 model year
Even though this was only its second year, the frame had to be changed for a substantially-altered version to suit the engine's new characteristics as a result of the greatly-enlarged flywheels. A new die-cast Schebler carburettor, of the same size, was fitted.

Brake and clutch pedals were now cadmium-plated. A revised kickstart mechanism made starting easier. From the middle of the model year onwards, a constant-mesh three-speed gearbox with reverse was available for the first time for sidecar-specification models. The new fishtail-silencer exhaust was supplied by Burgess and from now on the smaller plated items, previously nickel-plated, were now chromed.

Those customers who didn't like the single, larger, 7-inch headlamp fitted for this year could still order the old Dual Bullet version from H-D's accessories catalogue. A fire extinguisher was available ex-works, as was a first-aid kit and speedometer lighting. Potential Model V owners also had a new range of two-tone colour

1930 Model V sidecar outfit

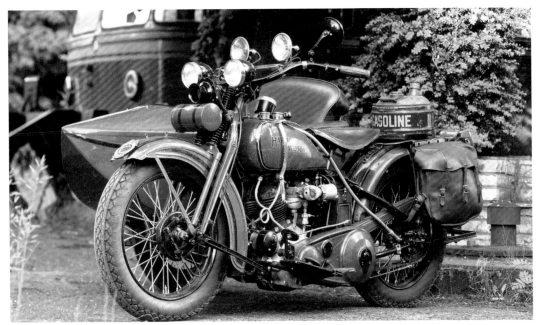
1930 Model V sidecar outfit

schemes to choose from for this year; Teak Red/Black, Maroon/Cream, White/Gold and Blue/Grey.

1932 model year
New cylinders giving improved airflow for better cooling and strengthened front forks were the most important new features for 1932. The exhaust was revised again for a slightly more elegant appearance and was finished in heat-resistant black; fuel tanks could be ordered with, instead of the traditional lettering on a plain background, a new curving design – the forerunner of the following year's Art Deco eagle design – highlighting the lettering.

1933 model year
This year saw the introduction for the first time of the famous Buddy Seat, an enormous pear-shaped saddle for two (very close) people that would become so popular in the '50s and '60s. Full chrome-plating of all decorative parts, including handlebars, chainguard, exhaust pipe and silencer, cost only 15 bucks extra. The Harley eagle, in the form of a kind of radiator mascot for the front mudguard, was introduced as an accessory this year, as was the running light for the front mudguard.

Towards the end of the model year the carburettor was changed from a Schebler, the long-standing supplier, to a similar instrument from Linkert, who would supply Harley for the next few decades.

The headlamp glass was no longer flat but convex; curved towards the outside. The Art Deco eagle decorating the petrol tank was incidentally a feature that was only available in 1933.

1934 model year
The only novelties worthy of note in the accessories catalogue were rear-view mirrors. Technical modifications were limited to the modified Straight-bore cylinders and aluminium-alloy pistons. Apart from that, new tank graphics – the flying diamond design – were used, the Airflow taillamp was fitted and the (for this year only) upswept High-Flo tailpipe exhaust.

1935 model year
Introduced late this year, the new models appearing in December rather than September, as usual, the most obvious new features were the toolbox which moved from under the headlamp to the rear right-hand side of the frame and the large-fishtail exhaust silencer. To go with the flying diamond tank graphics, prospective customers for Model Vs also had a new range of two-tone colour schemes to choose from for this year; Teak Red/Black, Venetian Blue/Silver, Verdant Green/Black, Egyptian Ivory/Regent Brown and Olive Green/Black. As with the Forty-Fives, Harley capitalized on the pent-up demand amongst those Americans who could afford motorcycles for accessorizing and customizing their machines by offering Groups of accessories; the private customer had the choice between the standard specification, the Solo Group of accessories and the Deluxe Solo Group. Even the police, sidecar-users and commercial customers were offered their own accessory Groups.

The idea was that the accessories most frequently selected by customers were combined into Groups that were offered at a price less than the total of the individual items. Thus the affluent motorcyclist, buying, for example, a VLH, could order the Deluxe Solo Group and for his 50 bucks would get the full chrome-plating package, crashbars, sidestand, 100 mph speedometer, steering damper, stop-/taillamp, saddlebags and other such goodies.

With the minimum of modification (the new models used exactly the same transmissions and cycle parts as the 74 cu. in./1217 cc. Model Vs; the capacity increase being achieved by stroking the existing engine by one-quarter of an inch), the first Eighty models appeared at the end of the model year.

1936 model year

The unbelievably-extensive catalogue of sidecars diminished slowly over the seven years of the Model Vs' life, so that by 1936, beside the usual chassis and bodies, only three complete combinations were still offered. All Package Trucks (for the Post Office and private carriers) and those sidecars built for the Big Twins were now manufactured by the Abresh Body Company of Milwaukee.

Cylinders and cylinder heads (with nine holding-down studs instead of seven, reworked combustion chambers and deeper cooling fins) were again revised for the last model year. The constant-mesh four-speed gearbox from the new ohv Model E was now available as an option. Strangely enough, the Model VMG (with Bosch magneto ignition) was presented once more in this last year and found more buyers than before.

1933 Model V: two-tone paint schemes and the Art Deco eagle design

Advertisement for 1933 Model V range

1934 Model V: the first version of the Art Deco flying-diamond design

1934 Model V

1934 Model V: St Bernard toolbox and High-Flo exhaust

1935 Model V sidecar outfit

1936 Model VL

1936 Model VD

1936 Model VL

1936 Model V sidecar outfit

Model V 1930 to 1933
Model VE 1933
Model VF 1933

Engine
Engine	Air-cooled 45° sv V-twin
Bore x stroke	3⁷⁄₁₆ x 4 (3.44 x 4) in.
Bore x stroke	87.31 x 101.6 mm
Capacity	74.24 cu. in. (1217 cc)
Power output (SAE):	
1930	27(20)/4000 rpm
1931-33	28(21)/4000 rpm
Compression ratio	4.0:1
Carburettor	1930-33: Schebler 1¼-inch
	1933½: Linkert 1¼-inch Type M-41
Ignition	Battery-and-coil
Exhaust system	Twin-tube silencer (1930)
	1 small-fishtail silencer (1931-33)

Transmission
Gearbox	3-speed
Gearchange	Hand change, on tank
Final drive	Chain

Cycle parts
Frame construction	Single-loop, curved front downtube
Front fork	Springers, exposed-springs, forged I-beam legs
Rear suspension	None
Front brake	Drum, via cable and handlebar lever
Rear brake	Drum, via rod and pedal

General information
Unladen weight	240 kg
Permissible total weight	410 kg
Front and rear tyres	4.00-19 (1930-33)
	4.40-19 (1930)
	4.50-19 (1931-33)
Wheels	2.15 x 19 wire-spoked
Maximum speed	127 km/h
Tank capacity	21.3 litres

Model VS 1930 to 1933
Model VSR 1931
Model VSE 1933
Model VFS 1933
with sidecars

Engine
Engine	Air-cooled 45° sv V-twin
Bore x stroke	3⁷⁄₁₆ x 4 (3.44 x 4) in.
Bore x stroke	87.31 x 101.6 mm
Capacity	74.24 cu. in. (1217 cc)
Power output (SAE)	25(18)/4000 rpm
Compression ratio	3.6:1
Carburettor	1930-33: Schebler 1-inch
	1933½: Linkert Type 1-inch M-16
Ignition	Battery-and-coil
Exhaust system	Twin-tube silencer (1930)
	1 small-fishtail silencer (1931-33)

Transmission
Gearbox	3-speed, optional with reverse (from 1931)
Gearchange	Hand change, on tank
Final drive	Chain

Cycle parts
Frame construction	Single-loop, curved front downtube
Front fork	Springers, exposed-springs, forged I-beam legs
Rear suspension	None
Front brake	Drum, via cable and handlebar lever
Rear brake	Drum, via rod and pedal

General information
Unladen weight	approx. 320 kg
Permissible total weight	approx. 600 kg
Front and rear tyres	4.00-19 (1930-33)
	4.40-19 (1930)
	4.50-19 (1931-33)
Wheels	2.15 x 19 wire-spoked
Maximum speed	100 km/h
Tank capacity	21.3 litres

Sidecar details are given at the end of this Chapter.

Model VM 1930
Model VMG 1930 to 1931

Engine
Engine	Air-cooled 45° sv V-twin
Bore x stroke	3⁷⁄₁₆ x 4 (3.44 x 4) in.
Bore x stroke	87.31 x 101.6 mm
Capacity	74.24 cu. in. (1217 cc)
Power output (SAE)	1930: 27(20)/4000 rpm
	1931: 28(21)/4000 rpm
Compression ratio	4.0:1
Carburettor	Schebler 1¼-inch
Ignition	Magneto
Exhaust system	Twin-tube silencer (1930)
	1 small-fishtail silencer (1931)

Transmission
Gearbox	3-speed
Gearchange	Hand change, on tank
Final drive	Chain

Cycle parts
Frame construction	Single-loop, curved front downtube
Front fork	Springers, exposed-springs, forged I-beam legs
Rear suspension	None
Front brake	Drum, via cable and handlebar lever
Rear brake	Drum, via rod and pedal

General information
Unladen weight	240 kg
Permissible total weight	410 kg
Front and rear tyres	4.00-19 (1930-31)
	4.40-19 (1930)
	4.50-19 (1931)
Wheels	2.15 x 19 wire-spoked
Maximum speed	127 km/h
Tank capacity	21.3 litres

Model VMS 1930 to 1931
with sidecar

Engine
Engine	Air-cooled 45° sv V-twin
Bore x stroke	3⁷⁄₁₆ x 4 (3.44 x 4) in.
Bore x stroke	87.31 x 101.6 mm
Capacity	74.24 cu. in. (1217 cc)
Power output (SAE)	25(18)/4000 rpm
Compression ratio	3.6:1
Carburettor	Schebler 1½-inch
Ignition	Battery-and-coil
Exhaust system	Twin-tube silencer (1930)
	1 small-fishtail silencer (1931)

Transmission
Gearbox	3-speed
Gearchange	Hand change, on tank
Final drive	Chain

Cycle parts
Frame construction	Single-loop, curved front downtube
Front fork	Springers, exposed-springs, forged I-beam legs
Rear suspension	None
Front brake	Drum, via cable and handlebar lever
Rear brake	Drum, via rod and pedal

General information
Unladen weight	approx. 320 kg
Permissible total weight	approx. 600 kg
Front and rear tyres	4.00-19 (1930-31)
	4.40-19 (1930)
	4.50-19 (1931)
Wheels	2.15 x 19 wire-spoked
Maximum speed	100 km/h
Tank capacity	21.3 litres

Sidecar details are given at the end of this Chapter.

Model VC 1930 to 1933
Model VCM 1930

Engine
Engine	Air-cooled 45° sv V-twin
Bore x stroke	3⁷⁄₁₆ x 4 (3.44 x 4) in.
Bore x stroke	87.31 x 101.6 mm
Capacity	74.24 cu. in. (1217 cc)
Power output (SAE)	25(18)/4000 rpm
Compression ratio	3.6:1
Carburettor	1930-33: Schebler 1-inch
	1933½: Linkert 1-inch Type M-16
Ignition	Model VC: Battery-and-coil
	Model VCM: Magneto
Exhaust system	Twin-tube silencer (1930)
	1 small-fishtail silencer (1931-33)

Transmission
Gearbox	3-speed
Gearchange	Hand change, on tank
Final drive	Chain

Cycle parts
Frame construction	Single-loop, curved front downtube
Front fork	Springers, exposed-springs, forged I-beam legs
Rear suspension	None
Front brake	Drum, via cable and handlebar lever
Rear brake	Drum, via rod and pedal

General information
Unladen weight	240 kg
Permissible total weight	410 kg
Front and rear tyres	4.00-19 (1930-33)
	optional 4.40-19 (1930)
	optional 4.50-19 (1931-33)
Wheels	2.15 x 19 wire-spoked
Maximum speed	120 km/h
Tank capacity	21.3 litres

Model VL 1930 to 1933
Model VLE 1933
Model VLM 1930 to 1931

Engine
Engine	Air-cooled 45° sv V-twin
Bore x stroke	3⁷⁄₁₆ x 4 (3.44 x 4) in.
Bore x stroke	87.31 x 101.6 mm
Capacity	74.24 cu. in. (1217 cc)
Power output (SAE)	1930-32: 30(22)/4000 rpm
	1933: 32(24)/4200 rpm
Compression ratio	4.5:1
Carburettor	1930-33: Schebler 1¼-inch
	1933½: Linkert 1¼-inch Type M-41
Ignition	Models VL/VLE: Battery-and-coil
	Model VLM: Magneto
Exhaust system	Twin-tube silencer (1930)
	1 small-fishtail silencer (1931-33)

Transmission
Gearbox	3-speed
Gearchange	Hand change, on tank
Final drive	Chain

Cycle parts
Frame construction	Single-loop, curved front downtube
Front fork	Springers, exposed-springs, forged I-beam legs
Rear suspension	None
Front brake	Drum, via cable and handlebar lever
Rear brake	Drum, via rod and pedal

General information
Unladen weight	240 kg
Permissible total weight	410 kg
Front and rear tyres	4.00-19 (1930-31)
	optional 4.40-19 (1930)
	optional 4.50-19 (1931-33)
Wheels	2.15 x 19 wire-spoked
Maximum speed	137 km/h
Tank capacity	21.3 litres

Sidecar details are given at the end of this Chapter.

Model VD 1934 to 1936
Model VFD 1934 to 1936
Model VMG 1936

Engine
Engine	Air-cooled 45° sv V-twin
Bore x stroke	3⁷⁄₁₆ x 4 (3.44 x 4) in.
Bore x stroke	87.31 x 101.6 mm
Capacity	74.24 cu. in. (1217 cc)
Power output (SAE)	28(21)/4000 rpm
Compression ratio	4.0:1
Carburettor	Linkert 1¼-inch Type M-21
Ignition	Models VD/VFD: Battery-and-coil Model VMG: Magneto
Exhaust system	High-Flo exhaust (1934) 1 Fishtail silencer large (1935-36)

Transmission
Gearbox	3-speed optional 4-speed (from 1936)
Gearchange	Hand change, on tank
Final drive	Chain

Cycle parts
Frame construction	Single-loop, curved front downtube
Front fork	Springers, exposed-springs, forged I-beam legs
Rear suspension	None
Front brake	Drum, via cable and handlebar lever
Rear brake	Drum, via rod and pedal

General information
Unladen weight	240 kg
Permissible total weight	410 kg
Front and rear tyres	4.00-19 (1934-36) 4.50-19 (1934-36) optional 4.00-18 (1935-36)
Wheels	2.15 x 19 wire-spoked optional 2.15 x 18 wire-spoked (1935-36)
Maximum speed	127 km/h
Tank capacity	21.3 litres

Model VDS 1934 to 1936
Model VFDS 1934 to 1936
with sidecars

Engine
Engine	Air-cooled 45° sv V-twin
Bore x stroke	3⁷⁄₁₆ x 4 (3.44 x 4) in.
Bore x stroke	87.31 x 101.6 mm
Capacity	74.24 cu. in. (1217 cc)
Power output (SAE)	25(18)/4000 rpm
Compression ratio	3.6:1
Carburettor	Linkert Type 1-inch M-16
Ignition	Battery-and-coil
Exhaust system	High-Flo exhaust (1934) 1 Fishtail silencer large (1935-36)

Transmission
Gearbox	3-speed, optional with reverse, optional 4-speed (from 1936)
Gearchange	Hand change, on tank
Final drive	Chain

Cycle parts
Frame construction	Single-loop, curved front downtube
Front fork	Springers, exposed-springs, forged I-beam legs
Rear suspension	None
Front brake	Drum, via cable and handlebar lever
Rear brake	Drum, via rod and pedal

General information
Unladen weight	approx. 320 kg
Permissible total weight	approx. 600 kg
Front and rear tyres	4.00-19 (1934-36) 4.50-19 (1934-36) optional 4.00-18 (1935-36)
Wheels	2.15 x 19 wire-spoked optional 2.15 x 18 wire-spoked (1935-36)
Maximum speed	100 km/h
Tank capacity	21.3 litres

Sidecar details are given at the end of this Chapter.

Model VLD 1933 to 1936

Engine
Engine	Air-cooled 45° sv V-twin
Bore x stroke	3⁷⁄₁₆ x 4 (3.44 x 4) in.
Bore x stroke	87.31 x 101.6 mm
Capacity	74.24 cu. in. (1217 cc)
Power output (SAE)	36(26)/4500 rpm
Compression ratio	5.0:1
Carburettor	Linkert 1¼-inch Type M-21
Ignition	Battery-and-coil
Exhaust system	1 small-fishtail silencer (1933) High-Flo exhaust (1934) 1 Fishtail silencer large (1935-36)

Transmission
Gearbox	3-speed optional 4-speed (from 1936)
Gearchange	Hand change, on tank
Final drive	Chain

Cycle parts
Frame construction	Single-loop, curved front downtube
Front fork	Springers, exposed-springs, forged I-beam legs
Rear suspension	None
Front brake	Drum, via cable and handlebar lever
Rear brake	Drum, via rod and pedal

General information
Unladen weight	240 kg
Permissible total weight	410 kg
Front and rear tyres	4.00-19 (1933-36) 4.50-19 (1933-36) optional 4.00-18 (1935-36)
Wheels	2.15 x 19 wire-spoked optional 2.15 x 18 wire-spoked (1935-36)
Maximum speed	140 km/h
Tank capacity	21.3 litres

Model VLH 1936

Engine
Engine	Air-cooled 45° sv V-twin
Bore x stroke	3⁷⁄₁₆ x 4¼ (3.44 x 4.25) in.
Bore x stroke	87.31 x 107.95 mm
Capacity	78.89 cu. in. (1293 cc)
Power output (SAE)	38(28)/4500 rpm
Compression ratio	5.5:1
Carburettor	Linkert 1¼-inch Type M-41L
Ignition	Battery-and-coil
Exhaust system	1 Fishtail silencer large

Transmission
Gearbox	3-speed optional 4-speed
Gearchange	Hand change, on tank
Final drive	Chain

Cycle parts
Frame construction	Single-loop, curved front downtube
Front fork	Springers, exposed-springs, forged I-beam legs
Rear suspension	None
Front brake	Drum, via cable and handlebar lever
Rear brake	Drum, via rod and pedal

General information
Unladen weight	240 kg
Permissible total weight	410 kg
Front and rear tyres	4.00-19 optional 4.50-19 optional 4.00-18
Wheels	2.15 x 19 wire-spoked optional 2.15 x 18 wire-spoked (1935-36)
Maximum speed	140 km/h
Tank capacity	21.3 litres

Sidecar details are given at the end of this Chapter.

Model VDDS 1935
Model VHS 1936
with sidecars

Engine
Engine	Air-cooled 45° sv V-twin
Bore x stroke	3⁷⁄₁₆ x 4¼ (3.44 x 4.25) in.
Bore x stroke	87.31 x 107.95 mm
Capacity	78.89 cu. in. (1293 cc)
Power output (SAE)	35(26)/4000 rpm
Compression ratio	5.0:1
Carburettor	Linkert 1¼-inch Type M-41L
Ignition	Battery-and-coil
Exhaust system	1 Fishtail silencer large

Transmission
Gearbox	3-speed optional 4-speed (from 1936)
Gearchange	Hand change, on tank
Final drive	Chain

Cycle parts
Frame construction	Single-loop, curved front downtube
Front fork	Springers, exposed-springs, forged I-beam legs
Rear suspension	None
Front brake	Drum, via cable and handlebar lever
Rear brake	Drum, via rod and pedal

General information
Unladen weight	approx. 320 kg
Permissible total weight	approx. 600 kg
Front and rear tyres	4.00-19
	4.50-19
	optional 4.00-18
Wheels	2.15 x 19 wire-spoked
	optional 2.15 x 18 wire-spoked (1935-36)
Maximum speed	120 km/h
Tank capacity	21.3 litres

Model VFHS 1936
with sidecar

Engine
Engine	Air-cooled 45° sv V-twin
Bore x stroke	3⁷⁄₁₆ x 4¼ (3.44 x 4.25) in.
Bore x stroke	87.31 x 107.95 mm
Capacity	78.89 cu. in. (1293 cc)
Power output (SAE)	35(26)/4000 rpm
Compression ratio	5.0:1
Carburettor	Linkert 1¼-inch Type M-41L
Ignition	Battery-and-coil
Exhaust system	1 Fishtail silencer large

Transmission
Gearbox	3-speed, optional with reverse, optional 4-speed
Gearchange	Hand change, on tank
Final drive	Chain

Cycle parts
Frame construction	Single-loop, curved front downtube
Front fork	Springers, exposed-springs, forged I-beam legs
Rear suspension	None
Front brake	Drum, via cable and handlebar lever
Rear brake	Drum, via rod and pedal

General information
Unladen weight	approx. 320 kg
Permissible total weight	approx. 600 kg
Front and rear tyres	4.00-19
	4.50-19
	optional 4.00-18
Wheels	2.15 x 19 wire-spoked
	optional 2.15 x 18 wire-spoked (1935-36)
Maximum speed	120 km/h
Tank capacity	21.3 litres

Sidecar details are given at the end of this Chapter.

Model VLDD 1935

Engine
Engine	Air-cooled 45° sv V-twin
Bore x stroke	3⁷⁄₁₆ x 4¼ (3.44 x 4.25) in.
Bore x stroke	87.31 x 107.95 mm
Capacity	78.89 cu. in. (1293 cc)
Power output (SAE)	38(28)/4500 rpm
Compression ratio	5.5:1
Carburettor	Linkert 1¼-inch Type M-41L
Ignition	Battery-and-coil
Exhaust system	1 Fishtail silencer large

Transmission
Gearbox	3-speed, optional with reverse,
Gearchange	Hand change, on tank
Final drive	Chain

Cycle parts
Frame construction	Single-loop, curved front downtube
Front fork	Exposed-spring Springer forks
Rear suspension	None
Front brake	Drum, via cable and handlebar lever
Rear brake	Drum, via rod and pedal

General information
Unladen weight	240 kg
Permissible total weight	410 kg
Front and rear tyres	4.00-19
	4.50-19
	optional 4.00-18
Wheels	2.15 x 19 wire-spoked or optional 2.15 x 18 wire-spoked
Maximum speed	140 km/h
Tank capacity	21.3 litres

Model VFH 1936

Engine
Engine	Air-cooled 45° sv V-twin
Bore x stroke	3⁷⁄₁₆ x 4¼ (3.44 x 4.25) in.
Bore x stroke	87.31 x 107.95 mm
Capacity	78.89 cu. in. (1293 cc)
Power output (SAE)	35(26)/4500 rpm
Compression ratio	5.0:1
Carburettor	Linkert 1¼-inch Type M-41L
Ignition	Battery-and-coil
Exhaust system	1 Fishtail silencer large

Transmission
Gearbox	3-speed optional 4-speed
Gearchange	Hand change, on tank
Final drive	Chain

Cycle parts
Frame construction	Single-loop, curved front downtube
Front fork	Springers, exposed-springs, forged I-beam legs
Rear suspension	None
Front brake	Drum, via cable and handlebar lever
Rear brake	Drum, via rod and pedal

General information
Unladen weight	240 kg
Permissible total weight	410 kg
Front and rear tyres	4.00-19
	4.50-19
	optional 4.00-18
Wheels	2.15 x 19 wire-spoked or optional 2.15 x 18 wire-spoked
Maximum speed	140 km/h
Tank capacity	21.3 litres

Sidecar details are given at the end of this Chapter.

Motorcycle catalogue prices ($ US)

Model	1930	1931	1932	1933	1934	1935	1936
V	340	340	320	310	-	-	-
VC	340	340	320	310	-	-	-
VD	-	-	-	-	310	320	320
VDS	-	-	-	-	310	320	320
VDDS	-	-	-	-	-	347	-
VF	-	-	-	310	-	-	-
VFS	-	-	-	310	-	-	-
VFD	-	-	-	-	310	-	-
VFDS	-	-	-	-	310	320	-
VFH	-	-	-	-	-	-	340
VFHS	-	-	-	-	-	-	340
VHS	-	-	-	-	-	-	340
VL	340	340	320	310	-	-	-
VLD	-	-	-	325	310	320	320
VLDD	-	-	-	-	-	347	-
VLE	-	-	-	310	-	-	-
VLH	-	-	-	-	-	-	340
VLM	340	340	-	-	-	-	-
VM	340	-	-	-	-	-	-
VMG	340	340	-	-	-	-	340
VMS	340	340	-	-	-	-	-
VS	340	340	320	310	-	-	-
VSE	-	-	-	310	-	-	-

Prices are motorcycles only, less sidecars.

Prices and availability of optional Accessory Groups for motorcycles ($ US)

Group	1935	1936
Solo Group #1	24.90	-
Standard Solo Group	-	28.00
Deluxe Solo	49.50	49.50
Police Group #1	49.90	53.00
Deluxe Police	92.50	90.50

Note that Standard means a specification higher than standard ex-works finish.

Prices and availability of optional Accessory Groups for sidecars ($ US)

Group	1935	1936
LT Sidecar Group #1	16.75	-
Standard Sidecar Group	-	15.75
LT Deluxe Group	39.90	39.50
Truck Deluxe Group	26.90	26.90

Note that Standard means a specification higher than standard ex-works finish.

Sidecar catalogue prices ($ US)

Model		1930	1931	1932	1933	1934	1935	1936
GM	Goulding Side Van with Cover	X	-	-	-	-	-	-
GMC	Chassis for GM	X	-	-	-	-	-	-
K	Single-seat sidecar	-	-	-	-	-	-	110
K	Body only for K	-	-	-	-	-	-	52
KC	Chassis only for K	-	-	-	-	-	-	70
LC	Chassis only for LT/LT Body	X	X	X	X	70	70	-
LT	Single-seat sidecar	X	X	X	X	105	110	-
LT	Body only for LT	-	X	X	X	48	52	-
M	Side Van	X	X	X	X	115	115	115
M	Body only for M	-	X	X	X	60	60	60
MC	Chassis for M, MO, MXP, MDC, MNP, MT	X	X	X	X	72	72	72
MDC	Body only for FC	-	X	X	-	-	-	-
MDC	Laundry Side Van	X	X	X	-	-	-	-
MNP	Body only for MNP	-	X	X	-	-	-	-
MNP	Newspaper delivery Side Van	X	X	X	-	-	-	-
MO	Package Truck without Cover	X	X	X	X	105	106	105
MO	Body only for MO	-	X	X	X	47	47	47
MT	Body only for MT	-	X	X	X	X	-	-
MT	Post Package Truck	X	X	X	X	X	-	-
MW	Body only for MW	-	X	X	X	X	-	-
MW	Side Van without Cover	X	X	X	-	-	-	-
MWC	Chassis only for MW and MWP	X	X	X	X	78	78	78
MWP	Body only for MW and MWP	-	X	X	X	-	-	-
MWP	Double compartment Side Van[1]	X	X	X	X	-	-	-
MXP	Body only for MXP	-	X	X	X	X	-	-
MXP	Double compartment Side Van[2]	X	X	X	X	X	-	-
Q	Chassis only for QT	X	X	X	X	-	-	-
QT	Body only for QT	-	X	X	X	-	-	-
QT	Double-adult sidecar	X	X	X	X	-	-	-

X = While the sidecar or part thereof listed was available, their prices are unknown for the years 1930-33.
[1] with 49.75-inch (1264 mm) track
[2] with 56.75-inch (1441 mm) track

Production figures

Model	1930	1931	1932	1933	1934	1935	1936
V	1960	825	478	233	-	-	-
VC[1]	1174	465	239	106	-	-	-
VCE[2]	-	-	-	3	-	-	-
VD	-	-	-	-	664	585	176
VDS	-	-	-	-	1029	1189	623
VDDS	-	-	-	-	-	N./Av.	-
VE	-	-	-	N./Av.	-	-	-
VF[3]	-	-	-	N./Av.	-	-	-
VFS	-	-	-	499	-	-	-
VFD[4]	-	-	-	-	5	N./Av.	N./Av.
VFDS	-	-	-	-	1330	327	600
VFH[5]	-	-	-	-	-	-	N./Av.
VFHS	-	-	-	-	-	-	35
VHS	-	-	-	-	-	-	305
VL	3246	3477	2684	886	-	-	-
VLD	-	-	-	780	4527	3963	1577
VLDD	-	-	-	-	-	179	-
VLE	-	-	-	N./Av.	-	-	-
VLH	-	-	-	-	-	-	2046
VLM	N./Av.	1	-	-	-	-	-
VM	1	-	-	-	-	-	-
VMG	19	17	1	-	-	-	118
VMS	13	1	-	-	-	-	-
VS	3612	1994	1233	164	-	-	-
VSE	-	-	-	N./Av.	-	-	-
VSR	-	10	-	-	-	-	-

[1] Includes Models VC and VCM
[2] Prototype
[3] Included in the Model VFS total
[4] For 1935 and 1936, included in the Model VFDS total
[5] Included in the Model VFHS total

The Model U Big Twins
1937 to 1948

Model description and technical information

One year after the launch of the sensational Model E, Harley followed up with the introduction of the new Model U range to replace the existing side-valve Big Twins. The Model Us shared the modified frame and cycle parts of the Model E and thus gained its good looks (the new two-part fuel tank and horseshoe-shaped oil tank under the saddle were virtually identical to those of the ohv machines), along with a duplex-loop frame with integral mounting eyes for sidecar attachments fitted in the twin downtubes. To fit a sidecar to these machines was now a simple matter of passing the sidecar's fittings through the eyes in the forged lugs and tightening the fasteners; there was no longer any need for clamped-up fittings that either worked loose or risked distorting and even breaking frame tubes. The front fork also got the improved chrome-molybdenum-steel oval-section tubular rigid (sprung) legs instead of the forged I-beam legs previously used, while Ride Control friction damping for the front fork springs was introduced as an extra-cost option. Models so equipped are clearly identified by the two slotted steel plates mounted on each side of the fork springs.

Nominally of the same capacity as the Model Vs, the new range was available in two sizes; the 74 cu. in./1209 cc. Model U and the 79 cu. in./1302 cc. Model UH. The Seventy-Four had been engineered by reducing its bore (allegedly so that it could share the same pistons as the ohv Model E) and extending its stroke, whereas the Eighty retained the same bore as both Model Vs (and all the Seventy-Fours going back to 1921) and gained its extra cubes by sharing the extended stroke of the Model U Seventy-Four. This ultra-undersquare bore-stroke ratio made for flexible engines with masses of torque and low-speed pulling power which gave the big

Model range

Model U	74 cu. in./1209 cc. sv V-twin, low compression	(1937-48)
Model US	74 cu. in./1209 cc. sv V-twin, low compression, sidecar-specification	(1937-48)
Model UL	74 cu. in./1209 cc. sv V-twin, high compression	(1937-48)
Model UMG	74 cu. in./1209 cc. sv V-twin, low compression, magneto ignition	(1937-40)
Model UH	80 cu. in./1302 cc. sv V-twin, low compression	(1937-41)
Model UHS	80 cu. in./1302 cc. sv V-twin, low compression, sidecar-specification	(1937-41)
Model ULH	80 cu. in./1302 cc. sv V-twin, high compression	(1937-41)

(Except Army contracts and competition machines)
UMG = Government use (e.g., Police or courier services)

1937 Model U – but the oil tank should be the same colour as the fuel tank and mudguards, and that's a 1941-on Rocket Fin silencer in chrome plate

1937 Model U sidecar outfit: Bronze Brown with Delphine Blue stripe with Gold edge

Flatheads strong and untiring performance over long rides. The power band extended for over 4,000 rpm before vibration became unbearable and power outputs varied between just over 32 horsepower to nearly 40 (the same as the ohv Model E, supposedly Harley's super-sports model) for the high-compression Model ULH. Given that they were of the same weight, this made the big side-valves virtually as fast as the (admittedly smaller-capacity) Model Es.

The long-overdue changeover to a recirculating dry-sump engine lubrication system instead of the previous total-loss system was a major step forward. The Model E's four-speed constant-mesh gearbox was available instead of the standard three-speed, a reverse gear being available for sidecar-specification machines.

The range of variations offered was sharply reduced; only six choices, three in each capacity. One was a low-compression sidecar-hauler, the other two differed only in compression ratio.

The policy of offering accessories in options Groups was naturally continued and extended for the new models. It was about this time that the factory began to put pressure on its dealers only to place orders for new machines with one of the options Groups; the 'standard' machine effectively no longer existed. Standard colours available were Teak Red with Black stripe and Gold edge, Bronze Brown with Blue stripe and Gold edge, Delphine Blue with Teak Red stripe and Gold edge, Police Silver with Black stripe and Gold edge and Olive Green with Black stripe and Gold edge. For this year only, the oil tank was included in the main paint scheme; only the wheel rims, frame and forks were black.

It is interesting that Harley, after the troubled early years of the Forty-Fives, the near-disaster of the Model Vs' introduction and the delayed introduction and subsequent teething troubles of the Model E, should choose to launch two complete new model ranges, the Model W and the Model U, in the same year. The Motor Company must have learned some lessons though;

1938 Model U sidecar outfit

while the Model Ws weren't exactly balls of fire and continued to have some weaknesses, they were at least reliable and dependable machines that by and large satisfied their customers. The Model Us, on the other hand were completely reliable, dependable and gave their owners every satisfaction from the start and continued to do so for many years. Valued, especially in far-flung export markets, for their slogging ability, their ease of maintenance and repair and their utter dependability, they were a complete success for H-D. They were probably only dropped in 1948 because of the increasing popularity of the ohv Big Twins, rather than because of any particular deficiency in the machines themselves.

Model development

1938 model year

In typical Harley fashion the frame was modified as early as its second year of production, with a revised steering head, fitted with a self-aligning cone on the lower bearing, heavier-gauge tubing and strengthening at

1939 Model U in 1936 Model E colours of Maroon and Nile Green

critical points. Furthermore the instrument panel was fitted with warning lights instead of gauges; red for low oil pressure and green for battery charging. The oil tank returned to its normal colour of black.

For the gearbox as well there were modifications; new ball bearings and improved third and fourth gears being the principal changes. The rear brake linings were shortened to prevent chattering, the backplate was reinforced and the rear brake shoes were interconnected by a two-piece cup bearing fitted to the pivot stud.

1938 was also the year in which spotlamps were offered for the first time as accessories, to be mounted on a bar bolted across the front forks. These spotlamps, later also called passing lamps (for overtaking) were to be fitted as standard equipment to the Electra-Glide from 1965 onwards and would be included in its optional FLH fairing. Thus for 70 years the Harley enthusiast has been happiest with three headlamps up front. In return, as it were, for keeping the price the same as in the previous year, the factory now insisted that its dealers only place orders for new machines with one of the options Groups.

1939 model year
For this year the intake manifold was extended by nearly three-quarters of an inch (20 mm) to move the carburettor a bit further away from the hot cylinder heads, and an asbestos gasket was fitted between carburettor and intake manifold to eliminate fuel vapourization and so improve engine running when hot and, particularly, when starting a hot engine. The carburettor also got a float chamber drain fitting. Further modifications were made to the gearbox and Neoprene-covered ignition HT leads were fitted.

On the styling front, the restyled Cats' Eye instrument panel appeared, as did the Boattail taillamp, the body of which was painted to match the mudguard, and stainless-steel mudguard tips. Colours available were: Black with Ivory tank panel, Teak Red with Black panel, and Airway Blue with White panel.

1940 model year
Even if war was raging in Europe, any American could still choose from all the Harley models and not experience any undue delay in delivery. The prevailing air of militarism was echoed, though, in the names of this year's paint colours; Black with Flight Red stripe, Flight Red with Black stripe, Squadron Grey with Bittersweet stripe and Clipper Blue with White stripe.

The hot news of the year was that the Eighty got deeply-finned aluminium-alloy cylinder heads as standard equipment; for the Seventy-Fours, they were only to be had as extra-cost options.

The timing cover was now finned and neutral was now restored to between first and second gears in the gate. Both halves of the petrol tank were now linked via a pipe, and all Harleys got chrome-plated metal tank badges in a teardrop style. D-shaped footboards were introduced and 5.00-16 tyres, with their substantially greater air volume giving much-improved ride comfort over the old fitment, were made available as options on all V-twin Harleys; by the following year they would completely replace the old 4.00-18 tyres.

Instead of the previous two options Groups, Harley now offered three. The cheapest, with the not exactly spectacular name of Utility Group, included useful equipment like reinforced 4-ply rating tyres, a steering damper, crashbars and a sidestand. Building on that was the Sport Solo Group, which added a trip odometer, an air cleaner, a running light on the front mudguard, chrome-plated rings on the wheels and a chrome-plated cover for the exhaust, not forgetting the coloured gearchange knob.

1940 Model ULH – aluminium-alloy cylinder heads as standard equipment. The biggest-ever Big Twin (until 1978!)

1941 model year
All big Harleys got new Rocket-Fin swallowtail-silencer exhausts and the battery was now earthed to the frame. The crankcases were redesigned, a more modern design of multi-plate clutch was used, a larger 7-inch (178 mm) air cleaner was fitted, the Airplane-Style speedometer (black face with large silver numbers) was introduced, the 16-inch wheels and tyres introduced the previous year were now standard (18-inch optional), and the steering head angle was now 29 degrees.

At the end of the model year the Eighty was discontinued; only introduced as a stop-gap in case the new ohv model should be further delayed or even unsuccessful, and kept on due to its initially strong sales, by 1941 it was superfluous to the Harley range. Its sales were falling off, assisted probably by the fact that it was significantly more expensive than the Seventy-Fours but principally by the fact that the ohv models, with their lubrication systems now completely efficient, were now accepted as proven and reliable machines and the sporting rider who valued the extra performance of the Eighty would rather be seen on the super-sporting ohv Model E than on an old Flathead. And Harley had just launched the ohv Seventy-Four.

1942 model year
It may be strange to British readers to think of civilian motorcycles being produced in 1942, but remember that the model year started in August or September, so production and sales were well under way before Pearl Harbor, naturally, changed everything. Throughout the war years, civilians were effectively barred from buying new motorcycles and technical changes were few; the only developments were in the materials used as substitutes for those resources which were entirely reserved for production of war materiel.

1943 model year
Chrome decoration was now frowned upon, everything being sacrificed to savings in material and to war production. What did not go to the Army went to the authorities such as the police forces. Since one was only allowed to buy a new Harley once one had obtained special permission from the authorities, only a few big Harleys found their way into civilian use; colours available were restricted to Grey and Silver. Seat padding was made with horsehair. Naturally, there were no Deluxe options.

1944 model year
In this year even carburettor bodies were delivered black-painted and tyres were made of the war-substitute synthetic rubber Type S-3 (the lowest-quality made).

1945 model year
With the end of the War in sight, some production restrictions were eased and the various options Groups, for sidecars as well as solos, slowly surfaced again. Apart from that, there were no changes.

1946 model year
The 1946 model year marked a slow return to full peacetime production. The majority of pre-war options were once again available, even the various chrome trim pieces, discontinued during the war years, being

gradually made available again. At first only Grey and Flight Red paint finishes were listed. Where the saddles of the wartime machines were padded with horsehair, latex mesh was now permitted to be used once more. For the first time, Monroe hydraulic dampers were available as optional extras for the front fork springs. The steering head angle was revised to 30 degrees.

1947 model year

The Special Solo Group, introduced in 1945 as the top-of-the-range luxury specification, nearly doubled in price from 55 to 100 dollars, but did include more options for the money. Post-war inflation caused the price of the Model Us, in base specification, to increase by some 28% from 427 to 545 dollars.

For this year only the Bull-Neck frame was fitted, with its oval-section steering head rather than the usual cylindrical forging. The gearchange was changed round again so that first was at the back of the gate and closest to the rider, then neutral, second and finally third (or fourth, depending on gearbox choice); gearchanging was still by hand, via the tank-mounted gate, and clutch operation by foot.

Colour choice was getting back to normal; Brilliant Black, Flight Red, Skyway Blue and Police Silver were available for the US, with Olive Green for export models. Amongst other minor modifications were the red Speed Ball tank badges, the Tombstone taillamp (so beloved in later years) and the new speedometer fitted to all 1947 Harleys; the speedometer needle was now red and the numbers on the speedometer dial were enlarged to improve its readability.

1948 model year

For their last year, the Model Us were fitted with the brand-new Wishbone frame. Now regarded as aged and outmoded, this was the end of the line for the big Flatheads and also the final year that the Springer front forks were fitted to Harley Big Twins – until 1988, that is.

1947 Model U sidecar outfit

1947 Model US with sidecar outfit, in Olive Green export trim

1947 Model U sidecar outfit

1948 Model U. Monroe hydraulic damper for the front fork springs lurking behind the horn

Model U 1937 to 1948
Model US 1937 to 1948

Engine
Engine	Air-cooled 45° sv V-twin
Bore x stroke	3⁵⁄₁₆ x 4²⁵⁄₃₂ (3.31 x 4.28) in.
Bore x stroke	84.14 x 108.74 mm
Capacity	73.79 cu. in. (1209 cc)
Power output (SAE)	32.5(24)/4200 rpm
Compression ratio	5.0:1
Carburettor	Linkert 1¼-inch Type M-51
Ignition	Battery-and-coil
Exhaust system	1 Fishtail silencer large (1937-40)
	1 Silencer with Schwalbenschwanz (1941-48)

Transmission
Gearbox	U: 3-speed optional 4-speed
	US: 3-speed with reverse optional 4-speed
Top gear ratio	3-speed: U 4.08:1; US 4.76:1
	4-speed: U 3.90:1; US 4.29:1
Gearchange	Hand change, on tank
Final drive	Chain

Cycle parts
Frame construction	1937-47: Duplex-loop with straight front downtubes
	1948: Duplex-loop with splayed front downtubes
Front fork	Exposed-spring Springer forks, oval-tube legs
Rear suspension	None
Front brake	Drum, via cable and handlebar lever
Rear brake	Drum, via rod and pedal

General information
Unladen weight	U: 240 kg; US: approx. 340 kg
Permissible total weight	U: 440 kg; US: approx. 600 kg
Front and rear tyres	4.00-18 (1937-40)
	5.00-16 (1940-48)
Wheels	2.15 x 18 wire-spoked
	3.00 x 16 wire-spoked
Maximum speed	U: 135 km/h; US: 115 km/h
Tank capacity	15.1 litres

Model UMG 1937 to 1940

Engine
Engine	Air-cooled 45° sv V-twin
Bore x stroke	3⁵⁄₁₆ x 4²⁵⁄₃₂ (3.31 x 4.28) in.
Bore x stroke	84.14 x 108.74 mm
Capacity	73.79 cu. in. (1209 cc)
Power output (SAE)	32.5(24)/4200 rpm
Compression ratio	5.0:1
Carburettor	Linkert 1¼-inch Type M-51
Ignition	Bosch magneto
Exhaust system	1 Fishtail silencer large

Transmission
Gearbox	3-speed optional 4-speed
Top gear ratio	3-speed: 4.08:1
	4-speed: 3.90:1
Gearchange	Hand change, on tank
Final drive	Chain

Cycle parts
Frame construction	Duplex-loop with straight front downtubes
Front fork	Exposed-spring Springer forks, oval-tube legs
Rear suspension	None
Front brake	Drum, via cable and handlebar lever
Rear brake	Drum, via rod and pedal

General information
Unladen weight	240 kg
Permissible total weight	440 kg
Front and rear tyres	4.00-18 (1937-40)
	5.00-16 (1940)
Wheels	2.15 x 18 wire-spoked
	3.00 x 16 wire-spoked
Maximum speed	135 km/h
Tank capacity	15.1 litres

Model UL 1937 to 1948

Engine
Engine	Air-cooled 45° sv V-twin
Bore x stroke	3⁷⁄₁₆ x 4⁹⁄₃₂ (3.31 x 4.28) in.
Bore x stroke	84.14 x 108.74 mm
Capacity	73.79 cu. in. (1209 cc)
Power output (SAE)	37.5(28)/4200 rpm
Compression ratio	5.5:1
Carburettor	Linkert 1¼-inch Type M-51
Ignition	Battery-and-coil
Exhaust system:	
1937-40	1 large fishtail silencer
1941-48	1 Rocket-Fin silencer

Transmission
Gearbox	3-speed (4-speed optional)
Top gear ratio	3-speed: 4.08:1
	4-speed: 3.90:1
Gearchange	Hand change, on tank
Final drive	Chain

Cycle parts
Frame construction:	
1937-47	Duplex-loop with straight front downtubes
1948	Duplex-loop with splayed front downtubes
Front fork	Exposed-spring Springer forks, oval-tube legs
Rear suspension	None
Front brake	Drum, via cable and handlebar lever
Rear brake	Drum, via rod and pedal

General information
Unladen weight	240 kg
Permissible total weight	440 kg
Front and rear tyres:	
1937-40	4.00-18
1940-48	5.00-16
Wheels:	
1937-40	2.15 x 18 wire-spoked
1940-48	3.00 x 16 wire-spoked
Maximum speed	135 km/h
Tank capacity	15.1 litres

Model ULH 1937 to 1941

Engine
Engine type	Air-cooled 45° sv V-twin
Bore x stroke	3⁷⁄₁₆ x 4⁹⁄₃₂ (3.44 x 4.28) in.
	87.31 x 108.74 mm
Capacity	79.47 cu. in. (1302 cc)
Power output (SAE)	39(29)/4200 rpm
Compression ratio	5.7:1
Carburettor	Linkert 1¼-inch Type M-51L
Ignition	Battery-and-coil
Exhaust system:	
1937-40	1 large fishtail silencer
1941	1 Rocket-Fin silencer

Transmission
Gearbox	3-speed (4-speed optional)
Top gear ratio	3-speed: 3.90:1
	4-speed: 3.73:1
Gearchange	Hand change, on tank
Final drive	Chain

Cycle parts
Frame construction	Duplex-loop with straight front downtubes
Front fork	Exposed-spring Springer forks, oval-tube legs
Rear suspension	None
Front brake	Drum, via cable and handlebar lever
Rear brake	Drum, via rod and pedal

General information
Unladen weight	240 kg
Permissible total weight	440 kg
Front and rear tyres:	
1937-40	4.00-18
1940-41	5.00-16
Wheels:	
1937-40	2.15 x 18 wire-spoked
1940-41	3.00 x 16 wire-spoked
Maximum speed	140 km/h
Tank capacity	15.1 litres

Model UH 1937 to 1941

Engine
Engine type	Air-cooled 45° sv V-twin
Bore x stroke	3⁷⁄₁₆ x 4⁹⁄₃₂ (3.44 x 4.28) in. 87.31 x 108.74 mm
Capacity	79.47 cu. in. (1302 cc)
Power output (SAE)	38.5(28)/4200 rpm
Compression ratio	5.2:1
Carburettor	Linkert 1¼-inch Type M-51L
Ignition	Battery-and-coil
Exhaust system:	
1937-40	1 large fishtail silencer
1941	1 Rocket-Fin silencer

Transmission
Gearbox	3-speed optional 4-speed
Top gear ratio	3-speed: 3.90:1; 4-speed: 3.73:1
Gearchange	Hand change, on tank
Final drive	Chain

Cycle parts
Frame construction	Duplex-loop with straight front downtubes
Front fork	Exposed-spring Springer forks, oval-tube legs
Rear suspension	None
Front brake	Drum, via cable and handlebar lever
Rear brake	Drum, via rod and pedal

General information
Unladen weight	240 kg
Permissible total weight	440 kg
Front and rear tyres	4.00-18 (1937-40) 5.00-16 (1940-41)
Wheels	2.15 x 18 wire-spoked (1937-40) 3.00 x 16 wire-spoked (1940-41)
Maximum speed	140 km/h
Tank capacity	15.1 litres

Model UHS 1937 to 1941

Engine
Engine type	Air-cooled 45° sv V-twin
Bore x stroke	3⁷⁄₁₆ x 4⁹⁄₃₂ (3.44 x 4.28) in.
Bore x stroke	87.31 x 108.74 mm
Capacity	79.47 cu. in. (1302 cc)
Power output (SAE)	38.5(28)/4200 rpm
Compression ratio	5.2:1
Carburettor	Linkert 1¼-inch Type M-51L
Ignition	Battery-and-coil
Exhaust system:	
1937-40	1 large fishtail silencer
1941	1 Rocket-Fin silencer

Transmission
Gearbox	3-speed with reverse, optional 4-speed
Top gear ratio	3-speed: 4.51:1; 4-speed: 4.29:1
Gearchange	Hand change, on tank
Final drive	Chain

Cycle parts
Frame construction	Duplex-loop with straight front downtubes
Front fork	Exposed-spring Springer forks, oval-tube legs
Rear suspension	None
Front brake	Drum, via cable and handlebar lever
Rear brake	Drum, via rod and pedal

General information
Unladen weight	UHS: approx. 340 kg
Permissible total weight	UHS: approx. 600 kg
Front and rear tyres	4.00-18 (1937-40) 5.00-16 (1940-41)
Wheels	2.15 x 18 wire-spoked (1937-40) 3.00 x 16 wire-spoked (1940-41)
Maximum speed	105 km/h
Tank capacity	15.1 litres

Production figures

Model	1937	1938	1939	1940	1941	1942	1943	1944	1945	1946	1947	1948
U	612	504	421	260	884	421	493	580	513	670	422	401
US	1080	1193	1327	1516	1888	978	1315	206	217	1052	1267	1006
UL	2861	1099	902	822	715	405	11	366	555	1800	1243	970
ULH	1513	579	384	672	420	-	-	-	-	-	-	-
UH	185	108	92	187	126	-	-	-	-	-	-	-
UHS	400	132	109	163	112	-	-	-	-	-	-	-
UMG	150	141	82	N./Av.	-	-	-	-	-	-	-	-

Catalogue price ($ US)

Model	1937	1938	1939	1940	1941	1942	1943	1944	1945	1946	1947	1948
U	395	395	395	385	385	385	385	385	427	427	545	590
US	395	395	395	385	385	385	385	385	427	427	545	590
UL	395	395	395	385	385	385	385	385	427	427	545	590
ULH	415	415	415	410	410	-	-	-	-	-	-	-
UH	415	415	415	410	410	-	-	-	-	-	-	-
UHS	415	415	415	410	410	-	-	-	-	-	-	-
UMG	395	395	395	385	-	-	-	-	-	-	-	-

Prices are for motorcycles without sidecars.

Catalogue price of sidecars ($ US)

Model	1937	1938	1939	1940	1941	1942	1943	1944	1945	1946	1947	1948
LE	125	125	125	125	125	125	125	125	125	139	139	173	185
LE Body	63	-	-	-	-	-	-	-	-	-	-	-	-
LEC	80	-	-	-	-	-	-	-	-	-	-	-	-
M	135	135	135	135	135	135	135	135	135	145	180	197
M Body	68	-	-	-	-	-	-	-	-	-	-	-
MC	80	80	80	-	-	80	80	80	86	86	110	118
MO	115	115	115	115	115	-	-	-	-	-	-	-
MWC	-	90	90	-	-	-	-	-	-	-	-	-

C. = Chassis only i.e., frame, wheel and fittings only (for aftermarket coach-built bodies).

Prices and availability of optional Accessory Groups for motorcycles ($ US)

Group	1937	1938	1939	1940	1941 1942	1943 1944	1945	1946	1947	1948
Standard Solo	21.75	16.70	15.50	-	-	-	-	-	-	-
Standard Commercial	20.00	-	-	-	-	-	-	-	-	-
Utility Solo	-	-	-	11.00	14.50	14.50	14.50	14.50	34.00	24.00
Sport Solo	-	-	-	22.50	27.00	-	-	-	-	59.25
Special Solo	-	-	-	-	-	-	44.50	55.00	100.00	92.00
Deluxe Solo	49.00	49.75	47.00	46.00	60.00	-	-	-	-	-
Standard Police	49.50	52.25	41.00	39.50	42.50	42.50	42.50	57.00	75.00	73.00
Deluxe Police	83.50	-	-	-	-	-	-	-	-	-

Note that Standard means a specification higher than standard ex-works finish.

Prices and availability of optional Accessory Groups for sidecars ($ US)

Group	1937	1938	1939	1940	1941 1942	1943 1944	1945	1946	1947	1948
Standard Commercial	20.00	14.25	14.00	-	-	-	-	-	-	-
Standard Sport	17.50	17.50	17.75	-	-	-	-	-	-	-
Standard Truck	-	14.25	14.00	-	-	-	-	-	-	-
Utility	-	-	-	8.50	12.00	12.00	12.00	12.00	31.50	33.50
Utility Truck	-	-	-	8.50	12.00	12.00	12.00	12.00	31.50	33.50
Deluxe Sport	44.00	44.00	47.00	50.00	54.00	-	49.00	49.00	85.00	90.50
Deluxe Truck	29.00	29.00	31.75	-	-	-	-	-	-	-

Note that Standard means a specification higher than standard ex-works finish.

The story from 1945

After the War had been won, many military vehicles remained behind on the battlefields of south-east Asia and Europe. For the majority an overhaul was not worthwhile and they were unceremoniously scrapped. Those vehicles that were still serviceable, however, saw further use by occupation troops and were then often left behind, as returning them to the US was not seen as sensible. In addition, not all the machines ordered under the War Department's still-current contracts had been issued, so the Army's department responsible for vehicles soon found itself with thousands of brand-new Harley WLAs on its hands which it actually no longer needed. Logically, the contract must be fulfilled, in spite of everything. So the War Department simply decided to get rid of the war-surplus machines by selling them on immediately. Thus even private individuals could take away one of these military machines (albeit without weaponry) for an initial price of 450 dollars (it was lowered, later on) without having to wait around unnecessarily in the dealers' showrooms.

Naturally this exercise cost the taxpayer quite a bit, having to cover the difference between the purchase price and the selling price. But Harley also lost money, since the Army's dumping of surplus motorcycles on the market substantially reduced the demand for new machines and significantly affected H-D's production. The roughly 29,000 machines produced for the Army in 1942 and 1943 compare with the bare 12,000-odd machines produced in 1945. Things were barely better in 1946, with some 15,000 produced and it was not until 1947 that things really began to pick up again with over 20,000 motorcycles being produced, but even then production numbers were nothing like during the war years.

In addition to this two more, less predictable, developments complicated things for the management at West Juneau Avenue. On the one hand the image of the motorcyclist was still not particularly good in the US, in spite of past attempts to exploit the use of motorcycles by big Hollywood stars by using photos of celebrities posed on their new Harleys in the Motor Company's advertising. On the Fourth of July 1947 participants in the annual Gypsy Tour sponsored by the American Motorcycle Association (AMA) converged on Hollister, in California, for a three-day meeting, the town was overwhelmed by the unexpectedly-large number of motorcyclists who attended and things began to get out of hand. The police were forced to intervene, some contemporary press reports claiming that tear gas had to be used to bring fighting between rival groups of motorcyclists under control, but most arrests were for public intoxication, reckless driving, and disturbing the peace. Press reporting wildly exaggerated actual events; a typical example being the famous – probably especially-posed, several days later – photograph of a drunken motorcyclist teetering on a 1940 Harley Model U surrounded by broken beer bottles. Taken by Barney Peterson, a photographer from the *San Francisco Chronicle* and appearing in *Life* magazine's July 21st 1947 edition, this photo served to reinforce the public's perception of motorcyclists and in the process did Harley's name no good at all. The Hollister 'Riot' became the inspiration of the 1953 Stanley Kramer film *The Wild One*, starring Marlon Brando and Lee Marvin as early motorcycle-riding hell-raisers (with Brando riding a Triumph and not a Harley) and added fuel to the anti-motorcycling fire. In the meanwhile, this affair and others like it reinforced the motorcyclists' belief that they were being treated as second-class citizens and unfortunate attempts by the AMA to disown the troublemakers as 'the one per cent deviant that tarnishes the public image of both motorcycles and motorcyclists' backfired. Outlaw gangs, calling themselves the 'one-percenters' soon emerged; the Hells' Angels were founded in 1948.

On the other hand, the GIs posted overseas had learned the delights of the powerful and reliable British motorcycles which were clearly so much faster and so much better-handling than the domestic products. So it should have been no surprise that the British motorcycle industry, recovering relatively quickly after 1945 and resuming mass production encouraged by the attractive export licence provisions and less restrictive material allocations for export awarded by the cash-strapped British Government for whom 'export or die' was the rule, should begin to sell in significant numbers in the US. This situation was made significantly more dangerous for Harley-Davidson when, on September 18th 1949, the pound sterling was devalued, making British imports 30% cheaper overnight.

The situation of motorcyclists and the status of motorcycles were thus very different in post-war America than in Europe. In Europe, where recovery from the ravages of war was only slowly achieved, a motorcycle was often the first and only means of transport for the middle-class family. In 'austerity' Britain, the acquisition of brand-new cars and motorcycles was discouraged by a Purchase Tax on new vehicles of the order of 27%, and in any case they were virtually unavailable, being reserved almost exclusively for export. Furthermore, continued stringent petrol rationing meant that only the smallest and most economical vehicles could be considered by anyone other than the very rich. While in Germany a top-quality motorcycle like a Zündapp KS 600 carried a price tag of some 2,550 Deutschmarks, an Opel Olympia cost some 6,785 Deutschmarks, and in England a BSA A7 cost some £183 against the £505 needed for an Austin A40 Devon. In the USA, however, the discrepancy was nowhere near as great; a 1949 FL cost some 750 dollars but the cheapest Chevy was available for only 1,339 bucks – not even double the cost. In the years to come this relationship developed even further to the disadvantage of the motorcycle. In addition, the large pool of second-hand machines meant that anyone who had the money for a new machine of a given size and/or specification could far more easily acquire a good used example of a bigger or higher-specification model. Apart from that, almost no-one used motorcycles for getting to and from work or for taking the family picnicking; the motorcycle's only use was for sport or as a hobby pursuit – or even for rioting.

Harley-Davidson adopted a three-pronged response to these problems. First, they staged a raid across the Atlantic in 1947 and obtained by way of reparations payments the blueprints for the German DKW RT125 design as a way of quickly and cheaply developing a machine for the entry-level sector of the motorcycle market, to sell at a price that the competition couldn't begin to match. BSA did the same in England, where the DKW was flying off the production lines as the BSA Bantam (and Yamaha used the design to create their 1955 YA-1). Harley's reproduction was called the Model 125 and sold more than 10,000 machines in its first year, 1948. This helped Harley establish a new production record of more than 31,000 motorcycles manufactured during the year. What is more, H-D used the dollars earned during the War to increase production capacity. In 1947, at Harley's dealers' annual convention, the delegates were put on a train and sent on a night-time ride to a 'secret destination', which turned out to be the former A. O. Smith wartime propeller-building plant; a 260,000-square foot one-storey factory building on Capitol Drive in Wauwatosa, a suburb west of Milwaukee that the Motor Company had bought in 1946 for 1.5 million dollars to help cope with anticipated post-war demand. Initially intended as a machining facility only,

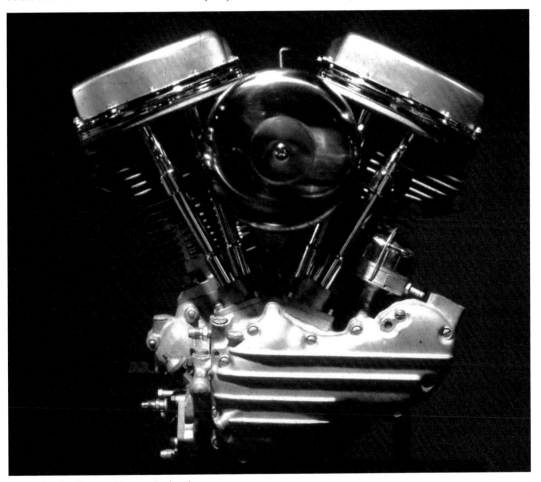

The engine for the next 18 years: Panhead

H-D seeking fame and glory: George Roeder and his 177 mph 250 Sprint-powered streamliner, Bonneville Salt Flats, 1965

The post-AMF H-D management team, in June 1981. Vaughn Beals on the right, with Willie G. Davidson behind.

The world-famous street sign outside the Harley factory: West Juneau Avenue

the new plant was soon producing and assembling engines and transmissions.

The second measure was the creation of a sporting image, not only by staging countless shows with audience appeal such as police displays, but also exclusive club activities with camping and barbecues, as well as competitions and games. The Harley fan could naturally read all about these events in regularly-appearing editions of *The Enthusiast*, the Motor Company's long-standing in-house publication. In addition advertising was intensified; for example in 1950 H-D advertised in the respected technical periodical *Popular Mechanics* and represented riding a Hydra-Glide as the coming experience. In addition the company judged the Hydra-Glide as 'as modern as a spaceship'.

Incidentally, right at the close of the same year, the then 69-year-old Arthur Davidson, the last of the Company's original Founding Fathers, died with his wife in a car accident in Milwaukee. Although border clashes had been escalating for some time, the Korean War broke out in late June 1950. US forces were engaged by July and major hostilities lasted until the armistice signed on July 27th 1953. Although a final batch of WLAs was delivered to the Army on the 29th February 1952, these were in fulfilment of an earlier contract; no further military contracts were forthcoming as a result of the Korean War, or subsequently.

Harley-Davidson's third measure was the introduction of the Model K, a pretty successor to the Model W whose deficiencies were already obvious and would be more so in the years to come. What was not so well advertised was that the Motor Company applied, unsuccessfully, to the government's Tariff Commission to raise duties on foreign imported motorcycles to 40% – a punitive 50% rise.

The last remaining competitor from inside the US, the legendary Indian Motocycle Company, had closed its gates in 1953; now Harley had only the foreign competition to fight.

However the Model 125 had proved to be a flash in the pan, as far as sales were concerned. Also, in spite of H-D's doing everything it could to resist the British competition, the marque's sales fell steadily throughout the four-year duration of the Korean War to under 10,000 machines in 1955. The Americans, in fact, seemed to have become particularly bored with motorcycles and over the next decade sales fluctuated between 10,000 and 15,000 motorcycles annually – once again, like the Thirties, too little to live on but too many to die – until recovery occurred, quite suddenly, in 1965.

The little DKW-based two-stroke was developed further, but represented to the majority of Americans nothing more than toys – an attitude fostered by many of Harley's own dealers, who simply didn't consider lightweight two-strokes to be proper motorcycles – and so most were treated with a corresponding lack of care and few survived in comparison with the bigger four-strokes. However, they did contribute to the factory's survival; for example the 4,900 large-capacity motorcycles and 1,000-odd Servi-Cars sold in 1955 would not have been enough on their own to ensure this. Most of the two-strokes anyway found their way into one of the many police forces across the country or into other aspects of official and commercial life such as parcel delivery services, couriers and newspaper deliveries. What they might have achieved, given proper development and promotion, was shown in 1956 when one Leroy Winters entered the Jack Pine Enduro (in that year a two-day, 500-mile event) and continued Harley's winning tradition (23 of 29 Jack Pines were won by H-D between 1923 and 1955), except that he did so riding a Model 165 (admittedly much modified, with a home-built swinging arm frame fitted eight years before the factory got around to it); the first time that a lightweight motorcycle had won this event, which was normally dominated by stripped-down and tuned-up Big Twins.

The Motor Company's management could still remember all too clearly the previous crises in 1920 and 1929-33 and knew well that single-minded concentration on motorcycle manufacture had not been the answer then, so they now tried to use the profits of wartime manufacture to give the factory wider-based support in peacetime; the new plant was one result of this policy. Another was the trimming of the range; production of the last side-valve Big Twins was stopped in 1948 in favour of the ohv models, these last now having developed into mature and reliable products. Similarly, the Model Ks replaced the war-tested but obsolete Model Ws in 1952 and were themselves to lead in the long run to the development of the Sportster range of 1957-on, which is still with us today. The Sporty, as the new model was immediately christened, finally had the up-to-date valvegear which enabled it to produce the engine speeds and power outputs capable of competing in a straight contest with the British imports.

However, a much more serious development then occurred which would turn the then comparatively-modest US motorcycle market completely upside-down. The nascent Japanese motorcycle industry began to export very small-capacity models to the US, the two-strokes in particular pointing up the 20-year gap in technology between themselves and Harley's DKW-based offerings, being superior both in terms of value for money and in power output. Thus the Japanese, selling in unprecedented numbers, ushered in the largest-ever boom in two-wheelers which gave the boys at H-D even more sleepless nights. Starting in 1959, Honda were the first company properly to organize imports into the USA. By clever marketing, the key to which was the slogan 'You meet the nicest people on a Honda', the Japanese promoted a complete change in the public perception of motorcyclists.

The DKW derivatives were ready to be pensioned off and certainly weren't about to win Harley any more prizes. Thanks to a complete lack of any other indigenous replacements, the Motor Company was forced to look out for a suitable partner and found one in the Italian motorcycle constructor Aermacchi. This former aircraft manufacturer had begun after the War to build small two- and four-stroke motorcycles, since post-war Italy was not allowed to build aircraft for many years. Motorcycles were something else, and these had been constantly developed by 1960 into strong, powerful and reliable machines (actually, Aermacchi only manufactured the engines, buying-in the cycle parts) which were light in weight, looked good, came from Europe (which in itself promised good sales) and came from an area in which, by American standards, labour costs were cheap. Half of the firm was taken over by 1960; at the beginning of 1961 William H. Davidson was appointed governor and Chairman in Italy.

As noted above, frames, pressed-steel components and all the rest were bought in by the company from outside suppliers. Unfortunately, the supply situation worsened progressively and substantially towards the middle of the decade, Italy then being in a state of turmoil with considerable industrial unrest. On top of this, inconsistent quality control caused increasingly-poor reliability. On the other hand the first joint model of the venture, the Model C Sprint, with its 246 cc./15 cu. in. four-stroke engine – nothing at all like the old-fashioned look of the DKW-derived two-strokes – was so exactly what the newcomer to motorcycling wanted and infinitely preferable to the Super 10. The scope of the co-operation with Aermacchi widened and by the beginning of the 1970s prospective customers could choose between a tiny 65 cc. (4 cu. in.) moped-like runabout, a 98 cc./6 cu. in. motorcycle and a range of 125s (7.5 cu. in.), 175s (11 cu. in.) and 250s (15 cu. in.), most of which were available in road and off-road style and all of which used two-stroke engines. The 248 cc./15 cu. in. and 342 cc./21 cu. in. four-stroke-engined models were sold in parallel, up to 1974. On top of that even pure racers were produced, with 250 (15 cu. in.) and 350 (21 cu. in.) two-stroke engines as well as the 342 cc./21 cu. in. four-stroke. So Harley-Davidson had the whole market covered, right up to the European middleweights. Above that was the Sportster range, the Super Glides (from 1971), and, naturally, at the top of the pile was the customary huge Electra-Glide.

In addition to motorcycles, in 1960 came the takeover of the Tomahawk Boat Company (for its experience with glass-fibre construction), which still today belongs to H-D. Their know-how was immediately put to good use in developing the hard Tour-Pak panniers which would be offered with the new Electra-Glide, as well as in the manufacture of Toppers, golf Caddys, Utilicars and snowmobiles. At bottom, the Motor Company was known for its Big Twins and the police forces were its best customers.

All these things taken together, the new models, and the new life breathed into the motorcycle market by the Japanese competition, caused sales figures to rise dramatically; 1965 saw the first real increase in nearly two decades, but in 1966 a new all-time production record was set of over 36,000 units.

In 1969 Harley-Davidson was taken over by the American Machine and Foundry corporation (AMF). Their top management saw the recovering motorcycle manufacturer as an excellent investment but had no closer interest in the end product – motorcycles. All processes were reorganized, all procedures checked for financial viability and, particularly, with a view to maximizing the profit-earning potential of each and every part of the Company.

Admittedly, AMF invested sufficiently in Harley-Davidson, but the increasingly-aggressive stance taken by the trades unions (who were not above actual sabotage on occasions) and the very cramped conditions pertaining in Milwaukee necessitated a more drastic solution to the problem.

Two views of the factory in Milwaukee

In 1972 AMF arranged the removal of the complete assembly line to an empty factory in York, in Pennsylvania. The Harley management remained with the engine and gearbox assembly lines in Milwaukee, Wisconsin, which is where it still is today. This separation did, however, create substantial communication problems from the beginning.

In spite of that, it was in the AMF era, in 1975, that the sales record fell of over 75,000 machines produced annually – a figure thought for a long time to be impossible to achieve. However this was done by abandoning the previously rigid rule that the only motorcycles built were those for which firm orders had been placed. Now the rule was production at any price, the merchandise to be thrown on to the market. Constant complaints and even recalls were the result, with accusations of dumping. Relationships between the Harley management and AMF's bosses

Two views inside the factory in York, Pennsylvania

became ever more strained, especially as the Harley board members had virtually no say in the direction of AMF's affairs.

In 1974 a strike, which lasted a whole 101 days, eventually paralysed all H-D's production plants. On top of that, even if the British motorcycle industry was virtually dead by this time, all the Japanese manufacturers now had a presence in the US motorcycle market, and they were now encroaching on H-D's home territory in the bigger capacity classes and producing machines with superior power outputs. The result was a steep drop in Harley sales – of the order of 25% – which could only be rectified in the succeeding year. Changing over to export markets was hardly possible, in view of the unfavourable exchange rates then pertaining and of factors like unsuitable models and lack of export infrastructure; for example it was not until the 1970s that an importer was established in West Germany. Problems were also emerging in the factories; some employees were stealing on a large scale. Long prison sentences followed well-publicised trials, but the damage was already done. By 1979 Harley-Davidson held only four per cent of the US motorcycle market.

Another factor was making itself apparent; since 1970 the US Government's Department of Transport (DOT) and Environmental Protection Authority (EPA), as well as various state authorities, notably California's, had burdened the manufacturers with constantly-intensified noise, exhaust and safety regulations for new models. Harley-Davidson, whose model range was already weak in terms of power outputs and based on old, noisy engine designs naturally suffered particularly badly in this respect. The results were declining power outputs and reduced torque development. In addition, the engines had to be modified to use unleaded petrol, which naturally incurred further development costs.

Even the numbers no longer added up; H-D hadn't done its homework properly – a slip into the red was only just avoided. Soon the question of the day was how long before AMF offloaded Harley-Davidson? Being a burden works both ways and in 1978 the Aermacchi factory in Italy was written-off and sold to Cagiva. In the meanwhile the Japanese competition had overtaken the Aermacchis both technically and on price, particularly on the race-tracks.

A man called Vaughn Beals, since 1975 Corporate Vice President and Group Executive of the AMF Motorcycle Products Group, got together a group of twelve Harley managers including Willie G. Davidson (the grandchild of one of the Founding Fathers, who had joined the Motor Company in 1963) and within a few weeks had raised the incredible sum of 75 million dollars in order to make AMF an offer for Harley-Davidson, thus creating privately-owned independent company from the former division of the publicly-owned AMF. Their enthusiasm as much as their business plan impressed enough banks to raise the loans.

So amid a fanfare of publicity on the 16th June 1981, Harley-Davidson was bought back from AMF. To mark this unique occasion the first motorcycle off the 'free' production line in York, was provided with a gilded oil dipstick. At last the Motor Company was again responsible for its own future.

Because of the quality-control problems engendered during the AMF era, a lot of time and effort had to be devoted to winning back the customers that had been lost, as well as winning new ones. By 1982 the output of new motorcycles had slumped to nearly half the 1975 figure and by 1983 the low point was reached; scarcely 30,000 machines were sold.

So Harley-Davidson scraped the bottom of the barrel again and came up with the same notion that had been tried in 1950 (and thus found themselves in company with the American car industry). An application was made to the US government for increased import duties to be levied on Japanese products to protect the domestic industry against the imminent supremacy of the Far East. This time they were partially successful; the then US president Ronald Reagan personally saw to it that additional import duties were levied on the Japanese for the next five years. In the meanwhile on the race-tracks, the prevailing mood of hostility to the Japanese was reflected in rule changes that kept the 'rice rockets' at a distance for a little while yet.

Although, except for a number of Sportsters, no military contracts for motorcycles were forthcoming from the late 1940s up to 1983, perhaps the survival of H-D was also due to the fact that, at the beginning of the 1980s Harley was again able to snatch a fat Army contract – for ammunition! Be that as it may, the reorganization slowly bore fruit and sales figures were on the increase. The authorities, above all the police, were still the most faithful Harley customers, but this core began to crumble in the 1980s, as more and more went over to Japanese machinery. There was an outcry in the Harley community when Wisconsin State police took delivery of its first Kawasakis in May 1982. More than 100 Harley riders united spontaneously to take part in a protest ride against this disgrace.

Under the new direction of Willie G. Davidson H-D returned to the tried and trusted methods of earlier times. One, perhaps the most important, is the creation of an artificial shortage implied in the maxim 'Always build one motorcycle fewer than you can sell' – this guarantees on the one hand that the customer always gets the motorcycle he or she wants, while on the other it stimulates demand (it also saves money and storage space by keeping stocks to a minimum). In addition many new models, new features and technical developments appeared, culminating in the dawn of the new age in 1983, with the launch of the Evolution engine.

But that's another story.

The post-war two-stroke singles

1948 to 1965

Model description and technical information

To the victor go the spoils. And so it was that DKW, the motorcycle-producing subsidiary of the German Auto Union vehicle manufacturer, already severely damaged in the aftermath of the War, was forced to 'lend' the blueprints of its 1939 RT125 design to the Americans. Thus emerged, through simple copying and conversion of metric to United States Customary System units of measurement (just as BSA did to produce the Bantam), the first two-stroke Harley, the Model 125 (whose engine numbers bore the S model designation that had hitherto only been used for racers). While the new model's metric roots were echoed in its name (a first for Harley, whose people were used to thinking in cubic inches) its bore and stroke were converted to American standard units and so the little Harley was by no means a replica of the original, still being produced in Germany; several minor differences are discernible. Harley's two-strokes, incidentally, were always called the Model 125, Model 165, Super 10, Pacer, Ranger, Scat or Bobcat – the Hummer was the low-cost 125 cc./7.6 cu. in.-engined model sold between February 1955 and 1959; while the other models are often nowadays referred to as 'Hummers', this is incorrect.

While it was very difficult for a single-cylinder two-stroke to capture the hearts of Harley dealers, the more than 10,000 sales of the new model since its appearing on the market in the first months of 1948 spoke for themselves. The 6-volt electrical system, including battery-and-coil ignition, was state-of-the-art in Europe at that time and no-one expected anything better of the machine in the USA. It was fine as a city runabout or for getting around in the suburbs, while some were used by pupils getting to high school and perhaps even by a few workers commuting to their nearby factories.

With simple rubber-band springing of the pressed-steel girder front fork, a modern (for a Harley) foot change gearbox and a maximum speed (sitting up) of some 47 mph (75 km/h), it was a good little bike and motorcycle-hungry teenagers lapped it up. There was one colour available for the 1948 Model 125 – Brilliant Black. Only one option was available – chrome-plated wheel rims, at $7.50. Naturally, sidecars were not envisaged for the smallest Harley.

Model range

Model 125
Model S	125 cc. (7.6 cu. in.)	(1948-52)

Model 165
Model ST	165 cc. (10.1 cu. in.)	(1953-59)
Model STU	165 cc. (10.1 cu. in.)	(1954-59)

Hummer
Model B	125 cc. (7.6 cu. in.)	(1955-59)

Super 10
Model BT	165 cc. (10.1 cu. in.)	(1960-61)
Model BTU	165 cc. (10.1 cu. in.)	(1960-61)

Pacer
Model BT	175 cc. (10.7 cu. in.)	(1962-65)
Model BTU	165 cc. (10.1 cu. in.)	(1962)
Model BTU	175 cc. (10.7 cu. in.)	(1963-64)

Ranger
Model BTF	165 cc. (10.1 cu. in.)	(1962)

Scat
Model BTH	175 cc. (10.7 cu. in.)	(1962-65)

Bobcat
Model BTH	175 cc. (10.7 cu. in.)	(1966)

Model development

1949 model year
The engines were finished in a heat-resistant silver colour, while the exhaust silencers were painted in a heat-resistant black.

1950 model year
The 1950 Model 125 was available in Sportsman Yellow, Flight Red and Riviera Blue as well as Brilliant Black. Strengthened gudgeon pins and a new saddle accompanied a modified generator as the sole alterations for this year.

1951 model year
For 1951 the pressed-steel girder front fork was finally replaced by a telescopic fork, just as the bigger Harleys had had since 1949, albeit in a much

1949: The Harley-DKW Model 125

1950 Model 125

1952 Model 125 in Rio Blue, with Tele-Glide front forks and CycleRay headlamp

simpler construction which was christened Tele-Glide to distinguish it from the Hydra-Glide suspension. New mudguards, a new lighting system with bigger 7-inch (178 mm) CycleRay headlamp and streamlined Bullet taillamp and a redesigned exhaust silencer – the original fishtail silencer being replaced by a more modern-looking tubular unit – completed the new features for this year.

1952 model year
The last year of the original Model 125 brought yet another new speedometer, a revised first gear ratio and an oil-lubricated front fork (replacing the previous year's grease-lubricated suspension). A waterproofed rear brake and several detail improvements rounded off the facelift. Standard colours available were Tropical Green, Persian Red and Rio Blue, with Marine Blue (metallic) available at extra cost.

1953 model year
Harley was increasing the engine size of its models generally and the smallest was no different; a bored-out 166 cc./10 cu. in. engine formed the basis of the more advanced Model 165 ST. The primary drive reduction ratio was revised, making the machines a little faster. Standard colours available were Pepper Red, Glacier Blue and Forest Green, with Glamour Green an extra-cost option.

1954 model year
The Model 165 could be ordered with one of two power outputs; the ST version producing just over five horsepower and the STU being fitted with a restriction in the carburettor to lower its output to less than five horsepower. This enabled the STU version to be sold in those States (principally in the mid-west) where 14-year-olds with a junior driver's licence were allowed to ride motorcycles of less than five horsepower (the Model 125, with its 3.5 horsepower, had always fallen into this category).

Instead of the previous 60-mph speedometer, a 70-mph unit was fitted. The gearchange was strengthened. All models were fitted with the brass Harley-Davidson Golden Anniversary 50th birthday medallion on the front mudguard. Colours available were Pepper Red, Glacier Blue, Daytona Ivory, Anniversary Yellow and Forest Green.

1955 model year
An easily changed exhaust and the return of the smaller 125 cc./7.6 cu. in. engine in the Model B Hummer were the high points of 1955. The Hummer, named after Dean Hummer, a Harley dealer in Omaha, Nebraska who in 1952, 1953 and 1954 sold more Model 125s and 165s than any other H-D dealer in the country, was introduced at a special dealers' convention in Milwaukee, in February 1955. It was intended as a low-cost entry-level machine and was 85 dollars cheaper (or nearly three-quarters of the price) of the Model 165. Cost reductions were effected by simply deleting the toolbox, the front brake and the speedometer, as well as any chrome-plating, and by using flywheel-magneto ignition and lighting and a bulb horn instead of the full battery electrical system of the Model 165. The smaller machine suffered by comparison with the 165 version, seeming rather

1957 Model 165

1958 Hummer

1959 Model 165

1958 Hummer (tank badges incorrect)

mean and miserly. Apart from that, there were few modifications other than new materials for certain components and naturally the new big V tank badge (for the 165 – the Hummer got transfers).

1956 model year
The sole new feature was the fitting of 18-inch wheels to the Model 165s.

1957 model year
For 1957 the Hummer finally got a front brake. The Model 165s got new speedometers and again all got new tank graphics.

1958 model year
Only new tank badges.

1959 model year
Along with yet more new tank badging, a version of the Buddy Seat (a big seat for two slim people) was available as an extra-cost option for all models. For the 165 models a trip odometer was introduced, and a return to a smaller headlamp unit. The Model 165 and the Hummer were both discontinued at the end of the model year.

1960 model year
Harley-Davidson introduced the Super 10 at the end of 1959 as the replacement for the Models 165 and Hummer in the single-cylinder two-stroke category. The 10 referred to the engine's capacity in cubic inches, a departure for the designations of these models which had hitherto been based on their capacities in metric units and a reflection of the fact that the engine was now considered at H-D as a purely-American product. Neither technically nor aesthetically was the Super 10 much of an advance on its predecessors and the entire range began to look a bit outdated from now on; the basic design by now, of course, being over twenty years old. The Super 10 was still fitted with the original DKW design's rigid frame, as would all small Harleys be until 1962, which didn't look too good against the new Japanese machines now beginning to appear on the market.

Nevertheless in 1960 more Super 10s were sold than were Model 165s in the year before (but only two-thirds of the combined sales of the previous year's two models). The redesigned engine, still based on the bored-out DKW design, was again available in over-five-horsepower and under-five-horsepower versions; the latter for those States still permitting 14-year-olds to ride on the public highway.

The cylinder heads were now of aluminium-alloy and the electrical system, based on the Hummer's flywheel-magneto ignition and lighting system, should be waterproof (allegedly).

1961 model year
Apart from the obligatory change of tank badging, the only new feature was the adoption of 16-inch wheels.

1962 model year
This year saw a marked widening of the model range. The Super 10s were replaced by the more powerful Pacer with a longer-stroke 175 cc./11 cu. in. engine which produced 6.5 horsepower. It had to wait until the following year for a restricted under-five-horsepower version to appear, so the 166 cc. U-version was retained for a year to fill the gap.

The new Ranger (166 cc./10 cu. in. engine) and Scat (like the Pacer, fitted with a 175 cc./11 cu. in. engine, but in this version producing 10 horsepower) were introduced. The name alone identifies the differences: the Ranger was an off-road motorcycle with full knobbly tyres and modified cycle parts giving high ground clearance. The high-level exhaust (to permit crossing shallow rivers and to assist the rider in getting the machine over fallen trees and other obstacles), Buckhorn handlebars and missing front mudguard, as well as the complete absence of lighting, horn and speedometer were all intended to suit it to its intended market; the hunter and fisherman who needed something small and light to get him to his chosen playground. The immense difference in the size of the 84-tooth final driven sprocket, compared with that of the normal road models, was necessary to be able to achieve the gear ratios required for such a low-powered engine to propel the machine and rider across difficult terrain. The Ranger was probably no great success and so was offered only in this year.

The Scat filled the middle ground between the pure off-road Ranger and the worthy but plain road-only Pacer and was in fact a forerunner of the trailbikes which were to be so popular a decade later. It was intended to be a road machine capable of some off-road work when required. Its less-aggressive trials-type tyres, high-level exhaust, abbreviated mudguards, Buckhorn handlebars and high-ground-clearance cycle parts gave it something in common with the Ranger, but its front mudguard and full road equipment (lighting, speedometer, horn, etc.) made it much better for the road; furthermore, it could be ordered with the Buddy Seat. Colours available were Tango Red and Skyline Blue, with Hi-Fi Red and Hi-Fi Purple available at extra cost on the Pacer and Scat.

1963 model year
For 1963 the Ranger had been discontinued and both the Pacer and Scat finally got the Glide-Ride swinging-arm rear suspension using twin laid-down undamped spring units located under the gearbox which anticipated the similar arrangement (in terms of location) fitted to the 1984-on Softail models. The rear mudguard was raised and extended to suit suspension travel. The machines retained their individual road/off-road identities, the new suspension being equally better for either in their particular field. The Pacer models also got the Scat's Buckhorn handlebars.

1966 Bobcat

1964 and 1965 model years
For the 1964 and 1965 model years the only differences were in the styling of the petrol tank. All existing models were discontinued at the end of the 1965 model year.

1966 model year
The small Harley two-strokes would survive only for this year, this sector of the market then being left to the Aermacchi models imported in the meantime in large numbers. The Pacer was renamed Bobcat for its last year; there were no other models. The chief characteristic of this quite modern-looking motorcycle was the one-piece fibre-glass tank cover and seat unit, which blended completely into the rear mudguard; again, a forerunner of developments in the next decade. Colours available were Holiday Red or Indigo Metallic, with Sparkling Burgundy as an extra-cost option.

Successor to the Scat was the Aermacchi Rapido MLS-125, which would be on sale from 1968.

Catalogue prices ($ US)

Model	1948	1949	1950	1951	1952	1953	1954	1955	1956	1957	1958	1959
125 S	325	325	325	365	365	-	-	-	-	-	-	-
165 ST	-	-	-	-	-	405	405	405	405	445	465	475
165 STU	-	-	-	-	-	-	405	405	405	445	465	475

Catalogue prices ($ US)

Model	1955	1956	1957	1958	1959	1960	1961	1962	1963	1964	1965	1966
Hummer	320	320	356	375	385	-	-	-	-	-	-	-
Super 10	-	-	-	-	-	455	465	-	-	-	-	-
Super 10 BTU	-	-	-	-	-	465	465	-	-	-	-	-
Pacer	-	-	-	-	-	-	-	465	485	495	505	-
Pacer BTU 165	-	-	-	-	-	-	-	465	-	-	-	-
Pacer BTU 175	-	-	-	-	-	-	-	-	485	495	-	-
Ranger	-	-	-	-	-	-	-	440	-	-	-	-
Scat	-	-	-	-	-	-	-	475	495	505	515	-
Bobcat	-	-	-	-	-	-	-	-	-	-	-	515

Production figures

Model	1948	1949	1950	1951	1952	1953	1954	1955	1956	1957	1958	1959
125 S	10117	7291	4708	5101	4576	-	-	-	-	-	-	-
165 ST*	-	-	-	-	-	4225	2835	2263	2219	2401	2445	2311

* Includes STU version from 1954-on

Production figures

Model	1955	1956	1957	1958	1959	1960	1961	1962	1963	1964	1965	1966
Hummer	1040	1384	1350	1677	1285	-	-	-	-	-	-	-
Super 10	-	-	-	-	-	2488	1587	-	-	-	-	-
Super 10 BTU	-	-	-	-	-	**	**	-	-	-	-	-
Pacer	-	-	-	-	-	-	-	1983	824	600	500	-
Pacer BTU 165	-	-	-	-	-	-	-	***	39	-	-	-
Pacer BTU 175	-	-	-	-	-	-	-	-	-	50	-	-
Ranger	-	-	-	-	-	-	-	***	-	-	-	-
Scat	-	-	-	-	-	-	-	***	877	800	750	-
Bobcat	-	-	-	-	-	-	-	-	-	-	-	1150

** Included with BT 165
*** Included with BT 175

Model S 1948 to 1952

Motor
Engine	Single-cylinder air-cooled piston-ported 2-stroke
Bore & stroke	2.06 x 2.28 in (52.39 x 57.94 mm)
Capacity	7.62 cu. in. (125 cc)
Power output (SAE) SAE	3.5 hp (2.6 kW)/4800 rpm
Compression ratio	6.6:1
Ignition	Battery-and-coil
Exhaust system:	
1948-50	Fishtail silencer
1951-52	Tubular silencer

Transmission
Gearbox	3-speed
Gearchange	Foot change
Final drive	Chain

Cycle parts
Frame	Single-loop
Front forks:	
1948-50	Girder fork, rubber-band springing
1951-52	Tele-Glide telescopic forks
Rear suspension	None (rigid)
Front brake	5-inch (127 mm) drum, via cable and handlebar lever
Rear brake	5-inch (127 mm) drum, via rod and pedal

General data
Dry weight	80 kg
Permissible total weight	160 kg
Tyres – front and rear	3.25-19
Wheels	2.00 x 19 wire-spoked
Top speed	75 km/h
Fuel tank capacity	6.6 litres

Model ST 1953 to 1959
Model STU 1954 to 1959

Motor
Engine	Single-cylinder air-cooled piston-ported 2-stroke
Bore & stroke	2.38 x 2.28 in (60.33 x 57.94 mm)
Capacity	10.11 cu. in. (166 cc)
Power output (SAE):	
ST	5.5 hp (4 kW)/5000 rpm
STU	4.7 hp (3.5 kW)/5000 rpm
Compression ratio	6.6:1
Ignition	Battery-and-coil
Exhaust system	Tubular silencer

Transmission
Gearbox	3-speed
Gearchange	Foot change
Final drive	Chain

Cycle parts
Frame	Single-loop
Front forks	Tele-Glide telescopic forks
Rear suspension	None (rigid)
Front brake	5-inch (127 mm) drum, via cable and handlebar lever
Rear brake	5-inch (127 mm) drum, via rod and pedal

General data
Dry weight	85 kg
Permissible total weight	160 kg
Tyres – front and rear	3.25-19 (1953-55)
	3.50-18 (1956-59)
Wheels	2.50 x 19 wire-spoked (1953-55)
	2.50 x 18 wire-spoked (1956-59)
Top speed	93 km/h (ST)
	75 km/h (STU)
Fuel tank capacity	6.6 litres

Model B 1955 to 1959 **Hummer**

Motor
Engine	Single-cylinder air-cooled piston-ported 2-stroke
Bore & stroke	2.06 x 2.28 in (52.39 x 57.94 mm)
Capacity	7.62 cu. in. (125 cc)
Power output (SAE)	4.8 hp (3.5 kW)/4800 rpm
Compression ratio	6.6:1
Ignition	Flywheel magneto
Exhaust system	Tubular silencer

Transmission
Gearbox	3-speed
Gearchange	Foot change
Final drive	Chain

Cycle parts
Frame	Single-loop
Front forks	Tele-Glide telescopic
Rear suspension	None (rigid)
Front brake:	
1955-56	None
1957-59	5-inch (127 mm) drum, via cable and handlebar lever
Rear brake	5-inch (127 mm) drum, via rod and pedal

General data
Dry weight	75 kg
Permissible total weight	160 kg
Tyres – front and rear:	
1955	3.25-19
1956-59	3.50-18
Wheels:	
1955	2.50 x 19 wire-spoked
1956-59	2.50 x 18 wire-spoked
Top speed	75 km/h
Fuel tank capacity	6.6 litres

Model BT 1960 to 1961 **Super 10**
Model BTU 1960 to 1961 **Super 10**
Model BTU 1962 **Pacer**

Motor
Engine	Single-cylinder air-cooled piston-ported 2-stroke
Bore & stroke	2.38 x 2.28 in (60.33 x 57.94 mm)
Capacity	10.11 cu. in. (166 cc)
Power output (SAE):	
BT	5.5 hp (4 kW)/5000 rpm
BTU	4.7 hp (3.5 kW)/5000 rpm
Compression ratio	6.6:1
Ignition	Flywheel magneto
Exhaust system	Tubular silencer

Transmission
Gearbox	3-speed
Gearchange	Foot change
Final drive	Chain

Cycle parts
Frame	Single-loop
Front forks	Tele-Glide telescopic
Rear suspension	None (rigid)
Front brake	5-inch (127 mm) drum, via cable and handlebar lever
Rear brake	5-inch (127 mm) drum, via rod and pedal

General data
Dry weight	85 kg
Permissible total weight	160 kg
Tyres – front and rear:	
1960	3.50-19
1961-62	3.50-16
Wheels:	
1960	2.50 x 19 wire-spoked
1961-62	2.50 x 16 wire-spoked
Top speed	90 km/h (BT)
	75 km/h (BTU)
Fuel tank capacity	6.6 litres

Model BT 1962 to 1965 Pacer

Motor
Engine	Single-cylinder air-cooled piston-ported 2-stroke
Bore & stroke	2.38 x 2.41 in. (60.33 x 61.12 mm)
Capacity	10.66 cu. in./175 cc
Power output (SAE)	6.5 hp (5kW)/5000 rpm
Compression ratio	7.63:1
Ignition	Flywheel magneto
Exhaust system	Tubular silencer

Transmission
Gearbox	3-speed
Gearchange	Foot change
Final drive	Chain

Cycle parts
Frame	Single-loop
Front forks	Tele-Glide telescopic
Rear suspension	None (rigid) (1962); Swinging-arm (1963-65)
Front brake	5-inch (127 mm) drum, via cable and handlebar lever
Rear brake	5-inch (127 mm) drum, via rod and pedal

General data
Dry weight	88 kg
Permissible total weight	160 kg
Tyres – front and rear	3.50-16
Wheels	2.50 x 16 wire-spoked
Top speed	100 km/h
Fuel tank capacity	6.6 litres

Model BTU 1963 to 1964 Pacer

Motor
Engine	Single-cylinder air-cooled piston-ported 2-stroke
Bore & stroke	2.38 x 2.41 in. (60.33 x 61.12 mm)
Capacity	10.66 cu. in./175 cc
Power output (SAE)	5 hp (3.7kW)/5000 rpm
Compression ratio	7.63:1
Ignition	Flywheel magneto
Exhaust system	Tubular silencer

Transmission
Gearbox	3-speed
Gearchange	Foot change
Final drive	Chain

Cycle parts
Frame	Single-loop
Front forks	Tele-Glide telescopic
Rear suspension	Swinging-arm
Front brake	5-inch (127 mm) drum, via cable and handlebar lever
Rear brake	5-inch (127 mm) drum, via rod and pedal

General data
Dry weight	85 kg
Permissible total weight	160 kg
Tyres – front and rear	3.50-16
Wheels	2.50 x 16 wire-spoked
Top speed	75 km/h
Fuel tank capacity	6.6 litres

Model BTH 1962 Scat

Motor
Engine	Single-cylinder air-cooled piston-ported 2-stroke
Bore & stroke	2.38 x 2.41 in. (60.33 x 61.12 mm)
Capacity	10.66 cu. in./175 cc
Power output (SAE)	10 hp (7 kW)/5000 rpm
Compression ratio	7.63:1
Ignition	Flywheel magneto
Exhaust system	High-level, right-hand side

Transmission
Gearbox	3-speed
Gearchange	Foot change
Final drive	Chain

Cycle parts
Frame	Single-loop
Front forks	Tele-Glide telescopic
Rear suspension	None (rigid)
Front brake	5-inch (127 mm) drum, via cable and handlebar lever
Rear brake	5-inch (127 mm) drum, via rod and pedal

General data
Dry weight	60 kg
Permissible total weight	160 kg
Tyres – front and rear	3.50-18
Wheels	2.50 x 18 wire-spoked
Top speed	approx. 100 km/h
Fuel tank capacity	6.6 litres

Model BTH 1963 to 1965 Scat

Motor
Engine	Single-cylinder air-cooled piston-ported 2-stroke
Bore & stroke	2.38 x 2.41 in. (60.33 x 61.12 mm)
Capacity	10.66 cu. in./175 cc
Power output (SAE)	10(7)/5000 rpm
Compression ratio	7.63:1
Ignition	Flywheel magneto
Exhaust system	High-level, right-hand side

Transmission
Gearbox	3-speed
Gearchange	Foot change
Final drive	Chain

Cycle parts
Frame	Single-loop
Front forks	Tele-Glide telescopic
Rear suspension	Swinging-arm
Front brake	5-inch (127 mm) drum, via cable and handlebar lever
Rear brake	5-inch (127 mm) drum, via rod and pedal

General data
Dry weight	60 kg
Permissible total weight	160 kg
Tyres – front and rear	3.50-18
Wheels	2.50 x 18 wire-spoked
Top speed	approx. 100 km/h
Fuel tank capacity	6.6 litres

Model BTF 1962 Ranger

Motor
Engine	Single-cylinder air-cooled piston-ported 2-stroke
Bore & stroke	2.38 x 2.28 in (60.33 x 57.94 mm)
Capacity	10.11 cu. in. (166 cc)
Power output (SAE)	6(4.4)/5000 rpm
Compression ratio	6.6:1
Ignition	Flywheel magneto
Exhaust system	High-level, right-hand side

Transmission
Gearbox	3-speed
Gearchange	Foot change
Final drive	Chain

Cycle parts
Frame	Single-loop
Front forks	Tele-Glide telescopic
Rear suspension	None (rigid)
Front brake	5-inch (127 mm) drum, via cable and handlebar lever
Rear brake	5-inch (127 mm) drum, via rod and pedal

General data
Dry weight	60 kg
Permissible total weight	160 kg
Tyres – front and rear	3.50-18
Wheels	2.50 x 18 wire-spoked
Top speed	approx. 50 km/h
Fuel tank capacity	6.6 litres

Model BTH 1966 Bobcat

Motor
Engine	Single-cylinder air-cooled piston-ported 2-stroke
Bore & stroke	2.38 x 2.41 in. (60.33 x 61.12 mm)
Capacity	10.66 cu. in./175 cc
Power output (SAE)	10(7)/5000 rpm
Compression ratio	7.63:1
Ignition	Flywheel magneto
Exhaust system	Tubular silencer

Transmission
Gearbox	3-speed
Gearchange	Foot change
Final drive	Chain

Cycle parts
Frame	Single-loop
Front forks	Tele-Glide telescopic
Rear suspension	Swinging-arm
Front brake	5-inch (127 mm) drum, via cable and handlebar lever
Rear brake	5-inch (127 mm) drum, via rod and pedal

General data
Dry weight	70 kg
Permissible total weight	160 kg
Tyres – front and rear	3.50-18
Wheels	2.50 x 18 wire-spoked
Top speed	approx. 105 km/h
Fuel tank capacity	7.1 litres

Topper Scooters
1960 to 1965

Model description and technical information

As Harley-Davidson took advantage of every opportunity to enter new markets in the difficult times of the 1950s, it is not surprising that Milwaukee would attempt to jump on the bandwagon of the relatively short-lived (in the USA) scooter craze started by the importing of Vespas and Lambrettas. Unfortunately, the Harley Topper arrived too late on the market and by 1965 it was all over.

The very simply-constructed, but heavy, scooter used an engine derived from the Model 165s but rubber-mounted and with the cylinder laid flat (but not fan-cooled) for this installation. For the first time in Harley's history a two-wheeler had bodywork, with built-in legshields; fibre-glass was used for the rear body and engine cover, while the frame, footboards, legshields and front mudguard were of pressed-steel. Designed as a two-seater, it was a product of the late-1950s American industrial school of design and its appearance was therefore distinctly angular and sharp-edged. There was only one colour to begin with: Strato Blue with White panels.

The Topper got simple rear suspension with two adjustable springs long before the other Harley two-strokes. Front suspension was leading-link, but with a single spring working under tension and a single hydraulic damper, both on the right-hand side only. Steel disc wheels were used, larger in diameter than the Italian scooters which had started the whole craze. Another special feature was the simple-to-use 'twist-and-go' Scootaway automatic transmission which used a V-belt drive running between two variable-diameter pulleys to provide automatic adjustment of engine speed and power to the load imposed, stepless changes in reduction ratio being achieved by the front pulley changing in diameter under the control of steel balls moving along cam tracks under centrifugal force, while the rear pulley consisted of two halves with tapered faces able to move towards or away from each other under the control of a pair of large coil springs. Final drive was, however, by chain; all transmission components were exposed underneath the bodywork. In addition, the 200 lb (90 kg) scooter was, for some unfathomable reason, fitted with a parking brake – a hook device to lock the left hand-operated front brake when required.

The Topper was started like a lawnmower or outboard motor – with a rope and a recoil starter. The seat had space underneath for schoolbooks, snacks or two-stroke oil.

Like the Model 165s and Super 10s, the Topper was offered in two versions: the Model A over-five-horsepower and the Model AU under-five-horsepower versions, the latter for those States still permitting 14-year-olds with a junior driver's licence to ride two-wheelers of less than five horsepower on the public highway. Sales figures are not particularly helpful, but it does not seem to have had too many takers.

The Topper was actually introduced in mid-May 1959, but genuine '59 models are probably the exception rather than the rule today; the 1960 model was introduced, with no changes, some time around August.

Model development

1961 model year
Two sidecars were introduced for this year, which either made the scooter a (presumably very slow) three-seater or enabled it to carry parcels and other goods for commercial users. The original low-compression engine was replaced by a high-compression version, the Model AH, which, if one can believe H-D's brochures, was at least 50% more powerful, going from 5.5 to 9 horsepower. The Model AU was of course left with its restricted output unchanged. The transmission received a major modification to the front pulley, going from an unenclosed 'dry' assembly to a sealed oil bath 'wet' assembly during the year. Colours available were Pepper Red, Granada Green or Strato Blue, all with Birch White panels, Hi-Fi Red being available at extra cost.

1962 to 1965 model years
Colours available for 1962 were Tango Red, Skyline Blue and Granada Green, either overall, or with Birch White panels. In the following years only badging was changed and minimal aesthetic (differing paint schemes – Tango or Fiesta Red, overall or with White panels, or Black and White) and technical modifications made. The Model AU disappeared from the catalogue at the end of the 1964 model year; the Model AH (Pacific Blue or Holiday Red, with Birch White panels) only lasted a year longer.

1960 Model A Topper – Strato Blue and White

1961 Topper – Granada Green and White

A Topper sidecar outfit

Model range

Scooters

Model A	Standard compression engine	(1960)
Model AH	Higher compression engine	(1961-65)
Model AU	Standard compression engine, restricted output	(1960-64)

All models have automatic transmission and are available with a sidecar.

Sidecars available

Model LA	Single-seater sidecar	(1961-63)
Model LM	Utility Box sidecar	(1961-63)

Catalogue prices ($ US)

Model	1960	1961	1962	1963	1964	1965
A	430	-	-	-	-	-
AH	-	445	445	460	470	480
AU	430	445	445	460	470	-

Prices are without sidecars, for which no prices are available.

A Topper with saddlebags

Production figures

Model	1960	1961	1962	1963	1964	1965
A	3801	-	-	-	-	-
AH	-	1341	***	972	800	500
AU	*	**	***	6	25	-

*	Included with A
**	Included with AH
***	No details available from Harley-Davidson

Model A 1960
Model AU 1961 to 1964

Motor
Engine	Single-cylinder air-cooled piston-ported 2-stroke
Bore & stroke	2.38 x 2.28 in (60.33 x 57.94 mm)
Capacity	10.11 cu. in. (166 cc)
Power output (SAE):	
A	5.5 hp (4 kW)/5000 rpm
AU	4.7 hp (3.5 kW)/5000 rpm
Compression ratio	6.6:1
Ignition	Flywheel magneto
Exhaust system	1 silencer

Transmission
Gearbox	Stepless automatic, ratios from 5.9 to 18.0
Gearchange	Automatic
Final drive	Chain

Cycle parts
Frame	Pressed-steel
Front forks	Leading link, single spring and single damper
Rear suspension	Swinging-arm
Front brake	5-inch (127 mm) drum, via cable and handlebar lever
Rear brake	5-inch (127 mm) drum, via rod and pedal

General data
Dry weight	90 kg
Permissible total weight	250 kg
Wheelbase	1308 mm
Tyres – front and rear	4.00-12
Wheels	2.50 x 12 steel disc
Top speed	approx. 95 km/h (A)
	approx. 80 km/h (AU)
Fuel tank capacity	6.4 litres

Model AH 1961 to 1965

Motor
Engine	Single-cylinder air-cooled piston-ported 2-stroke
Bore & stroke	2.38 x 2.28 in (60.33 x 57.94 mm)
Capacity	10.11 cu. in. (166 cc)
Power output (SAE)	9 hp (7 kW)/5500 rpm
Compression ratio	8.0:1
Ignition	Flywheel magneto
Exhaust system	1 silencer

Transmission
Transmission	Stepless automatic, ratios from 5.9 to 18.0
Gearchange	Automatic
Final drive	Chain

Cycle parts
Frame	Pressed-steel
Front forks	Leading link, single spring and single damper
Rear suspension	Swinging-arm
Front brake	5-inch (127 mm) drum, via cable and handlebar lever
Rear brake	5-inch (127 mm) drum, via rod and pedal

General data
Dry weight	90 kg
Permissible total weight	250 kg
Wheelbase	1308 mm
Tyres – front and rear	4.00-12
Wheels	2.50 x 12 steel disc
Top speed	approx. 115 km/h
Fuel tank capacity	6.4 litres

Knucklehead: Models E and F Big Twins

1936 to 1947

Model description and technical information

During the Depression of the early Thirties, an equally-difficult time for all motorcycle manufacturers, Harley decided that attack was the best form of defence and undertook development of their next generation of motorcycles, even though they had only just replaced their ancient ioe machines with the side-valve-engined Model Vs. Thus, in only seven model years, Milwaukee took the two most important steps that would bring it into the modern motorcycling world. The new and up-to-date overhead-valve Model E stood head and shoulders above the rest in a world of mundane side-valves. With its 60 cu. in./989 cc. capacity (almost universally rounded up to identify it as the Sixty-One), it slotted in exactly between the small side-valve Forty-Fives and the Big Twin range of 74 cu. in./1209 cc. Model Us and 80 cu. in./1302 cc. Model UHs – which would gradually be supplanted by the ohv models between 1936 and 1948.

First proposed in 1931 by William S. Harley and referred to as the 'sump oiler', development of a new ohv model with the weight of a Forty-Five but the power of a Seventy-Four was agreed by the board. By early 1934 the first complete prototype had been assembled and was being tested, but then the project ran into trouble with the development of the new recirculating oil lubrication system. Originally intended for launch in the autumn as a 1935 model, the ohv was held back (even though the launch of the 1935 model range was delayed from the usual September until December) and an ultra-Big side-valve Twin, the Eighty, whose extra power was also aimed at the sidecar-driver and to head off opposition from Indian, was hurriedly prepared and on sale by the following summer. Development was reportedly hindered by the fact that anti-Depression legislation then banned companies from requiring employees to work overtime

1936 Model EL in Venetian Blue with Croydon Cream mudguard panels and wheel rims

and insisted on the employment of additional workers (not always experienced men) instead. By now President Walter Davidson had had enough and insisted on the new model's being on sale by June 1936 in order to recoup some of its development costs. Although it was not yet fully ready, the Model E was shown to dealers at their annual convention on November 25th 1935 and new machines started to reach the dealers by January-February 1936; in spite of only having half the normal sales year in which to make its mark, the Model E still managed to sell 1,704 units in 1936 (although it should be noted that some dispute this figure, believing that sales were higher – 1,836, 1,926, or even 2,000 machines). While it was typical of Harley to take the cautious approach and to carry on selling the old model alongside the new, in this case it was a good thing as the Model Vs' teething troubles were as nothing to those of the Model Es, and the latter's development costs were much greater. But all this was in the future for the new motorcycle.

The main innovation of the new model, apart from its valvegear, was its sophisticated dry-sump recirculating lubrication system which was supplied from the now-famous 0.8-Imp. gallon (3.8 litre) horseshoe-shaped oil tank located behind the seat tube. This reduced oil consumption from the 184 to 211 miles per litre to be expected of a machine using the total-loss lubrication system to a figure between 211 and 422 miles per litre.

Adjustable pushrods and the characteristic rocker housings on the cylinder heads with their four chrome-plated metal covers (soon replaced by large chrome-plated threaded plugs), for which it would later be christened the 'Knucklehead', are the obvious points of identification for an engine which was otherwise well-designed, for all the unfortunate problems, large and small, caused by its over-hasty launch on to the market. Perpetual oil leaks and other mechanical faults dealt a heavy blow to the reputation of the early ohv models and it wasn't for some years, even up to 1941, until Harley had the engine's problems truly sorted.

It wasn't just the valvegear and the lubrication system which were new; a new four-speed constant-mesh gearbox was also fitted, all previous Big Twin gearboxes being sliding-mesh. Gearchanging was still by hand via a tank-mounted gate – one couldn't expect customers to take in too many innovations at once – but the selector mechanism was well-designed and executed for much smoother and quicker gearchanging.

The frame was different from that of any other Harley in that it was a duplex-loop design with twin downtubes – even the new Model Vs only had single-loop frames – which made it much stiffer for improved handling. It was constructed of high-quality chrome-molybdenum steel and was also beautifully integrated into the overall styling of the motorcycle. However, although it is a strange thing to say in the context of a 515 lb (234 kg) motorcycle, it was light in construction and proved to be a bit too light in use, especially when sidecar fixings were clamped to the downtubes; the

1936 Model EL in Maroon with Nile Green mudguard panels and wheel rims

Model range

Model E	60 cu. in./989 cc. standard-compression engine	(1936-37 and 1942-47)
Model ES	60 cu. in./989 cc. standard-compression engine, sidecar-specification	(1936-41 and 1945-47)
Model EL	60 cu. in./989 cc. high-compression engine	(1936-47)
Model F	74 cu. in./1207 cc. standard-compression engine	(1942-47)
Model FS	74 cu. in./1207 cc. standard-compression engine, sidecar-specification	(1941 and 1945-47)
Model FL	74 cu. in./1207 cc. high-compression engine	(1941-47)

stress put on the slim downtubes could cause them to crack. The front fork got improved chrome-molybdenum-steel oval-section tubular rigid (sprung) legs instead of the forged I-beam legs previously used, but the four massive volute springs remained, controlled for the first time as an extra-cost factory option by friction damping. The Ride Control, previously only seen on racers and available only from accessory suppliers, consisted of two slotted steel plates mounted on each side of the fork springs and set by tightening down an adjusting knob on the side of the assembly.

In terms of wheels and tyres, standard equipment was now the smaller 18-inch wheels with 4.00-section tyres which not only improved the ride and handling but were better suited to the more modern styling of the machine. Brakes were internally-expanding drums, 7¼-inch front and rear in the interests of reduced weight, but again were a little too weak – even by the poor standards of the day – for the machine's near-100 mph (160 km/h) performance.

New also were the two-part petrol tank, the famous instrument panel with speedometer, oil pressure gauge, ammeter, ignition switch and speedometer lighting switch lying centrally between the two tank halves and being much closer to the rider's field of view than the old speedometer. The speedometer itself had a white dial and, half-way through the model year, was fitted with a trip odometer. Front crashbars were standard equipment and the toolbox was fastened to a bracket on the right-hand rear frame tubes. The exhaust silencer terminated in the large-fishtail pattern common to the other 1935 models, even if the downpipes were (obviously) slightly different.

In the matter of colours, the Model E showed that Harley had finally left the 'you can have any colour you like as long as it's Olive Green' days of the Twenties well and truly behind them. Admittedly alternative colours had been offered since 1927 and two-tone paint schemes had even been available to prospective owners since 1930, but in 1935-6 the message to dealers was, effectively, 'Do not lose a sale because of a mere trifle like the colour scheme' (no doubt inspired by the fact that H-D's greatest rival, Indian, had been in the hands of a member of the du Pont family since 1930 and, with the resources of the du Pont de Nemours chemical empire behind them were now offering Fours, Chiefs and Scouts in a new and attractive range of du Pont paint colours). Truly modern colour schemes were offered as standard: Sherwood Green with Silver mudguard panels and wheel rims, Teak Red with Black mudguard panels and wheel rims, Dusk Gray with Royal Buff mudguard panels and wheel rims, Venetian Blue with Croydon Cream mudguard panels and wheel rims and the startling Maroon with Nile Green mudguard panels and wheel rims. However, in a circular to dealers, H-D offered to supply whatever colour scheme the customer wanted, in order to make the sale; order forms were modified to enable dealers to give details of the customer's demands.

From the very beginning three versions were offered: the high-compression Model EL, which was not available ex-works with a sidecar, and the medium-compression Model E, while the Model ES was fitted with the same pistons as the Model E but also had compression plates to lower its compression ratio even further for sidecar use.

During what remained of the Model E's first production year, changes were introduced in a feverish campaign of recalls and despatch of modified parts resembling the Model Vs' debacle, to address the machine's various shortcomings. The frame was strengthened (on the right-hand side only, to cope with the stresses of sidecar attachment, which was causing frames to crack), the kickstart was improved and the timing cover was modified no less than three times during the year. There were even three different types of oil tanks, and the cylinders, including the cylinder heads, had to be modified; the latter, particularly, received a whole series of changes and updates to the arrangements for sealing and enclosing the rockers and valve springs and for dealing with the return of oil from the top half of the engine to the sump.

Almost entirely unconcerned with the threat of war in Europe, the Model E was constantly developed and matured with the passing of the years into a thoroughly reliable machine which echoed the success of the Model Us and gained Harley a sound reputation for dependability. The fact that few were ordered for war service by the US Army does not reflect badly on the Model E; it merely illustrates the value offered to the government by the Harley Forty-Fives which took the lion's share of the contracts.

1937 Model EL with chrome-plated large fishtail silencer and saddlebags

Model development

1937 model year
Apart from several modifications to improve the reliability and oil-tightness of the engine, a 120-mph speedometer was fitted and a redesigned and strengthened frame with integral mounting eyes for sidecar attachments and improved support for the kickstart end of the gearbox. The rear brake drum, was now of stamped steel, increased in diameter to 8 inches (203 mm) and fitted with a reinforced backplate and improved operating linkage. At first fitted with longer brake shoe linings, problems with brake chatter and self-energising rear brakes caused further modifications throughout the year which resulted in the linings being shortened. This year's colours for the fuel and oil tanks plus mudguards were Bronze Brown with Delphine Blue stripe with Yellow edge, Teak Red with Black stripe with Gold edge, Delphine Blue with Teak Red stripe and Gold edge, while police machines were Police Silver with Black stripe with Gold edge. Wheel rims were Black and the offer of custom colours to the customer's design was still available.

1938 model year
For 1938 there were again a host of modifications and improvements, chief amongst which were truly effective full enclosures for the rocker arms and valves, but oil feed and return lines were revised, the oil tank breather modified and oil-sealing generally improved. The rear brake linings were shortened to prevent chattering, the backplate was reinforced and the rear brake shoes were interconnected by a two-piece cup bearing fitted to the pivot stud. New ball bearings and improved third and fourth gears were fitted to the gearbox and the clutch operating mechanism was improved. The instrument panel was fitted with warning lights instead of gauges; red for low oil pressure and green for battery charging. Numerous further small detail changes made everyday life easier and contributed to the machine's suitability for daily use.

The medium-compression engine now used the same pistons as the high-compression engine, with compression plates being fitted under the cylinder barrel to reduce the compression ratio to the desired level for sidecar use. The low-compression option was no longer available from now until 1942; solo motorcycles would automatically be delivered as high-compression Model ELs. Paint colours were: Hollywood Green with Gold stripe and Black edge, Teak Red with Black stripe and Gold edge, Silver Tan with Sunshine Blue stripe and Gold edge, Police Silver with Black stripe, Venetian Blue with White stripe and Burnt Orange edge and, for export only, Olive Green with Black stripe and Gold edge. These applied to fuel tanks and mudguards only, wheel rims and oil tanks being painted black.

1939 model year
Although Model E production was by now well under way, the modifications still flowed thick and fast. New pistons and rings, stiffer valve springs and revised valvegear were fitted, as was a modified clutch and an improved oil pump which operated at nearly twice the speed for greater flow and higher oil pressure;

1937 Model EL: Teak Red with Black stripe and Gold edge (including the oil tank, for this year only) and Speedster bars

unfortunately this caused over-oiling at low engine speed and so a lighter bypass valve spring had to be fitted later in the year. The carburettor also got a float chamber drain fitting. The gearchange pattern was altered, neutral now being found between second and third. This, however, only lasted a year before neutral was restored to between first and second gears in the gate. While the four-speed gearbox was still standard, two three-speed transmissions were introduced to special order, one of them with a reverse gear for sidecar-specification machines.

New front fork springs and a frame with a revised steering head, fitted with a self-aligning cone on the lower bearing, brought significant improvements in handling. On the styling front, the restyled Cats' Eye instrument panel appeared, as did the Boattail taillamp, the body of which was painted to match the mudguard, and stainless-steel mudguard tips. Colours available were: Black with Ivory tank panel, Teak Red with Black panel, and Airway Blue with White panel.

1940 model year
Yet again there were several mechanical improvements, such as modified cylinders and piston rings, the re-revised gearchange and all models now used the larger 1½-inch (38 mm) carburettor. The two halves of the petrol tank were now linked via a pipe so it was no longer necessary to fill each half separately. The front fork legs were heat-treated, an additional process which made them stiffer and less prone to flexing. This year's paint colours were: Black with Flight Red stripe, Flight Red with Black stripe, Squadron Grey with Bittersweet stripe and Clipper Blue with White stripe. The optional twin seat now had a special latex covering and chrome-plated metal tank badges in a teardrop style were fitted. D-shaped footboards were introduced and 5.00-16 tyres on wider 3-inch (76.2 mm) rims, with their substantially greater air volume giving much-improved ride comfort over the old fitment, were made available as options; they rapidly became very popular.

1941 model year
Even while the Second World War was raging, life went on as normal in the United States and the 60 cu. in./989 cc. Model E got a big sister, the 74 cu. in./1207 cc. Model F, the two being intended to sell side-by-side and the Seventy-Four receiving the same technical improvements that the Sixty had received the previous year. The piston rings were again revised, as were the crankcase and flywheels (the Model E being fitted with the larger and heavier components needed to create the Model F's longer-stroke engine), while the timing cover was now finned. The bigger machine was also fitted with different pistons to give the low-compression version, reserved in the first year of production for sidecar machines but from 1942-on standard fitment for solos, so that sidecars could be ordered with either engine, and the more powerful high-compression Model FL version. The Knucklehead's lubrication problems were finally and completely ended, five years after its introduction, with the introduction of a bypass valve controlled by a centrifugal governor that provided maximum lubrication to the engine at high speed. At low speeds, it directed the oil back to the supply tank, bypassing the timing gear case. The frame was amended to give a steering head angle of 29 degrees (previously 28) and increased trail to better suit the fatter 16-inch tyres and wheels which were now standard

equipment, the 18-inch former standard sizes now being optional.

A more modern and stronger design of multi-plate clutch was used, a larger 7-inch (178 mm) air cleaner was fitted, and the battery was now earthed to the frame. The Airplane-Style speedometer (black face with large silver numbers) was introduced to all Harleys this year, as was the new Rocket-Fin swallowtail-silencer exhaust.

Frames and cycle parts, fixtures and fittings, were the same as those on the smaller model so that only the Harley *cognoscenti* can tell the difference between a Model E Sixty and a Model F Seventy-Four of the same year; even the tank badges were the same for both from 1941 through to 1946, and then again for 1947 and 1948. Colours available were the same for both, for 1941 and 1942: Brilliant Black, Flight Red, Cruiser Green and Skyway Blue (all over – no two-tone paint schemes for these years), with Police Silver for police machines and Olive Green for any exports.

1942 model year
It may be strange to British readers to think of civilian motorcycles being produced in 1942, but remember that the model year started in August or September, so production and sales were well under way before Pearl Harbor, naturally, changed everything. Since one was only allowed to buy a new Harley once one had obtained special permission from the authorities, only a few big Harleys found their way into civilian use; the machines that were produced were reserved for the armed forces and for the authorities such as the police forces. Materials deemed essential for the war effort were strictly rationed; no more chrome was allowed, for example, for the plating of decorative items of trim. Colours available (all over – no two-tone paint schemes for these years), initially at least, were the same as in 1941: Brilliant Black, Flight Red, Cruiser Green and Skyway Blue, with Police Silver for police motorcycles and Olive Green for any exports.

The low-compression option was again available for solo motorcycles; the best-quality petrol was reserved for military use and civilians had to make do with very poor-quality fuel. Apart from that, no changes worthy of the name were made.

1943 model year
On the grounds of economizing on materials the front mudguard running light was no longer fitted, and even trim items of stainless steel were deleted. The air cleaner housing was painted black and saddles were padded with horsehair; colours available were restricted to Grey and Silver. Synthetic materials were urgently required for armaments production.

1944 model year
For 1944 even carburettor bodies were delivered black-painted and tyres were made of the war-substitute synthetic rubber Type S-3 (the lowest-quality made).

1945 model year
No changes. In September Monroe hydraulic dampers were available as optional extras for the front fork springs, replacing the Ride Control friction-damping device.

1946 model year
The frame was changed again, the steering head angle being revised to 30 degrees for improved straight-line stability. Where the saddles of the wartime machines were padded with horsehair, with the end of the war the use of latex mesh, as pre-war, was now permitted once more. The motorcycles

1939 Model EL: Black with Ivory tank panel

gradually regained the glitter lost during the war years as the various chrome trim pieces gradually became available again. Although, as usual with Harley-Davidson, precise information is not available, such improvements being introduced as and when on the production lines, a modified front fork was fitted some time towards the end of the model year.

1947 model year
Colour choice was getting back to normal; Brilliant Black, Flight Red, Skyway Blue and Police Silver were available for the US, with Olive Green for export models. Post-war inflation caused the price of the Model Es, in base specification, to increase by some 27% from 464 to 590 dollars. Amongst other minor modifications were the new handlebar grips, red Speed Ball tank badges, the famous Tombstone taillamp and the new speedometer fitted to all 1947 Harleys; the speedometer needle was now red and the numbers on the speedometer dial were enlarged to improve its readability.

The frame was changed three times during the model year, the best-known version being the Bull-Neck frame, with its oval-section steering head rather than the usual cylindrical forging.

This was the last year for the Knucklehead; a more modern unit with aluminium-alloy cylinder heads and hydraulic tappets was being developed which required a new frame. Effectively, the 1948 Models E and F were completely new motorcycles which only had their front fork construction in common with the original 1936 model.

1947 Model EL in Skyway Blue

Motorcycle catalogue prices ($ US)

Model	1936	1937	1938 1939	1940	1941	1942 1943 1944	1945 1946	1947
E	380	435	-	-	-	425	464	590
ES	380	435	435	430	425	-	464	590
EL	380	435	435	430	425	425	464	590
ELF	-	-	-	-	-	-	-	-
ELS	-	-	-	-	-	-	-	-
F	-	-	-	-	-	465	465	605
FS	-	-	-	-	465	-	465	605
FL	-	-	-	-	465	465	465	605

Prices are for motorcycles only, without sidecars.

Sidecar catalogue prices ($ US)

Model	1936	1937	1938 1939	1940 1941	1942 1943 1944	1945 1946	1947
LE	-	125	125	125	125	139	173
LE Body	-	63	-	-	-	-	-
LEC Chassis	-	80	-	-	-	-	-
K (single-seater)	110	-	-	-	-	-	-
K Body	52	-	-	-	-	-	-
KC Chassis	70	-	-	-	-	-	-
M	115	135	135	135	135	135	180
M Body	60	68	-	-	-	-	-
MC Chassis	72	80	80	-	80	86	110
MO	105	115	115	115	-	-	-
MO Body	47	-	-	-	-	-	-
MWC Chassis	78	-	90	-	-	-	-

Chassis = Frame, wheel and fittings only (for aftermarket coach-built bodies).

Prices and availability of optional Accessory Groups for motorcycles ($ US)

Group	1936	1937	1938	1939	1940	1941	1942	1943	1944
Standard Solo	14.00	21.75	16.70	15.50	-	-	-	-	-
Standard Commercial	-	20.00	14.25	14.00	-	-	-	-	-
Utility Solo	-	-	-	-	11.00	14.50	14.50	14.50	14.50
Sport Solo	-	-	-	-	22.50	27.00	27.00	-	-
Deluxe Solo	34.50	49.00	49.75	47.00	46.00	60.00	60.00	-	-
Standard Police	44.00	49.50	52.25	41.00	39.50	42.50	42.50	42.50	42.50
Deluxe Police	80.50	83.50	-	-	-	-	-	-	-

Group	1945	1946	1947
Utility Solo	14.50	14.50	34.00
Sport Solo	-	-	-
Deluxe Solo	44.50	55.00	100.00
Standard Police	42.50	57.00	75.00

Note that Standard means a specification higher than standard ex-works finish.

Prices and availability of optional Accessory Groups for sidecars ($ US)

Group	1936	1937	1938	1939	1940	1941	1942	1943	1944
Standard Sport	15.75	17.50	17.50	17.75	-	-	-	-	-
Standard Truck	-	-	14.25	14.00	-	-	-	-	-
Utility	-	-	-	-	8.50	12.00	12.00	12.00	12.00
Utility Truck	-	-	-	-	8.50	12.00	12.00	12.00	12.00
Deluxe Sport	39.50	44.00	44.00	47.00	50.00	54.00	54.00	-	-
Deluxe Truck	26.90	29.00	29.00	31.75	-	-	-	-	-

Group	1945	1946	1947
Utility	12.00	12.00	31.50
Utility Truck	12.00	12.00	31.50
Deluxe Sport	49.00	49.00	85.00

Note that Standard means a specification higher than standard ex-works finish.

Production figures

Model	1936	1937	1938	1939	1940	1941	1942	1943	1944	1945	1946	1947
E	152	126	-	-	-	-	***	***	***	***	***	***
ES	26	70	189	214	176	261	164	105	180	282	244	237
EL ***	1526	1829	2289	2695	3893	2280	620	53	116	398	2098	4117
F and FS	-	-	-	-	-	156	107	12	67	131	418	334
FL	-	-	-	-	-	2452	799	33	172	619	3986	6893

*** Where a separate figure is not given for Model Es, they are included in the total figure of all Model ELs.

Model E 1936 to 1937

Motor
Engine	Air cooled 45° ohv V-twin
Bore and stroke	3 5/16 x 3 1/2 (3.31 x 3.5) in.
Bore and stroke	84.14 x 88.9 mm
Capacity	60.33 cu. in. (989 cc)
Power output (SAE)	37 hp (27 kW)/4800 rpm
Compression ratio	6.5:1
Carburettor	Linkert 1 1/4-inch Type M-5
Ignition	Battery-and-coil
Exhaust system	Large fishtail silencer

Transmission
Gearbox	4-speed
Overall reduction ratio	3.73:1
Gearchange	Hand change, on tank
Final drive	Chain

Cycle parts
Frame	Duplex-loop, twin downtube
Front forks	Springers – exposed-springs, oval-tube legs
Rear suspension	None (rigid)
Front brake	7 1/4-inch dia. drum, via cable and handlebar lever
Rear brake	7 1/4-inch dia. drum, via rod and pedal

General data
Dry weight	240 kg
Permissible total weight	440 kg
Tyres – front and rear	4.00-18
Wheels front and rear	2.15 x 18 wire-spoked
Top speed	140 km/h
Fuel tank capacity	14.2 litres

Model ES 1936 to 1941

Motor
Engine	Air cooled 45° ohv V-twin
Bore and stroke	3 5/16 x 3 1/2 (3.31 x 3.5) in.
Bore and stroke	84.14 x 88.9 mm
Capacity	60.33 cu. in. (989 cc)
Power output (SAE):	34 hp (25 kW)/4800 (1936-37)
	37 hp (27 kW)/4800 SAE (1938-39)
	40 hp (29 kW)/4800 SAE (1940-41)
Compression ratio:	
1936-37	5.6:1
1938-41	6.5:1
Carburettor:	
1936-39	Linkert 1 1/4-inch Type M-5
1940	Linkert 1 1/4-inch Type M-25
1941	Linkert 1 1/4-inch Type M-35
Ignition	Battery-and-coil
Exhaust system:	
1937-40	Large fishtail silencer
1941	Rocket-Fin silencer

Transmission
Gearbox	4-speed (optional 3-speed with reverse)
Overall reduction ratio	3-speed: 4.75:1; 4-speed: 4.29:1
Gearchange	Hand change, on tank
Final drive	Chain

Cycle parts
Frame	Duplex-loop, twin downtube
Front forks	Springers – exposed-springs, oval-tube legs
Rear suspension	None (rigid)
Front brake	Drum, via cable and handlebar lever
Rear brake	Drum, via rod and pedal

General data
Dry weight	approx. 350 kg
Permissible total weight	approx. 600 kg
Tyres – front and rear	4.00-18 (1936-40)
	5.00-16 (1940-41)
Wheels front and rear	2.15 x 18 wire-spoked (1936-40)
	3.00 x 16 wire-spoked (1940-41)
Top speed	approx. 110 km/h
Fuel tank capacity	14.2 litres

Model E 1942 to 1950

Motor
Engine	Air cooled 45° ohv V-twin
Bore and stroke	3 5/16 x 3 1/2 (3.31 x 3.5) in.
Bore and stroke	84.14 x 88.9 mm
Capacity	60.33 cu. in. (989 cc)
Power output (SAE)	40 hp (29 kW)/4800 SAE (1942-1949)
	44(32)/4800 SAE (1950)
Compression ratio	6.5:1
Carburettor	Linkert 1 1/2-inch Type M-35 (1942-48)
	Linkert 1 1/2-inch Type M-36 (1949-50)
	Linkert 1 1/2-inch Type M-61 (1950)
Ignition	Battery-and-coil
Exhaust system	Rocket-Fin silencer (1942-49)
	Tubular silencer (1950)

Transmission
Gearbox	4-speed (optional 3-speed with reverse)
Overall reduction ratio	3-speed: 3.90:1; 4-speed: 3.73:1
Gearchange	Hand change, on tank
Final drive	Chain

Cycle parts
Frame	1942-47: Twin-loop, with straight front downtubes
	1948-50: Twin-loop, with splayed front downtubes
Front forks	Springers – exposed-springs, oval-tube legs
Rear suspension	None (rigid)
Front brake	Drum, via cable and handlebar lever
Rear brake	Drum, via rod and pedal

General data
Dry weight	240 kg
Permissible total weight	440 kg
Tyres – front and rear	5.00-16 optional 4.00-18 or 4.50-18
Wheels front and rear	3.00 x 16 wire-spoked optional 2.15 x 18
Top speed	140 km/h
Fuel tank capacity	14.2 litres

Model ES 1945 to 1950

Motor
Type	Knucklehead (1945-47)
	Panhead (1948-50)
Engine	Air cooled 45° ohv V-twin
Bore and stroke	3 5/16 x 3 1/2 (3.31 x 3.5) in.
Bore and stroke	84.14 x 88.9 mm
Capacity	60.33 cu. in. (989 cc)
Power output (SAE)	40 hp (29 kW)/4800 SAE (1945-1949)
	44(32)/4800 SAE (1950)
Compression ratio	6.5:1
Carburettor	Linkert 1 1/2-inch Type M-35 (1945-48)
	Linkert 1 1/2-inch Type M-36 (1949-50)
	Linkert 1 1/2-inch Type M-61 (1950)
Ignition	Battery-and-coil
Exhaust system	Rocket-Fin silencer (1945-49)
	Tubular silencer (1950)

Transmission
Gearbox	4-speed (optional 3-speed)
Overall reduction ratio	3-speed: 3.90:1; 4-speed: 3.73:1
Gearchange	Hand change, on tank
Final drive	Chain

Cycle parts
Frame	1945-47: Twin-loop, with straight front downtubes
	1948-50: Twin-loop, with splayed front downtubes
Front forks	Springers – exposed-springs, oval-tube legs
Rear suspension	None (rigid)
Front brake	Drum, via cable and handlebar lever
Rear brake	Drum, via rod and pedal

General data
Dry weight	approx. 350 kg
Permissible total weight	approx. 600 kg
Tyres – front and rear	5.00-16 optional 4.00-18 or 4.50-18
Wheels front and rear	3.00 x 16 wire-spoked optional 2.15 x 18
Top speed	approx. 100 km/h
Fuel tank capacity	14.2 litres

Model EL 1936 to 1952
Model ELP 1949
Model ELF 1952

Motor

Type	Knucklehead (1936-47) Panhead (1948-52)
Engine	Air cooled 45° ohv V-twin
Bore and stroke	3⁵⁄₁₆ x 3½ (3.31 x 3.5) in.
Bore and stroke	84.14 x 88.9 mm
Capacity	60.33 cu. in. (989 cc)
Power output (SAE)	40 hp (29 kW)/4800 SAE (1936-37)
	43(32)/4800 SAE (1938-39)
	46(34)/4800 SAE (1940-49)
	50(37)/4800 SAE (1950-52)
Compression ratio	6.5:1 (1936-37)
	7.0:1 (1938-52)
Carburettor	Linkert 1½-inch Type M-5 (1936-38)
	Linkert 1½-inch Type M-25 (1940)
	Linkert 1½-inch Type M-35 (1941-48)
	Linkert 1½-inch Type M-36 (1949-50)
	Linkert 1½-inch Type M-61 (1950-52)
Ignition	Battery-and-coil
Exhaust system	Large fishtail silencer (1936-40)
	Rocket-Fin silencer (1941-49)
	Tubular silencer (1950-52)

Transmission

Gearbox	4-speed (optional 3-speed)
Overall reduction ratio	3-speed: 3.90:1
	4-speed: 3.73:1 (1936-48)
	3.90:1 (1949-52)
Gearchange	Hand change, on tank, from 1952 optional Foot change
Final drive	Chain

Cycle parts

Frame	1936-47: Twin-loop, with straight front downtubes
	1948-52: Twin-loop, with splayed front downtubes
Front forks	Springers – exposed-springs, oval-tube legs (1941-48)
	Hydraulically-damped telescopic (1949-52)
Rear suspension	None (rigid)
Front brake	Drum, via cable and handlebar lever
Rear brake	Drum, via rod and pedal

General data

Dry weight	240 kg
Permissible total weight	440 kg
Tyres – front and rear	4.00-18 (1936-52)
	optional 4.50-18 (1936-52)
	optional 5.00-16 (1940-52)
Wheels front and rear	2.15 x 18 wire-spoked (1936-40)
	optional 3.00 x 16 wire-spoked (1940-52)
Top speed	140 km/h
Fuel tank capacity	14.2 litres

Model ELS 1951 to 1952

Motor
Type	Panhead
Engine	Air cooled 45° ohv V-twin
Bore and stroke	3⁵⁄₁₆ x 3½ (3.31 x 3.5) in.
Bore and stroke	84.14 x 88.9 mm
Capacity	60.33 cu. in. (989 cc)
Power output (SAE)	50(37)/4800 SAE (1950-52)
Compression ratio	7.0:1
Carburettor	Linkert 1½-inch Type M-61
Ignition	Battery-and-coil
Exhaust system	Rocket-Fin silencer (1945-49) Tubular silencer

Transmission
Gearbox	4-speed (optional 3-speed with reverse)
Overall reduction ratio	3-speed: 4.76:1 4-speed: 4.29:1
Gearchange	Hand change, on tank from 1952 optional Foot change
Final drive	Chain

Cycle parts
Frame	Twin-loop, with splayed front downtubes
Front forks	Springers – exposed-springs, oval-tube legs
Rear suspension	None (rigid)
Front brake	Drum, via cable and handlebar lever
Rear brake	Drum, via rod and pedal

General data
Dry weight	approx. 350 kg
Permissible total weight	approx. 600 kg
Tyres – front and rear	4.00-18(1936-52) optional 4.50-18 (1936-52) optional 5.00-16 (1940-52)
Wheels front and rear	2.15 x 18 wire-spoked (1936-40) optional 3.00 x 16 wire-spoked (1940-52)
Top speed	approx. 110 km/h
Fuel tank capacity	14.2 litres

1941 Model FL: Rocket-Fin silencer, 5.00-16 wheels and tyres

1942 Model FL. Either the accessories were added post-war, or this one was built before Pearl Harbor was attacked

1945 Model FL. Some chrome pieces returning, but wartime austerity is still evident

1946 Model FL with sidecar. This one still has Ride Control friction damping for the front fork springs

1947 Model FL in Brilliant Black

1947 Model FL: The last of the Knuckleheads, but not quite the last of the Big Twin Springers

Model F 1942 to 1948
Model FS 1941
Model FS 1945 to 1948

Motor
Engine	Air cooled 45° ohv V-twin
Bore and stroke	3⁷⁄₁₆ x 3³¹⁄₃₂ (3.44 x 3.97) in.
Bore and stroke	87.31 x 100.81 mm
Capacity	73.66 cu. in. (1207 cc)
Power output (SAE)	46(34)/5000 rpm
Compression ratio	6.6:1
Carburettor	Linkert 1½-inch Type M-25 (1941)
	Linkert 1½-inch Type M-75 (1941)
	Linkert 1½-inch Type M-35 (1942-48)
Ignition	Battery-and-coil
Exhaust system	Rocket-Fin silencer

Transmission
Gearbox	4-speed (optional 3-speed)
Overall reduction ratio	F: 3.73:1
	FS: 4.29
	(3-speed with reverse)
	FS: 4.08 (4-speed)
Gearchange	Hand change, on tank
Final drive	Chain

Cycle parts
Frame	1941-47: Twin-loop, with straight front downtubes
	1948: Twin-loop, with splayed front downtubes
Front forks	Springers – exposed-springs, oval-tube legs
Rear suspension	None (rigid)
Front brake	Drum, via cable and handlebar lever
Rear brake	Drum, via rod and pedal

General data
Dry weight	240 kg
Permissible total weight	440 kg
Tyres – front and rear	5.00-16
	optional 4.00-18 or 4.50-18
Wheels front and rear	3.00 x 16 wire-spoked
	optional 2.15 x 18
Top speed	150 km/h
	(FS: approx. 115 km/h)
Fuel tank capacity	14.2 litres

Model FL 1941 to 1948

Motor
Engine	Air cooled 45° ohv V-twin
Bore and stroke	3⁷⁄₁₆ x 3³¹⁄₃₂ (3.44 x 3.97) in.
Bore and stroke	87.31 x 100.81 mm
Capacity	73.66 cu. in. (1207 cc)
Power output (SAE)	48(36)/5000 rpm
Compression ratio	7.0:1
Carburettor	Linkert 1½-inch Type M-25 (1941)
	Linkert 1½-inch Type M-75 (1941)
	Linkert 1½-inch Type M-35 (1942-48)
Ignition	Battery-and-coil
Exhaust system	Rocket-Fin silencer

Transmission
Gearbox	4-speed (optional 3-speed)
Overall reduction ratio	3.73:1
Gearchange	Hand change, on tank
Final drive	Chain

Cycle parts
Frame	1941-47: Twin-loop, with straight front downtubes
	1948: Twin-loop, with splayed front downtubes
Front forks	Springers – exposed-springs, oval-tube legs
Rear suspension	None (rigid)
Front brake	Drum, via cable and handlebar lever
Rear brake	Drum, via rod and pedal

General data
Dry weight	240 kg
Permissible total weight	440 kg
Tyres – front and rear	5.00-16
	optional 4.00-18 or 4.50-18
Wheels front and rear	3.00 x 16 wire-spoked
	optional 2.15 x 18
Top speed	150 km/h
Fuel tank capacity	14.2 litres

Panhead & Hydra-Glide: Models E and F Big Twins

1948 to 1957

Model description and technical information

Made in 60- and 74-cubic inch (989 and 1207 cc.) capacities, the Panhead, essentially, was a new top end, featuring aluminium-alloy cylinder heads and hydraulic tappets for cooler running and reduced maintenance, on the existing Knucklehead bottom end. It is instantly identifiable thanks to its chrome-plated pressed-steel valve/rocker covers (thought to resemble cooking pans) and it is from around this time that the names Knucklehead and Panhead came into use to identify the two engines. The less obvious introduction of hydraulic tappets for automatic adjustment of valve clearances was a technology used for the first time on a motorcycle but which had been used first on a mass-production engine by the luxury car manufacturer Cadillac, in 1938.

The taller engine required a new frame; christened the Wishbone frame thanks to its widely-splayed front downtubes, it was completely new and all-welded and could, for the first time, be fitted with a steering lock in those markets which required it. The front forks were the previous models' Springers, with their characteristic massive volute springs and chrome-molybdenum-steel oval-section tubular rigid (sprung) legs, and either Ride Control friction damping or Monroe hydraulic dampers. The new models were available in Brilliant Black, Flight Red or Azure Blue, with Police Silver for police machines.

Model development

1949 model year
The discontinuation of the Model U side-valves left the ohv models as the only remaining Harley Big Twins. A trimming of the range had seemed urgently required, especially as, from 1948, Harley was venturing into a completely new sector of the market with its small two-strokes. After much experimentation Harley-Davidson finally succeeded in developing a hydraulically-damped telescopic front fork suitable for all the Big Twins. The enclosed stanchions and sliders contributed substantially to the modern appearance of the machines; Harley-Davidson were so proud of their new Hydra-Glide front fork that, for the first time, they officially named their flagship model after this one sales feature. With the new engine and frame introduced the previous year, even if the latter still had a rigid rear end, these completed the comprehensive modernization of Harley's flagship model range.

The 1949 Big Twins were fitted with Harley's first hydraulically-damped telescopic front fork, the Hydra-Glide, which led to both Model E and F ranges being known by that name. However, not all 1949 Panheads were Hydra-Glides; the EP, ELP, FP and FLP models (EPS

Model range

Model	Specification	Years
Model E	60 cu. in./989 cc. standard-compression engine	(1948-50)
Model ES	60 cu. in./989 cc. standard-compression engine, sidecar-specification	(1948-50)
Model EL	60 cu. in./989 cc. high-compression engine	(1948-52)
Model ELF	60 cu. in./989 cc. high-compression engine	(1952)
Model ELP	60 cu. in./989 cc. high-compression engine, Springer front forks	(1949)
Model ELS	60 cu. in./989 cc. high-compression engine, sidecar-specification	(1951-52)
Model F	74 cu. in./1207 cc. standard-compression engine	(1948-50)
Model FS	74 cu. in./1207 cc. standard-compression engine, sidecar-specification	(1948-50)
Model FL	74 cu. in./1207 cc. medium-compression engine	(1948-57)
Model FLP	74 cu. in./1207 cc. medium-compression engine, Springer forks	(1949)
Model FLF	74 cu. in./1207 cc. medium-compression engine, foot change	(1952-57)
Model FLS	74 cu. in./1207 cc. medium-compression engine, sidecar-specification	(1951-52)
Model FLE	74 cu. in./1207 cc. high-compression engine	(1953-56) *
Model FLEF	74 cu. in./1207 cc. high-compression engine, foot change	(1953-56) *
Model FLH	74 cu. in./1207 cc. high-compression engine	(1955-57)
Model FLHF	74 cu. in./1207 cc. high-compression engine, foot change	(1955-57)

(All models with an F suffix have foot gearchange. All others still have the tank-mounted hand change.)
* = Traffic Combination low-compression engine for police use only.

and ELPS or FPS and FLPS, if sidecar-specification) indicate motorcycles with Panhead engines and Springer forks which sold, at a price which is not now known, to the diehards still not convinced of the superiority of the telescopic fork over the tried-and-trusted Springer and to the sidecar owner. Since the factory did not yet recommend the Hydra-Glide front fork as suitable for sidecar use, and officially pointed prospective sidecar owners at Springer-fork Panheads, there should have been no Model ESs or FSs sold in 1949; but in fact the records show that some 667 may have been, while some 585 Springer-equipped solo Panheads were sold. There is still much debate about the accuracy of Harley's sales figures, in this year as well as others.

The big Harleys could still be ordered either with the three-speed gearbox or with the more modern four-speed transmission. For sidecar-specification machines there was even a gearbox with three forward speeds and reverse. The frame had new Timken tapered-roller steering head bearings. The unsprung rear end was still around, but here at least H-D was in tune with the times; even the British motorcycles coming across the Atlantic still had rigid frames at this period.

The front brake was increased in size to 8 inches (203 mm), with a cast-iron drum and an unpolished aluminium-alloy backplate concealing an enclosed actuating arm operated via a larger cable, all of which, the factory claimed, made the new assembly 34% more effective than the previous brake (which wasn't saying much, given that the new model was verging on 600 pounds (272 kg) kerb weight and the Springer fork's front brake had been one of the weakest on the market).

Just as with American automobiles from 1938 onwards, it was now customary to fit special headlamps; the cars had their Sealed Beam units, Harley had its larger 8-inch (203 mm)-diameter Sealed Ray unit, in which headlamp glass, reflector and bulb were all one component in their housing, so that if a single filament blew, the whole headlamp had to be renewed. On top of that there were new deep-skirted and seamless Air Flow mudguards, pressed from a single piece of 20-gauge steel and fitted with stainless-steel trim pieces at the tips, and new handlebar grips (chrome-plated rubber-mounted handlebars were optional instead of the solidly-mounted and black-painted items fitted as standard). A new silicon-based heat-resistant paint finish was used on the cylinder barrels, exhaust pipes and silencers; the cylinders were finished in silver to match the aluminium-alloy heads, but while the exhaust pipes were also silver, the silencers were painted black (chrome-plated covers for the pipes and a stainless-steel cover were available as part of the Sport Solo and Deluxe Solo options Groups). Brilliant Black, Burgundy and Peacock Blue were the standard colour choices, with Metallic Congo Green available at extra cost, and Police Silver for police motorcycles. Stainless steel was optional for the front fork upper covers (standard finish was black-painted steel), but standard for the valve/rocker covers (from now until 1965), the air cleaner cover, the new mudguard tip trim

1948 Model FL: Enter the Panhead, in Flight Red

pieces, the timer unit (contact breaker housing) cap and the generator cover.

In nearly every one of the succeeding model years, minor changes were made to frames, decorative trim pieces and emblems which will not be given particular mention.

1950 model year
New cylinder heads with bigger intake ports made possible a 10% power increase. The low-compression Models E and F disappeared at the end of the model year.

Even the exhaust began to show the signs of the ending of material shortages and restrictions. The fishtail silencer evolved into the Mellow-Tone exhaust – a cigar-shaped silencer with a rounded tailpipe, the standard finish of which was silicon-based heat-resistant black paint, but which could be ordered fully chrome-plated as an optional extra and would be standard equipment for many years on all big Harleys. The Hydra-Glide front forks were redesigned to permit the trail to be adjusted for sidecar use; this meant that the Springer fork was now completely obsolete – until 1988. The front fork upper covers now had the name Hydra-Glide on them. The standard colour choices were Brilliant Black, Ruby Red and Riviera Blue with Police Silver on police machines and Metallic Green, White, Flight Red and Azure Blue available at extra cost.

1951 model year
A few modifications to the valvegear and new chrome-plated piston rings were the most important mechanical changes. The tank badges changed, naturally, with all Harleys of this year, to the underlined script badge, and, in spite of the onset of the Korean War, there were even further chrome-plated items including the exhaust pipes and, as an option, a chrome-plated Hydra-Glide emblem was available for the front mudguard, up to 1954. The standard colour choices were Brilliant Black, Persian Red and Rio Blue with Police Silver on police machines and Metallic Green, White, and Metallic Blue available at extra cost.

1952 model year
In the last year of production of the Sixty, all big Harleys received a completely new development; as of then, all customers who so wished could forgo the up-to-now standard offering of foot clutch and tank-mounted hand gearchange in favour of foot change and hand-operated clutch – still the standard layout today, but not immediately popular then; although the offer made the machines considerably easier to manage, it was not taken up by many riders. Motorcycles ordered for the authorities, particularly the police forces, stayed with the hand change, which remained on the options list until the first half of the 1970s.

The new foot change models were identified by the suffix F on the order forms, so that, for example, the Model EL became the ELF. To reduce the effort needed at the handlebar lever to operate a clutch that had been designed for foot control, an ingenious force-amplifying

1949 Model FL: Panhead engine, Hydra-Glide telescopic front forks and Air Flow mudguards in Peacock Blue

device called the Mousetrap was fitted to the frame front downtubes. Naturally, other transmission modifications were also required.

Even the frame was once more slightly altered, with inverted U-shaped toolbox brackets welded on the right-hand lower rear frame tubes, and the silencer was quieter. Later in the year came the rotating exhaust valves which were designed so as to rotate slightly when opened, thus reducing wear. The standard colour choices were Brilliant Black, Persian Red, Rio Blue and Tropical Green, with Police Silver on police machines and Bronco Brown Metallic, White, and Marine Blue Metallic available at extra cost.

In spite of still-respectable sales figures – nearly 1,000 Model Es being sold in 1952 – it was decided by H-D to stop producing the Sixty and to reduce costs by concentrating on the Seventy-Four which had always outsold the smaller-engined machine and recently had sold up to four or even six times as well. From 1953 onwards, the Model E vanished from the catalogue.

1953 model year

After the discontinuation of the Model E the Seventy-Four engine was now the biggest powerplant in the Harley range. The police machines were fitted with low-compression Traffic Combination engines which made them easier to manage at walking speeds and thereby facilitated one of the police motorcyclist's principal tasks, the monitoring of parking offenders. Normally a task for which the Servi-Cars were best suited, in the USA at that time this still involved marking the tyres of parked cars with chalk. On all engines, the hydraulic tappets were relocated at the bottom ends of the pushrods, closer to the oil pump and thereby ensuring more constant oil pressure; this necessitated a redesign of the entire valvegear.

On top of this, the speedometer was given another colour change and the outsize Buddy Seat, available for those who liked cosy togetherness, was now finished in imitation leather. Standard colours available were Pepper Red, Glacier Blue, Forest Green and Brilliant Black, with Police Silver on police machines and Cavalier Brown, White and Glamour Green available at extra cost.

1954 model year

Over the course of this year three different frames were fitted, all of them still with rigid rear ends. The last incarnation of these reverted to the straight front downtubes used on the 1936-47 models and so was christened the Straight-Leg frame; it lasted until 1957.

The handlebar grips were further improved and a quicker-action throttle twistgrip fitted, the Jubilee Trumpet horn was fitted and all models were fitted with the brass Harley-Davidson Golden Anniversary 50th birthday medallion on the front mudguard. Colour schemes were: Black, Pepper Red, Glacier Blue, Daytona Ivory, Anniversary Yellow or Forest Green, with Police White or Police Silver for police machines as standard. At no extra cost, Daytona Ivory tanks could be had with

1951 Model FL: Persian Red, new tank badge and Buddy Seat

1951 Model FL: Persian Red and chrome-plated exhaust with Mellow-Tone silencer

either Forest Green or Glacier Blue or Pepper Red mudguards, and either Pepper Red or Glacier Blue or Forest Green tanks with Daytona Ivory mudguards. For the first time there was an optional exhaust system with two silencers, one on the left and one on the right.

1955 model year
1955 saw the introduction of the Super Sport model in addition to the existing models. Simply known as the Model FLH, the new machine had an engine with an 8:1 compression ratio, which meant that a production Harley exceeded 60 horsepower for the first time; this gave it a top speed of some 99 mph (160 km/h). In addition a host of minor improvements increased the practical value of the machines right across the model range. While all models got the new big V tank badge, FL and FLH models were also fitted with the brass V medallion on the front mudguard. Standard colours available were Pepper Red, Atomic Blue, Anniversary Yellow, Aztec Brown and Brilliant Black; Silver and White were available for police machines. Glamorous Hollywood Green or combinations of colours other than Hollywood Green were available at extra cost.

1956 model year
The Panhead was already eight years old and was again carefully revised and developed; in this case by the adoption of a new higher-lift cam profile for improved gas flow on the FLH. The speedometer was redesigned with italic Day-Glo green numbers and marks, with gold painted centre. Standard colour schemes were Pepper Red with White tank panels and Red mudguards, Black with Champion Yellow tank panels and Black mudguards, Atomic Blue with Champion Yellow tank panels and Blue mudguards and Flamboyant Metallic Green with White tank panels at extra cost.

1957 model year
This would be the last year for the famous rigid-framed Harleys. Minimal technical and cosmetic alterations were made; steel alloy valve guides were fitted, and stronger valve springs. The tank badge was a round Lucite disc; this and the new front mudguard tip were H-D's first use of plastic for badges. Standard colour schemes this year were: Pepper Red with Black tank panels and Pepper Red mudguards, Black with Pepper Red tank panels and Black mudguards, Skyline Blue with Birch White tank panels and Blue mudguards, Birch White with Black tank panels and Birch White mudguards, Police Silver, Metallic Midnight Blue (tank and mudguards), with Birch White tank panels. At no extra cost were: Birch White, solid or black panels and matching mudguards. Production ended in the summer of 1957, although this did not mean the end of the line for the Panhead engine.

1955 Model FLH: big V tank badge and Jubilee Trumpet horn

1957 Model FLH: Metallic Midnight Blue with Birch White tank panels and new round plastic tank badge

1957 Model FLH in Police Silver

Motorcycle catalogue prices ($ US)

Model	1948	1949	1950	1951	1952	1953	1954	1955	1956	1957
E	635	735	735	-	-	-	-	-	-	-
ES	635	735	735	-	-	-	-	-	-	-
EL	635	735	735	885	955	-	-	-	-	-
ELF	-	-	-	-	955	-	-	-	-	-
ELS	-	-	-	885	955	-	-	-	-	-
F	650	750	750	-	-	-	-	-	-	-
FS	650	750	750	-	-	-	-	-	-	-
FL	650	750	750	900	970	1000	1015	1015	1055	1167
FLF	-	-	-	-	970	1000	1015	1015	1055	1167
FLS	-	-	-	900	970	-	-	-	-	-
FLE	-	-	-	-	-	1000	1015	1015	1055	-
FLEF	-	-	-	-	-	1000	1015	1015	1055	-
FLH	-	-	-	-	-	-	-	1083	1123	1243
FLHF	-	-	-	-	-	-	-	1083	1123	1243

These prices are for motorcycles only, without sidecars.

Sidecar catalogue prices ($ US)

Model	1948	1949	1950	1951	1952	1953	1954	1955	1956	1957
LE	185	205	205	240	260	260	270	270	270	304
M	197	215	215	245	270	270	280	280	280	314
MC Chassis	118	130	130	155	170	170	170	170	170	193

Chassis = Frame, wheel and fittings only (for aftermarket coach-built bodies).

Price and availability of optional Accessory Groups for motorcycles ($ US)

Group	1948	1949	1950	1951	1952	1953	1954	1955	1956	1957
Utility Solo	24.00	24.75	19.95	25.50	-	-	-	-	-	-
Standard Solo	-	-	-	-	28.45	28.45	31.50	31.50	-	-
Sport Solo	59.25	75.50	53.70	66.50	-	-	-	-	-	-
Deluxe Solo	92.00	122.00	95.50	115.00	73.50	83.30	88.00	88.00	-	-
Standard Police	73.00	87.00	67.50	78.75	85.50	85.50	90.00	90.00	65.00	65.00
Chrome Finish	-	-	-	-	-	-	-	-	54.00	64.90
Road Cruiser	-	-	-	-	-	-	-	-	108.00	117.35
King of the Highway	-	-	-	-	-	-	-	-	191.50	192.00

Note that Standard means a specification higher than standard ex-works finish.

Price and availability of optional Accessory Groups for sidecars ($ US)

Group	1948	1949	1950	1951	1952	1953	1954	1955	1956	1957
Utility	33.50	-	-	-	-	-	-	-	-	-
Utility Truck	33.50	-	-	-	-	-	-	-	-	-
Deluxe Sport	90.50	102.00	102.45	122.10	136.60	136.60	129.00	129.00	127.00	119.00

Production figures

Model	1948	1949	1950	1951	1952	1953	1954	1955	1956	1957
E	****	****	****	-	-	-	-	-	-	-
ES	198	177	268	-	-	-	-	-	-	-
EL ****	4321	3419	2046	1532	918†	-	-	-	-	-
ELF	-	-	-	-	918†	-	-	-	-	-
ELP	-	99	-	-	-	-	-	-	-	-
ELS	-	-	-	76	42	-	-	-	-	-
F+FS	334	490	544	-	-	-	-	-	-	-
FL+FLF *	8071	8014	7407	6560	5554	-	-	-	-	-
FLS	-	-	-	135	186	-	-	-	-	-
FLP	-	486	-	-	-	-	-	-	-	-
All FLs **	-	-	-	-	-	1986	-	-	-	-
All FLFs ***	-	-	-	-	-	3351	-	-	-	-
All FLs **/***	-	-	-	-	-	-	4757	-	-	-
FL	-	-	-	-	-	-	-	953	856	1579
FLF	-	-	-	-	-	-	-	2013	1578	1259
FLE	-	-	-	-	-	-	-	853	671	-
FLEF	-	-	-	-	-	-	-	220	162	-
FLH	-	-	-	-	-	-	-	63	224	164
FLHF	-	-	-	-	-	-	-	1040	2315	2614

* From 1952 includes FLF.   ** From 1953 includes FLE.   *** From 1953 includes FLEF.
**** Where a separarate figure is not given for Model Es, they are included in the total figure of all Model ELs.
† The figure of 918 includes both Models EL and ELF.

Model F 1949 to 1950
Model FS 1949 to 1950

Motor

Type	Panhead
Engine	Air cooled 45° ohv V-twin
Bore and stroke	3⁷⁄₁₆ x 3³¹⁄₃₂ (3.44 x 3.97) in.
Bore and stroke	87.31 x 100.81 mm
Capacity	73.66 cu. in. (1207 cc)
Power output (SAE)	46(34)/5000 SAE (1949)
	50(37)/5200 SAE (1950)
Compression ratio	6.6:1
Carburettor	Linkert 1½-inch
	Type M-36 (1949)
	Linkert 1½-inch
	Type M-45 (1949)
	Linkert 1½-inch
	Type M-74 (1950)
Ignition	Battery-and-coil
Exhaust system	Rocket-Fin silencer (1949)
	Tubular silencer (1950)

Transmission

Gearbox	4-speed (optional 3-speed)
Overall reduction ratio	3.73:1 (3-/4-speed Solo)
	4.29 (3-speed sidecar)
	4.08 (4-speed sidecar)
Gearchange	Hand change on tank
Final drive	Chain

Cycle parts

Frame	Twin-loop, with splayed front downtubes
Front forks	Telescopic, hydraulically-damped
Rear suspension	None (rigid)
Front brake	Drum, via cable and handlebar lever
Rear brake	Drum, via rod and pedal

General data

Dry weight	240 kg
Permissible total weight	440 kg
Wheelbase	1511 mm
Tyres – front and rear	5.00-16
	optional 4.00-18 or 4.50-18
Wheels front and rear	3.00 x 16 wire-spoked
	optional 2.15 x 18
Top speed	150 km/h
	(FS: approx. 115 km/h)
Fuel tank capacity	14.2 litres

Model FL 1949 to 1957
Model FLP 1949
Model FLS 1951 to 1952
Model FLF 1952 to 1957

Motor

Type	Panhead
Engine	Air cooled 45° ohv V-twin
Bore and stroke	3⁷⁄₁₆ x 3³¹⁄₃₂ (3.44 x 3.97) in.
Bore and stroke	87.31 x 100.81 mm
Capacity	73.66 cu. in. (1207 cc)
Power output (SAE)	48(36)/5000 SAE (1949)
	53(39)/5200 SAE (1950-54)
	55(40)/5400 SAE (1955-57)
Compression ratio	7.0:1 (1949-54)
	7.25:1 (1955-57)
Carburettor	Linkert 1½-inch Type M-36 (1949)
	Linkert 1½-inch Type M-45 (1949)
	Linkert 1½-inch Type M-74 (1950)
	Linkert 1½-inch Type M-74B (1950-57)
Ignition	Battery-and-coil
Exhaust system	Rocket-Fin silencer (1949)
	Tubular silencer (1951-57)
	optional 2 tubular silencers (1954-57)

Transmission

Gearbox	4-speed (optional 3-speed)
Overall reduction ratio	3.73:1 (3-/4-speed Solo)
	4.29 (3-speed sidecar) (1949-54)
	4.50 (3-speed sidecar) (1955-57)
	4.08 (4-speed sidecar) (1949-57)
Gearchange	Foot change (from 52), optional hand change on tank
Final drive	Chain

Cycle parts

Frame	1949-54: Twin-loop, with splayed front downtubes
	1955-57: Twin-loop, with straight front downtubes
Front forks: Hydra-Glide models	Telescopic, hydraulically-damped
P-suffix models	Springers – exposed-springs, oval-tube legs
Rear suspension	None (rigid)
Front brake	Drum, via cable and handlebar lever
Rear brake	Drum, via rod and pedal

General data

Dry weight	240 kg
Permissible total weight	440 kg
Wheelbase	1511 mm
Tyres – front and rear	5.00S16
	optional 4.00S-18 or 4.50S-18
	(to 1952)
Wheels front and rear	3.00 x 16 wire-spoked
	optional 2.15 x 18
	(to 1952)
Top speed	155 km/h
	(FLS: approx. 120 km/h)
Fuel tank capacity	14.2 litres

Model FLE 1953 to 1956
Model FLEF 1953 to 1956

Motor
Type	Panhead
Engine	Air cooled 45° ohv V-twin
Bore and stroke	3⁷⁄₁₆ x 3³¹⁄₃₂ (3.44 x 3.97) in.
Bore and stroke	87.31 x 100.81 mm
Capacity	73.66 cu. in. (1207 cc)
Power output (SAE)	53(39)/5200 SAE (1953-54)
	60(44)/5400 SAE (1955-56)
Compression ratio	7.0:1 (1953-54)
	8.0:1 (1955-56)
Carburettor	Linkert 1½-inch Type M-74B
Ignition	Battery-and-coil
Exhaust system	Tubular silencer, optional 2 tubular silencers

Transmission
Gearbox	4-speed (optional 3-speed)
Overall reduction ratio	3.57:1 (Solo)
	3.90 optional 4.08 (sidecar)
Gearchange	Foot change, optional hand change on tank
Final drive	Chain

Cycle parts
Frame	1953-55 Twin-loop, with splayed front downtubes, 1955-57 as FLH
Front forks	Telescopic, hydraulically-damped
Rear suspension	None (rigid)
Front brake	Drum, via cable and handlebar lever
Rear brake	Drum, via rod and pedal

General data
Dry weight	240 kg
Permissible total weight	440 kg
Wheelbase	1511 mm
Tyres – front and rear	5.00S16
Wheels front and rear	3.00 x 16 wire-spoked
Top speed	160 km/h (with sidecar: approx. 125 km/h)
Fuel tank capacity	14.2 litres

Model FLH 1955 to 1957
Model FLHF 1955 to 1957

Motor
Type	Panhead
Engine	Air cooled 45° ohv V-twin
Bore and stroke	3⁷⁄₁₆ x 3³¹⁄₃₂ (3.44 x 3.97) in.
Bore and stroke	87.31 x 100.81 mm
Capacity	73.66 cu. in. (1207 cc)
Power output (SAE)	60(44)/5400 rpm
Compression ratio	7.0:1 (1953-54)
	8.0:1
Carburettor	Linkert 1½-inch Type M-74B
Ignition	Battery-and-coil
Exhaust system	Tubular silencer, optional 2 tubular silencers

Transmission
Gearbox	4-speed (optional 3-speed)
Overall reduction ratio	3.57:1 (Solo)
	3.90 optional 4.08 (sidecar)
Gearchange	Foot change, optional hand change on tank
Final drive	Chain

Cycle parts
Frame	Twin-loop, with straight front downtubes
Front forks	Telescopic, hydraulically-damped
Rear suspension	None (rigid)
Front brake	Drum, via cable and handlebar lever
Rear brake	Drum, via rod and pedal

General data
Dry weight	240 kg
Permissible total weight	440 kg
Wheelbase	1511 mm
Tyres – front and rear	5.00S16
Wheels front and rear	3.00 x 16 wire-spoked
Top speed	160 km/h (with sidecar: approx. 125 km/h)
Fuel tank capacity	14.2 litres

Duo-Glide.
The Model F Big Twins

1958 to 1964
Model description and technical information

Autumn 1957 marked a quantum leap forward in terms of rear suspension technology for the Harley Big Twins with the introduction of swinging-arm rear suspension. The frame, christened the Step-Down frame thanks to the pronounced drop from the suspension unit top mountings to the top tubes, used two coil-sprung hydraulically-damped rear suspension units which naturally led to the creation of the Duo-Glide name and would remain unchanged in service up to 1964.

Intensive development of the Panhead engine produced a power increase of the order of 10%, which made the FLH of that year capable of a 102-mph (165 km/h) top speed. Outwardly, the engine had changed a lot, but the well-known valve/rocker covers were kept. In addition the rear brake was now hydraulically-operated, as was the stop-lamp switch.

The four-speed gearbox was standard equipment, but those who really wanted it could still order the old three-speed unit. As previously, H-D offered a three-speed-plus-reverse transmission for the sidecar-specification models.

Like many other ancillary items, the oil tank had had to be redesigned to fit into the new frame. The fat 5.00-16 wheels and tyres were retained, but new low-profile 5.10S16 tyres were available at extra cost. Two-tone paint schemes were standard for this year, in the following combinations: Calypso Red/Birch White, Black/Birch White, Skyline Blue/Birch White and Sabre Gray Metallic/Birch white. The first colour was applied to the tank top and mudguard sides, while the second was applied to the tank bottom and mudguard tops. Any solid standard colour could be ordered at no extra cost for those who wanted more choice. Police machines were available in Police Silver and Police Birch White.

1958 Model FLH

Model range

Models available

Model FL	74 cu. in./1207 cc. standard-compression engine	(1958-64)
Model FLF	74 cu. in./1207 cc. standard-compression engine, foot change	(1958-64)
Model FLH	74 cu. in./1207 cc. high-compression engine	(1958-64)
Model FLHF	74 cu. in./1207 cc. high-compression engine, foot change	(1958-64)

(All models with an F model code suffix have foot-change gearboxes. All others have hand-change.)

Optional Accessory Groups available

Chrome Finish (Group #F-1)	(1958-64)
Road Cruiser (Group #F-2)	(1958-62)
Road Cruiser (Group #F-2A)	(1959-61)
Road Cruiser (Group #F-3)	(1958-62)
Road Cruiser (Group #F-3A)	(1959-60)
Road Cruiser (Group #F-4)	(1958-62)
Road Cruiser (Group #F-4A)	(1959-60)
King of the Highway (Group #F-5)	(1958-64)
King of the Highway (Group #F-5A)	(1959-61)
King of the Highway (Group #F-6)	(1958-64)
King of the Highway (Group #F-6A)	(1959-61)
King of the Highway (Group #F-7)	(1958-64)
King of the Highway (Group #F-7A)	(1959-61)
Standard Police (Group #FP-1)	(1958-61 and 1963-64)
Standard Police (Group #FP-2)	(1958-61 and 1963-64)
Standard Police (Group #FP-3)	(1960-62)

Sidecars available

Model LE	Right-hand fitting sidecar	(1958-64)
Model LEC	Chassis only for fitting aftermarket coach-built bodies	(1962-64)

Optional Accessory Groups available for sidecars

Deluxe (Group #SC-1)	(1958-61)

After 1961 optional Accessory Groups were no longer listed for sidecars, although the individual items remained available.

Model development

1959 model year
For the Duo-Glide's second model year there were only a few alterations. A green neutral warning lamp appeared for the first time. The Arrow-Flite (red ball with chrome arrow) tank badge was fitted and the front mudguard tip is again metal, now chrome-plated. Footboards were painted black, instead of just being Parkerized. The centre stand was an option. That year's colour schemes were: Calypso Red, Black, Skyline Blue, Hi-Fi Red or Hi-Fi Turquoise tank top and mudguards with Birch White side panels. Any solid standard colour without tank panels was available at no extra cost. Even the police machines had a choice this year: Police Silver tank top and mudguards with Birch White side panels or Police Birch White.

1960 model year
The outstanding modification for the 1960 model year was a larger rear brake. On top of that came rear suspension units with revised damping, Stellite-tipped exhaust valves, rail-type oil control rings, heavier valve springs and an altered reduction ratio for the kickstart drive gears. Standard colours available were Black or Skyline Blue with Birch White tank panels, while Hi-Fi Red, Hi-Fi Green or Hi-Fi Blue with Birch White tank panels were available at extra cost; again, any solid standard colour without tank panels was available at no extra cost. Police machines could be ordered in Police Silver with Birch White tank panels or Police Birch White.

1961 model year
For the first time since 1940 a fishtail-silencer exhaust was available again, but now only as an optional extra and in the form of a slightly less flamboyant chrome-plated end-piece for the standard tubular silencer, rather than as a complete unit, as before. The GunSight four-pointed star tank badge appeared. The major change for the year was the introduction of the Single Fire ignition system in which two sets of contact breaker points, each with its own condenser and opened by a single-lobe cam, triggered two ignition HT coils. The left flywheel was marked later in the year for timing the rear cylinder separately, and the pushrods were redesigned with a longer upper

1958 Model FL: Optional twin-silencer exhaust system fitted

section and shorter pressed-in lower section. The generator's voltage regulator was now mounted externally, and would remain so up to the 1969 models.

The optional twin-silencer exhaust system was fitted with a Resonance Tube linking the two parts of the system. Standard colours available were unchanged apart from the substitution of Pepper Red for Skyline Blue.

1962 model year
Those who liked fishtail exhausts could now choose a complete fishtail-ended silencer of somewhat more flowing design. The Tombstone-style speedometer was introduced, with brushed aluminium tombstone-shaped centre and three round warning lamps replacing the two rectangular ones, and various other brushed aluminium components. Spotlights, also called passing lamps (for overtaking), were available as accessories, to be mounted on each side of the main headlamp. Standard colours available were again unchanged apart from the substitution of Tango Red for Pepper Red and Hi-Fi Purple for Hi-Fi Green.

1963 model year
The Panhead had now reached its 15th birthday and was again revised. External oil lines improved the supply of oil to and return from the rockers, the oil passages in the cylinders now being deleted, and earned these models the nickname of Outside-Oilers. Standard colours available were again unchanged apart from the addition of Horizon Metallic Blue. The tank badge was changed to a new design based on H-D's bar-and-shield logo.

The rear brake drum was widened, with 1¾-inch (44.5 mm) wide shoes giving some 40% greater contact area than the previous brake, and the wheel cylinder piston diameters revised so that less pedal pressure was required. It has to be said, however, that the drum-braked Harleys were never renowned for outstanding stopping power.

1964 model year
Harley introduced the two-key system; one key served for the ignition, while the other was for the steering lock that was now standard equipment in many markets. The sidestand was provided with a wider foot to support the machine better on starting. Standard colours available were Black or Fiesta Red with Birch White tank panels, while Hi-Fi Red and Hi-Fi Blue with Birch White tank panels were available at extra cost; the choice for police machines was the same as in 1960.

Three views of 1960 Model FLHs: Hi-Fi Red and Birch White and Hi-Fi Blue and Birch White, with Arrow-Flite tank badge

1961 Model FL: Hi-Fi Green and Birch White, with GunSight tank badge and fishtail end-piece for exhaust silencer

1962 Model FLH: New fishtail exhaust silencer

1963 Model FL: New design of tank badge

1964 Model FLH: Someone's spent a lot of money at the chrome-plater's!

1964 Model FL

Motorcycle catalogue prices ($ US)

Model	1958	1959	1960	1961	1962	1963	1964
FL	1255	1280	1310	1335	1335	1360	1385
FLF	1255	1280	1310	1335	1335	1360	1385
FLH	1320	1345	1375	1400	1400	1425	1450
FLHF	1320	1345	1375	1400	1400	1425	1450

These prices are for motorcycles only, without sidecars.

Sidecar catalogue prices ($ US)

Model	1958	1959	1960	1961	1962	1963	1964
LE	324	324	360	372	372	387	400
LEC Chassis	-	-	-	-	239	252	257

Production figures

Model	1958	1959	1960	1961	1962	1963	1964
FL	1591	1201	-	-	-	1096	-
FLF	1299	1222	-	-	-	950	-
FLH	195	121	-	-	-	100	-
FLHF	2953	3223	-	-	-	2100	-
All Fs	-	-	5967	4927	5184	-	-
All FLs	-	-	-	-	-	-	2775
All FLHs	-	-	-	-	-	-	2725
Total	6038	5767	5967	4927	5184	4246	5500

Model FL 1958 to 1964
Model FLF 1958 to 1964

Motor
Type	Panhead
Engine	Air cooled 45° ohv V-twin
Bore and stroke	$3^{7}/_{16}$ x $3^{31}/_{32}$ (3.44 x 3.97) in.
Bore and stroke	87.31 x 100.81 mm
Capacity	73.66 cu. in. (1207 cc)
Power output (SAE)	57(42)/5200 rpm
Compression ratio	7.25
Carburettor	Linkert 1½-inch Type M-74B
Ignition	Battery-and-coil
Exhaust system	Tubular silencer, optional 2 tubular silencers or 2 silencers with fishtails

Transmission
Gearbox	Solo: 4-speed (optional 3-speed) sidecar: 3-speed with reverse
Overall reduction ratio	3.00 (1) 2.71 (1) 1.82 (2) 1.50 (2) 1.23 (3) 1.00 (3) 1.00 (4) (R 2.66)
Secondary reduction ratio	2.32 2.32
Gearchange	Foot change, optional hand change on tank
Final drive	Chain

Cycle parts
Frame	Twin-loop, with straight front downtubes
Front forks	Telescopic, hydraulically-damped
Rear suspension	Swinging arm
Front brake	Drum, via cable and handlebar lever
Rear brake	Drum, hydraulically-operated

General data
Dry weight	240 kg
Permissible total weight	440 kg
Tyres – front and rear	5.00S16 optional 5.10S16
Wheels front and rear	3.00 x 16 wire-spoked
Top speed	155 km/h (with sidecar: approx. 120 km/h)
Fuel tank capacity	13.2 litres

Model FLH 1958 to 1964
Model FLHF 1958 to 1964

Motor
Type	Panhead
Engine	Air cooled 45° ohv V-twin
Bore and stroke	$3^{7}/_{16}$ x $3^{31}/_{32}$ (3.44 x 3.97) in.
Bore and stroke	87.31 x 100.81 mm
Capacity	73.66 cu. in. (1207 cc)
Power output (SAE)	66(49)/5600 rpm
Compression ratio	8.0:1
Carburettor	Linkert 1½-inch Type M-74B
Ignition	Battery-and-coil
Exhaust system	Tubular silencer, optional 2 tubular silencers or 2 silencers with fishtails

Transmission
Gearbox	Solo: 4-speed (optional 3-speed) sidecar: 3-speed with reverse
Overall reduction ratio	3.00 (1) 2.71 (1) 1.82 (2) 1.50 (2) 1.23 (3) 1.00 (3) 1.00 (4) (R 2.66)
Secondary reduction ratio	2.32 2.32
Gearchange	Foot change, optional hand change on tank
Final drive	Chain

Cycle parts
Frame	Twin-loop, with straight front downtubes
Front forks	Telescopic, hydraulically-damped
Rear suspension	Swinging arm
Front brake	Drum, via cable and handlebar lever
Rear brake	Drum, hydraulically-operated

General data
Dry weight	240 kg
Permissible total weight	440 kg
Tyres – front and rear	5.00S16 optional 5.10S16
Wheels front and rear	3.00 x 16 wire-spoked
Top speed	165 km/h (with sidecar: approx. 130 km/h)
Fuel tank capacity	13.2 litres

Electra-Glide.
The Model F Big Twins
1965 to 1984

Model description and technical information

The Duo-Glide and Sportster formed the backbone of Harley's range for the first half of the Sixties, although the bought-in Aermacchi singles also did well. Cautious development of the Model F range was the order of the day, until the 1965 and 1966 model years arrived and three major technical advances were introduced in quick succession. First, the biggest Harley was treated to a new frame which was less complex and so easier and cheaper to manufacture but retained the former round-section swinging arm. Second, the long-awaited 12-volt electrical system finally arrived, which made possible the fitting of an electric starter and so enabled the new model to be christened Electra-Glide. The kickstart actually remained available to special order up to 1972. Although all Electra-Glides were electric start-equipped, the factory sought to make this crystal clear by identifying them with the suffix letter B on order forms and similar paperwork (e.g. FLHB); the B does not feature in the engine number's model designations.

A year later it was decided in Milwaukee finally to pension off the good old Panhead and to bring in the Shovelhead engine, which was visually similar to the smaller and very successful unit fitted to the Sportsters since 1957. It was still an ohv Seventy-Four based on previous experience, but with a thorough redesign from the ground up. The name came about because of the prominent rocker boxes (not just valve/rocker covers), that were thought to resemble the shape of a shovel.

Besides the new cylinders and Power Pac (*sic*) aluminium-alloy cylinder heads, the crankcase right-hand half was completely remodelled. In addition, H-D finally turned away from Linkert, who had supplied them with carburettors since 1933, and from then on ordered their carburettors from Tillotson. The air cleaner was fitted with a round cover and so was

1965 Model FLH: Last of the Panheads, in Hi-Fi Red with Birch White tank panels

Model range

Seventy-Fours (74 cu. in./1207 cc.) models 1965 to 1969
Model FLB	Standard compression	(1965-69)
Model FLFB	Standard compression, foot gearchange	(1965-69)
Model FLHB	Higher compression	(1965-69)
Model FLHFB	Higher compression, foot gearchange	(1965-69)

(All models with an F suffix have foot gearchange, all others still have hand change. The B stands for Big Battery, which was necessary on the introduction of 12-volt electrics and the electric starter.)

Seventy-Fours (74 cu. in./1207 cc.) or '1200' models 1970 to 1980
Model FL	Standard compression	(1970)
Model FLH	Higher compression	(1970-80)
Model FLH Shrine	Higher compression	(1979-80)
Model FLH Liberty Edition	Higher compression	(1976)
Model FLH Annivesary Edition	Higher compression	(1978)
Model FLHS	Higher compression	(1977)

Liberty Edition (on the 200th anniversary of the American Declaration of Independence).

Eighties (82 cu. in./1340cc.) or '80' models 1978 to 1984
Model FLH	Higher compression	(1978-84)
Model FLHC	Higher compression	(1980-82)
Model FLH Heritage	Higher compression	(1981-82)
Model FLHX	Higher compression	(1984)
Model FLH Shrine	Higher compression	(1979-84)
Model FLHS	Higher compression	(1980-83)
Model FLHS	Higher compression	(1984)

Police motorcycles, 1970 to 1984
From 1970, the factory sales records once more distinguish separate police sales, for the first time since 1949. Apart from the civilian Shrine version, police models are covered in Volume 2 and so their details are not given here.

Model FLP	Low compression, foot gearchange	(1970-72)
Model FLP(F)	Low compression, foot gearchange	(1970-77)
Model FLP-1200	Low compression	(1978)
Model FLH-1200	Higher compression	(1979)
Model FL-80	Low compression	(1978-80)
Model FLH-80	Higher compression	(1979-80)
Model FLHP	Higher compression	(1981)
Model FLHP	Higher compression, Standard	(1982-84)
Model FLHP	Higher compression, Deluxe	(1982-84)

Optional Accessory Groups available for motorcycles
Chrome Finish (Group #F-1)	(1965-66)
King of the Highway (Group #F-5)	(1965-66)
King of the Highway (Group #F-6)	(1965-66)
King of the Highway (Group #F-7)	(1965-66)
Standard Police (Group #FP-1)	(1965-66)
Standard Police (Group #FP-2)	(1965-66)
Chrome Group	(1967-69)
King of the Highway Group	(1967-69)
Police Group	(1967-69)

Sidecars available
Model LE	Right-hand fitting sidecar (Electra Glide)	(1965-80)
Model LEC	Chassis only (for coach-built bodies)	(1965-69)
Model CLE	Right-hand fitting sidecar (Electra Glide)	(1980-84)
Model CLE	Right-hand fitting sidecar (Electra Glide Classic)	(1980-84)
Model MLE	Right-hand fitting sidecar (only for export to Mexico)	(1980)

1965 Model FLH: Last of the Panheads, in Hi-Fi Blue with Birch White tank panels

christened the Ham Can by owners. Further detail improvements were fitted such as an automatic ignition advance/retard mechanism in the timer unit (contact breaker housing), a modified speedometer drive, and fuel lines of petrol-resistant synthetic materials. The left-hand fuel tank half was no longer fitted with a breather assembly.

The previous four-speed gearbox was retained. The timing cover was now plain and unfinned, and carried internal oilways. The clutch drum now carried the electric starter's driven gear. The electric start was not universally welcomed by the enthusiast; only 'real men' could start Big Harleys and there were some who relished the chance to show off their prowess in this way. The only remaining V-twin with a kickstart was the XLCH Sportster, and it wasn't until the Seventies that a Big Harley would again be available with a kickstart, in the form of the Super Glide.

A large, unpolished aluminium-alloy casting connected engine and gearbox on the left-hand side, H-D thus managing the feat of sealing both assemblies at the same time. The oil tank had to be redesigned and the toolbox was deleted, for the first time, to make room for the much-bigger battery and the fuel tank now held over four Imp. gallons (19 litres).

New handlebar grips permitted easier manipulation of clutch and front brake controls. Incidentally, the Electra-Glide's optional FLH fairing, incidentally, was always fork-mounted and turned with the steering.

The bar-and-shield logo tank badge was retained for 1965, but changed to a new flattened-diamond design for 1966 onwards. Standard colours in the first year were Black or Holiday Red with Birch White tank panels, while Hi-Fi Red or Hi-Fi Blue with Birch White tank panels were available at extra cost; the choice for police machines was unchanged since 1960.

Model development

1967 model year
After so many new features had been introduced in the previous two years, for 1967 only the rear brake was modified and the wheel hubs and their covers improved. Turnout Mufflers – chrome-plated end-pieces for the standard tubular silencer which directed exhaust gases down and out to the left and right – were available as optional extras.

1968 model year
New features for this year concerned only the oil pump, now made of aluminium-alloy in the interests of weight-saving, and the new instrument panel. The revised clutch made riding easier. Sidecar bodies were now made of fibre-glass.

1969 model year
The front brake backplate, hitherto mounted on the

1967 Model FLH: The Shovelhead, complete with Turnout Mufflers

1965 Model FLH

1969 Model FLH

1969 Model FLH

1971 Model FLH: The Alternator Shovelhead with contact-breaker points in the cone-shaped timing cover

1971 Model FLH

left-hand fork leg, was now mounted on the right-hand leg and the operating cam was redesigned. A new stop-lamp switch was used.

1970 model year
For the first time since the Electra-Glide was launched in 1965, major changes were made. The engine received redesigned crankcase halves to enable it to take an alternator; the ensuing changes being sufficiently radical for the Alternator Shovelheads to be identified as a second-generation of the earlier Generator Shovelheads. A starter-generator replaced the previous generator equipment. The contact breakers and ignition advance assembly was located in the timing cover, the previous timer unit (contact breaker housing) now redundant and deleted. Thus the ignition timing was derived much more directly from the camshaft, with less risk of inaccuracy due to the backlash inherent in a long gear train.

The oil tank was also redesigned and provided with a more-easily removed dipstick for checking the level.

The exhaust downpipes were now chrome-plated; no more chrome-plated covers on painted serrated pipes. The previously round-section swinging arm was now of somewhat narrower square-section tubing.

The letter B, formerly a component of the order code (e.g. FLHB) was now no longer used; model designations reverted to their pre-1965 equivalents, the FLHB for example becoming the FLH again. The standard-model FL with the medium-compression engine now became a police-only option.

1971 model year
The link-up with the carburettor-manufacturer Tillotson didn't last long. After only five years, due to the unsatisfactory reliability of the Tillotson instruments, another supplier had to be found and the well-known firm of Bendix-Zenith were chosen.

1972 model year
The sensation of 1972 was the long-overdue introduction of disc brakes. However only the Electra-Glide was so equipped in the first year and then only at the front. New AMF/Harley-Davidson tank graphics decorated all Harleys for this year.

1973 model year
After the extremely good experience with the front disc brakes, these were now fitted at the rear as well. In addition, H-D started fitting all their V-twins with front forks made by the Japanese manufacturer, Kayaba; these were not only cost-effective, but their quality was of a high standard as well. The kickstart finally disappeared from the special-order section of the catalogue.

1974 model year
The alliance with Bendix-Zenith, which had promised so much three years before, was dissolved. Now the Japanese manufacturer Keihin started to supply carburettors and carried on doing so to the end of the millennium.

Next to a redesigned saddle, an anti-theft warning

1977 Model FLH

system was available as an optional extra for the first time.

1975 model year
The only technical innovation worth mentioning was the self-closing throttle twistgrip with twin cables and a return spring – a novelty for Harley-Davidson. This was the year in which Harley-Davidson achieved its best-ever sales record, with over 75,000 machines sold.

However the quality-control problems engendered by this policy of production at any price led to such a reaction in the marketplace that sales figures dropped rapidly immediately afterwards and kept dropping in the years to follow; only with the buy-back of the Company and the improvements in quality made after that was the situation restored.

1976 model year
The inspiration for the first of many special models came from the celebrations of the 200th anniversary of the American Declaration of Independence; the FLH-1200 Liberty Edition with a special metal-flake black paint finish and a huge Liberty Edition graphic featuring the American Eagle taking up most of the top front surface of the fairing. The Stone Age hand gearchange was finally deleted from the special-order section of the catalogue. A simple fact, but one with extensive consequences for model designations. Hitherto the machines with foot change had been identified by their F suffix letter – e.g. FLHF – while the hand-change machines had no suffix – e.g. FLH. Now that all models were in effect foot change, the F suffix disappeared, the FLHF becoming the FLH. So from this year on, FLH indicates a machine with foot gearchange, while before that it indicates a machine with hand gearchange.

1977 model year
Another year, another new saddle – or in this case, a dualseat. Further changes concerned the revised gearbox and the crankcase breather assembly, plus modifications to the hydraulic valve clearance adjustment and needle-roller-bearing rocker arms.

Following on from the Liberty Edition special model came the Electra-Glide Sport (FLHS), distinguished by its whitewall tyres as standard equipment.

1978 model year
In time for Harley-Davidson's 75th birthday, the Company brought out the FLH-1200 Anniversary Edition, with gold-painted cast aluminium-alloy wheels and the appropriate emblems as well as a special paint finish in Midnight Black with gold trim.

On the technical side, Milwaukee increased the Model F range by launching the larger-capacity 82 cu. in. (1340 cc.) engine. The bigger Shovelhead engine did not produce a great deal more outright power than the Seventy-Fours, but it made up for the loss that the smaller-engined models were suffering due to the ever-tighter anti-noise and emissions legislation then coming into force, and it produced its

A police model FL of 1978-80

power at lower engine speeds, which suited the market the Model Fs were now aimed at – the touring rider. Harley identified the two engines by marking the ever-larger rectangular air cleaner housing on the right-hand side of the engine. The Seventy-Fours abandoned their roots in American standard units and became the '1200's, while, in an uncharacteristic fit of modesty, Harley rounded-down the capacity of the bigger models' engines to make them '80's.

Apart from the larger air cleaners and more restrictive exhaust silencers now mandated by new legislation, Harley also introduced electronic ignition for more accurate and maintenance-free ignition timing and better combustion. The valves and valve guides were strengthened in the course of the model year and the camshaft was revised. Cast aluminium-alloy wheels, replacing the previous wire-spoked wheels, were available as optional extras.

1979 model year

Harley-Davidson launched the Electra-Glide Classic (Model FLHC), an, as it were, luxury version of the standard model which can actually be considered as a forerunner of the Tour Glide since it was fitted as standard with the Tour Pak (*sic*) equipment package. The principal items in this extensive collection of accessories were the front-fork-mounted fairing, hard panniers, crashbars and the top case. A sidecar, trimmed and painted (Tan and Creme, with hand-applied brown pinstriping) to match – the combination being designed as a co-ordinated entity by Willie G. Davidson himself – was also available. the sidecar's wheel matched those of the motorcycle – black-painted cast aluminium-alloy wheels with polished spokes and rims – to complete the ensemble.

The new police versions (Model FLHP), now also with the larger 82 cu. in. (1340 cc.) engine, were, strangely enough, also available for non-police customers to buy, albeit without the siren and the blue flashing lamps. There were also the Chain (final drive by chain) and Shrine (including full chain enclosure) variants. The latter was a limited edition with other features, principally concerned with enhancing the appearance of the machine, that was conceived following demand from the Shriners (a US charity organization) who staged elaborate shows featuring co-ordinated display riding by teams of motorcyclists all mounted on identical models that were polished and presented to a very high standard indeed). Hence the fully-enclosed final drive chain case; the riders did not want oil and grease splatters marring their perfect displays.

1980 model year

The introduction of the Tour Glide with five-speed gearbox and rubber-mounted engine/transmission marks a further milestone in Harley development; this model range will be covered in detail in a later volume.

1979 Model FLH-80

The existing model range with four-speed gearboxes and rigidly-mounted engine/transmission remained available.

The Electra-Glide Classic received an additional 'Classic' emblem bolted on to the front mudguard and the Motorola V-Fire electronic ignition now had electronic advance and retard.

The former special edition of 1977, the Model FLHS Electra-Glide Sport, returned to form a part of the regular range of models, but now as an 80, with the larger-capacity engine.

1981 model year

The ever-tightening grip of legislation curbing vehicle emissions and pollution is shown by the need to adopt an even more sophisticated 2nd-generation electronic ignition system; V-Fire II. So that the engines could use the lead-free fuel that was about to be introduced, their compression ratios had to be reduced to 7.4:1; strangely, this does not seem to be reflected in a corresponding drop in the Motor Company's published power outputs for its biggest engines. The valve guides were extended and better valve stem seals fitted. The rear brake caliper was now supplied by Girling, the manufacturer of braking systems and components.

The Harley tradition that had held since 1913 reaffirmed itself; as soon as the Motor Company tried to broaden its appeal and increase sales by introducing a larger-engined version of a model that was selling well, customers soon migrated to the bigger unit and sales of the still perfectly-acceptable small-capacity model fell off to the point that it was discontinued as being no longer viable. This now happened to the Seventy-Fours, aided no doubt in this case by the fact that the Eighties weren't significantly more expensive. There had been a Seventy-Four, usually, but not always, Harley's flagship model, in the range since 1921; now their place was taken by the Eighty.

The flow of H-D's special-edition models continued with the introduction of the Heritage Electra-Glide, based on the Model FLH. With its colour scheme (Olive Green and Orange) harking back to the Harleys of the Twenties and early Thirties, a windscreen, and tasselled, cowboy-style, saddle and saddlebags, it represented an evocation of the Motor Company's past that was one of its main sales themes at that time, in the aftermath of the buy-back from AMF.

1982 model year

The valves and their springs were again improved on all big Harleys, but apart from that, there were no changes worthy of the name made on the Electra-Glide for this year.

1983 model year

The introduction of the much more up-to-date five-speed Model FLHT Electra-Glide, with its rubber-mounted engine/transmission, was the major new feature for this year, and, even in its first year, the five-speeder outsold those old rigid-mount four-speed Electra-Glides which carried on as well (the Models FLHC and Heritage were discontinued). However, one feature in particular made the last rigid-mount four-speeders good machines; the addition of toothed-belt final drive for the final two years of production.

1984 model year

The parting shot of the old four-speeders was the revision of the Model FLHS Electra-Glide Sport and the introduction of the Model FLHX Electra-Glide Deluxe alongside the standard Model FLH.

The Sport was fitted with normal footrests, like those of the FX range, instead of the footboards which had become so completely a part of the identity of the Electra-Glide models. In addition it got the FX models' exhaust, with both silencers on the right-hand side and as a result looked slimmer and lighter than the standard Electra-Glide. The Deluxe, also known as the Last Edition model, was fitted once more with wire-spoked wheels, an exclusive black-and-white colour scheme and full touring equipment.

After 1984 the old rigid-mount four-speeders were finally discontinued and the five-speed Model FLT Tour Glide, sold in parallel for the last two years, took over the role and name of the Electra-Glide.

Catalogue prices ($ US)

Panhead and Generator-Shovelhead models

Model	1965	1966	1967	1968	1969
FLB	1530	1545	1735	1735	1885
FLFB	1530	1545	1735	1735	1885
FLHB	1595	1610	1800	1800	1900
FLHFB	1595	1610	1800	1800	1900

Prices are for motorcycles only, without sidecars.

1981 Model FLH Heritage Electra-Glide

1981 Model FLHC Electra-Glide Classic

1983 Model FLHS Electra-Glide Sport

1983 Model FLH

Production figures

Model	1965	1966	1967	1968	1969	1970	1971	1972	1973	1974	1975
FLP	-	-	-	-	-	-	-	-	-	791	900
FLH	-	-	-	-	-	-	-	-	-	5166	7400
FLHF	-	-	-	-	-	-	-	-	-	1310	1535
All FLs	2130	2175	2150	1718	1800	1706	1200	1600	1025	-	-
All FLHs	4800	5625	5600	5354	5500	5909	5475	8100	7750	-	-
Total	6930	7800	7750	7072	7300	7615	6675	9700	8775	7267	9835

Production figures

Model	1976	1977	1978	1979	1980	1981	1982	1983	1984
FLH-1200	11891	8691	4761	2612	1111	-	-	-	-
FLHP-1200	-	-	-	596	528	-	-	-	-
FLH Anniversary	-	-	2120	-	-	-	-	-	-
FLH-80	-	-	2525	3429	1625	2131	1491	1272	1983
FLHP-80	-	-	-	84	391	402	1832	878	1092
FLHC	-	-	-	4368	2480	1472	1284	-	-
FLHC sidecar outfits	-	-	-	-	353	463	152	*	-
FLHS	-	535	-	-	-	-	-	-	-
FLHS-80	-	-	-	-	914	1062	948	985	499
FLH-80 Heritage	-	-	-	-	-	784	313	-	-
FLHX-80	-	-	-	-	-	-	-	-	1258
Total	11891	9226	9406	11442	7512	6003	6185	3469	4832

Production figures for other police models will be given in Volume 2.
* Included with FLHC figure.

Model FLB 1965
Model FLFB 1965

Motor
Type	Panhead
Engine	Air cooled 45° ohv V-twin
Bore and stroke	3⁷⁄₁₆ x 3³¹⁄₃₂ (3.44 x 3.97) in.
Bore and stroke	87.31 x 100.81 mm
Capacity	73.66 cu. in. (1207 cc)
Power output (SAE)	57(42)/5200 rpm
Torque (Nm)	84/3200 rpm
Compression ratio	7.25
Carburettor	Linkert 1½-inch Type M-74B
Ignition	Battery-and-coil
Electrical system	12-volt
Starter	Electric starter
Exhaust system	2 into 1, 1 cylindrical silencer, optional 2 into 2, 2 silencers, optional with Fishtail

Transmission
Gearbox	Solo: 4-speed (optional 3-speed) sidecar: 3-speed with reverse
Overall reduction ratio	3.00 (1) 2.71 (1) 1.82 (2) 1.50 (2) 1.23 (3) 1.00 (3) 1.00 (4) (R 2.66)
Secondary reduction ratio	2.32 2.32
Gearchange	Foot change (FLFB) Hand change, on tank (FLB)
Final drive	Chain

Cycle parts
Frame	Twin-loop, with straight front downtubes
Front forks	Telescopic, hydraulically-damped
Rear suspension	Swinging arm
Front brake	Drum, via cable and handlebar lever
Rear brake	Drum, hydraulically-operated

General data
Dry weight	320 kg
Permissible total weight	492 kg
Wheelbase	1524 mm
Tyres – front and rear	5.00S16 optional 5.10S16
Wheels front and rear	3.00 x 16 wire-spoked
Top speed	155 km/h (with sidecar: approx. 120 km/h)
Fuel tank capacity	18.9 litres

Model FLHB 1965
Model FLHFB 1965

Motor
Type	Panhead
Engine	Air cooled 45° ohv V-twin
Bore and stroke	3⁷⁄₁₆ x 3³¹⁄₃₂ (3.44 x 3.97) in.
Bore and stroke	87.31 x 100.81 mm
Capacity	73.66 cu. in. (1207 cc)
Power output (SAE)	66(49)/5600 rpm
Torque (Nm)	88/3200 rpm
Compression ratio	8.0
Carburettor	Linkert 1½-inch Type M-74B
Ignition	Battery-and-coil
Electrical system	12-volt
Starter	Electric starter
Exhaust system	2 into 1, 1 cylindrical silencer, optional 2 into 2, 2 silencers, optional with Fishtail

Transmission
Gearbox	Solo: 4-speed (optional 3-speed) sidecar: 3-speed with reverse
Overall reduction ratio	3.00 (1) 2.71 (1) 1.82 (2) 1.50 (2) 1.23 (3) 1.00 (3) 1.00 (4) (R 2.66)
Secondary reduction ratio	2.32 2.32
Gearchange	Foot change (FLHFB) Hand change, on tank (FLHB)
Final drive	Chain

Cycle parts
Frame	Twin-loop, with straight front downtubes
Front forks	Telescopic, hydraulically-damped
Rear suspension	Swinging arm
Front brake	Drum, via cable and handlebar lever
Rear brake	Drum, hydraulically-operated

General data
Dry weight	320 kg
Permissible total weight	492 kg
Wheelbase	1524 mm
Tyres – front and rear	5.00S16 optional 5.10S16
Wheels front and rear	3.00 x 16 wire-spoked
Top speed	165 km/h (with sidecar: approx. 125 km/h)
Fuel tank capacity	18.9 litres

Model FLB 1966 to 1969
Model FLFB 1966 to 1969

Motor
Type	Generator Shovelhead
Engine	Air cooled 45° ohv V-twin
Bore and stroke	3⁷⁄₁₆ x 3³¹⁄₃₂ (3.44 x 3.97) in.
Bore and stroke	87.31 x 100.81 mm
Capacity	73.66 cu. in. (1207 cc)
Power output (SAE)	57(42)/5200 rpm
Torque (Nm)	84/3200 rpm
Compression ratio	8.0
Carburettor	Tillotson
Ignition	Battery-and-coil
Electrical system	12-volt
Starter	Electric starter
Exhaust system	2 into 1, 1 cylindrical silencer, optional 2 into 2, 2 silencers, optional with Fishtail

Transmission
Gearbox	Solo: 4-speed (optional 3-speed) Sidecar: 3-speed with reverse
Overall reduction ratio	3.00 (1) 2.71 (1) 1.82 (2) 1.50 (2) 1.23 (3) 1.00 (3) 1.00 (4) (R 2.66)
Secondary reduction ratio	2.32 2.32
Gearchange	Foot change (FLFB) Hand change, on tank (FLB)
Final drive	Chain

Cycle parts
Frame	Twin-loop, with straight front downtubes
Front forks	Telescopic, hydraulically-damped
Rear suspension	Swinging arm
Front brake	Drum, via cable and handlebar lever
Rear brake	Drum, hydraulically-operated

General data
Dry weight	320 kg
Permissible total weight	492 kg
Wheelbase	1524 mm
Tyres – front and rear	5.00S16 optional 5.10S16
Wheels front and rear	3.00 x 16 wire-spoked
Top speed	155 km/h (with sidecar: approx. 120 km/h)
Fuel tank capacity	18.9 litres

Model FLHB 1966 to 1969
Model FLHFB 1966 to 1969

Motor
Type	Generator Shovelhead
Engine	Air cooled 45° ohv V-twin
Bore and stroke	3⁷⁄₁₆ x 3³¹⁄₃₂ (3.44 x 3.97) in.
Bore and stroke	87.31 x 100.81 mm
Capacity	73.66 cu. in. (1207 cc)
Power output (SAE)	66(49)/5600 rpm
Torque (Nm)	88/3200 rpm
Compression ratio	8.0
Carburettor	Tillotson
Ignition	Battery-and-coil
Electrical system	12-volt
Starter	Electric starter
Exhaust system	2 into 1, 1 cylindrical silencer, optional 2 into 2, 2 silencers, optional with Fishtail

Transmission
Gearbox	Solo: 4-speed (optional 3-speed) Sidecar: 3-speed with reverse
Overall reduction ratio	3.00 (1) 2.71 (1) 1.82 (2) 1.50 (2) 1.23 (3) 1.00 (3) 1.00 (4) (R 2.66)
Secondary reduction ratio	2.32 2.32
Gearchange	Foot change (FLHFB) Hand change, on tank (FLHB)
Final drive	Chain

Cycle parts
Frame	Twin-loop, with straight front downtubes
Front forks	Telescopic, hydraulically-damped
Rear suspension	Swinging arm
Front brake	Drum, via cable and handlebar lever
Rear brake	Drum, hydraulically-operated

General data
Dry weight	320 kg
Permissible total weight	492 kg
Wheelbase	1524 mm
Tyres – front and rear	5.00S16 optional 5.10S16
Wheels front and rear	3.00 x 16 wire-spoked
Top speed	165 km/h (with sidecar: approx. 125 km/h)
Fuel tank capacity	18.9 litres

Model FL 1970
Model FLH 1970 to 1972
Model FLHF 1970 to 1972

Motor
Type	Alternator Shovelhead
Engine	Air cooled 45° ohv V-twin
Bore and stroke	3⁷⁄₁₆ x 3³¹⁄₃₂ (3.44 x 3.97) in.
Bore and stroke	87.31 x 100.81 mm
Capacity	73.66 cu. in. (1207 cc)
Power output (SAE)	FL: 57(42)/5200 rpm
	FLH: 58(43)/5150 DIN
	FLH: 62(46)/5400 rpm
	FL: 87/3600 rpm
	FLH: 93/3500 DIN
Compression ratio	FL: 7.25
	FLH: 8.0
Carburettor	Tillotson (1970)
	Bendix-Zenith (1971-72)
Ignition	Battery-and-coil
Electrical system	12-volt
Starter	Electric starter
Exhaust system	2 into 2, 2 cylindrical silencers

Transmission
Gearbox	Solo: 4-speed (optional 3-speed)
	sidecar: 3-speed with reverse
Overall reduction ratio	3.00 (1) 2.71 (1)
	1.82 (2) 1.50 (2)
	1.23 (3) 1.00 (3)
	1.00 (4) (R 2.66)
Secondary reduction ratio	2.32 2.32
Gearchange	Foot change (FLHF)
	Hand change, on tank (FLH)
Final drive	Chain

Cycle parts
Frame	Twin-loop, with straight front downtubes
Front forks	Telescopic, hydraulically-damped
Rear suspension	Swinging arm
Front brake	Drum, via cable and handlebar lever
Rear brake	Drum, hydraulically-operated

General data
Dry weight	320 kg
Permissible total weight	492 kg
Wheelbase	1524 mm
Tyres – front and rear	5.00S16 optional 5.10S16
Wheels front and rear	3.00 x 16 wire-spoked
Top speed	156 km/h
	(with sidecar: approx. 125 km/h)
Fuel tank capacity	18.9 litres

Model FLH 1973 to 1975

Motor
Type	Alternator Shovelhead
Engine	Air cooled 45° ohv V-twin
Bore and stroke	3⁷⁄₁₆ x 3³¹⁄₃₂ (3.44 x 3.97) in.
Bore and stroke	87.31 x 100.81 mm
Capacity	73.66 cu. in. (1207 cc)
Power output (SAE)	62(46)/5400 rpm
	58(43)/5150 DIN
Torque (Nm)	95/3900 rpm
	93/3500 DIN
Compression ratio	8.0
Carburettor	Bendix-Zenith
Ignition	Battery-and-coil
Electrical system	12-volt
Starter	Electric starter
Exhaust system	2 into 2, 2 cylindrical silencers

Transmission
Gearbox	Solo: 4-speed (optional 3-speed)
	sidecar: 3-speed with reverse
Overall reduction ratio	3.00 (1) 2.71 (1)
	1.82 (2) 1.50 (2)
	1.23 (3) 1.00 (3)
	1.00 (4) (R 2.66)
Secondary reduction ratio	2.32 2.32
Gearchange	Foot change
Final drive	Chain

Cycle parts
Frame	Twin-loop, with straight front downtubes
Front forks	Telescopic, hydraulically-damped
Rear suspension	Square-section swinging arm
Front brake	Hand-operated hydraulic disc
Rear brake	Drum, hydraulically-operated (1973)
	Disc (1974-75)

General data
Dry weight	320 kg
Permissible total weight	492 kg
Wheelbase	1549 mm
Tyres – front and rear	5.10S16
Wheels front and rear	3.00 x 16 wire-spoked
Top speed	156 km/h
	(with sidecar: approx. 125 km/h)
Fuel tank capacity	18.9 litres optional 13.2 litres

Model FLH 1976 to 1977
Model FLHS 1977

Motor
Type	Alternator Shovelhead
Engine	Air cooled 45° ohv V-twin
Bore and stroke	3⁷⁄₁₆ x 3³¹⁄₃₂ (3.44 x 3.97) in.
Bore and stroke	87.31 x 100.81 mm
Capacity	73.66 cu. in. (1207 cc)
Power output (SAE)	60(44)/5200 rpm
	58(43)/5150 DIN
Torque (Nm)	93/3500 DIN
Compression ratio	8.0
Carburettor	Keihin B80B (SAE)
Ignition	Battery-and-coil
Electrical system	12-volt
Starter	Electric starter
Exhaust system	2 into 2, 2 cylindrical silencers

Transmission
Gearbox	4-speed (sidecar with reverse)
Overall reduction ratio	3.00 (1)
	1.82 (2)
	1.23 (3)
	1.00 (4)
	(R 2.66)
Secondary reduction ratio	2.32
Gearchange	Foot change
Final drive	Chain

Cycle parts
Frame	Twin-loop, with straight front downtubes
Front forks	Telescopic, hydraulically-damped
Rear suspension	Square-section swinging arm
Front brake	Hand-operated hydraulic disc
Rear brake	Foot-operated hydraulic disc

General data
Dry weight	320 kg
Permissible total weight	492 kg
Wheelbase	1549 mm
Tyres – front and rear	5.00S16 optional 5.10S16
Wheels front and rear	3.00 x 16 wire-spoked
Top speed	156 km/h (with sidecar: approx.125 km/h)
Fuel tank capacity	18.9 litres optional 13.2 litres

Model FLH 1978

Motor
Type	Alternator Shovelhead
Engine	Air cooled 45° ohv V-twin
Bore and stroke	3⁷⁄₁₆ x 3³¹⁄₃₂ (3.44 x 3.97) in.
Bore and stroke	87.31 x 100.81 mm
Capacity	73.66 cu. in. (1207 cc)
Power output (SAE)	60(44)/5200 rpm
	58(43)/5150 DIN
Torque (Nm)	93/3500 DIN
Compression ratio	8.0
Carburettor	Keihin B80B (SAE)
Ignition	Electronic V-Fire
Electrical system	12-volt
Starter	Electric starter
Exhaust system	2 into 2, 2 cylindrical silencers

Transmission
Gearbox	4-speed (sidecar with reverse)
Overall reduction ratio	3.00 (1)
	1.82 (2)
	1.23 (3)
	1.00 (4)
	(R 2.66)
Secondary reduction ratio	2.32
Gearchange	Foot change
Final drive	Chain

Cycle parts
Frame	Twin-loop, with straight front downtubes
Front forks	Telescopic, hydraulically-damped
Rear suspension	Square-section swinging arm
Front brake	Hand-operated hydraulic disc
Rear brake	Foot-operated hydraulic disc

General data
Dry weight	320 kg
Permissible total weight	492 kg
Wheelbase	1549 mm
Tyres – front and rear	MT90S16
Wheels front and rear	3.00 x 16 wire-spoked
Top speed	156 km/h (with sidecar: approx. 125 km/h)
Fuel tank capacity	18.9 litres optional 13.2 litres

Model FLH 1979
Model FLH Shrine 1979

Motor
Type	Alternator Shovelhead
Engine	Air cooled 45° ohv V-twin
Bore and stroke	3⁷⁄₁₆ x 3³¹⁄₃₂ (3.44 x 3.97) in.
Bore and stroke	87.31 x 100.81 mm
Capacity	73.66 cu. in. (1207 cc)
Power output (SAE)	60(44)/5200 SAE (1978-79)
	58(43)/5150 DIN (1978)
	58(43)/5300 DIN (1979)
Torque (Nm)	87.5/3550 DIN (1978)
	93/3500 DIN (1979)
Compression ratio	8.0
Carburettor	Keihin B80B (SAE) 1978-79
	Keihin B80A (DIN) 1978
	Keihin B81A (DIN) 1979
Ignition	Electronic V-Fire
Electrical system	12-volt
Starter	Electric starter
Exhaust system	2 into 2, 2 cylindrical silencers

Transmission
Gearbox	4-speed (sidecar with reverse)
Overall reduction ratio	3.00 (1)
	1.82 (2)
	1.23 (3)
	1.00 (4)
	(R 2.66)
Secondary reduction ratio	2.32
Gearchange	Foot change
Final drive	Chain

Cycle parts
Frame	Twin-loop, with straight front downtubes
Front forks	Telescopic, hydraulically-damped
Rear suspension	Square-section swinging arm
Front brake	Hand-operated hydraulic disc
Rear brake	Foot-operated hydraulic disc

General data
Dry weight	320 kg
Permissible total weight	492 kg
Wheelbase	1562 mm
Tyres – front and rear	MT90S16
Wheels front and rear	3.00 x 16 cast-alloy
Top speed	156 km/h (with sidecar: approx. 125 km/h)
Fuel tank capacity	18.9 litres (1978)
	19.85 litres (1979)

Model FLH 1980
Model FLH Shrine 1980

Motor
Type	Alternator Shovelhead
Engine	Air cooled 45° ohv V-twin
Bore and stroke	3⁷⁄₁₆ x 3³¹⁄₃₂ (3.44 x 3.97) in.
Bore and stroke	87.31 x 100.81 mm
Capacity	73.66 cu. in. (1207 cc)
Power output (SAE)	60(44)/5200 rpm
	58(43)/5300 DIN
Torque (Nm)	93/3500 DIN
Compression ratio	8.0
Carburettor	Keihin B80D (SAE)
	Keihin B81A (DIN)
Ignition	Electronic V-Fire
Electrical system	12-volt
Starter	Electric starter
Exhaust system	2 into 2, 2 cylindrical silencers

Transmission
Gearbox	4-speed (sidecar with reverse)
Overall reduction ratio	3.00 (1)
	1.82 (2)
	1.23 (3)
	1.00 (4)
	(R 2.66)
Secondary reduction ratio	2.32
Gearchange	Foot change
Final drive	Chain

Cycle parts
Frame	Twin-loop, with straight front downtubes
Front forks	Telescopic, hydraulically-damped
Rear suspension	Square-section swinging arm
Front brake	Hand-operated hydraulic disc
Rear brake	Foot-operated hydraulic disc

General data
Dry weight	345 kg
Permissible total weight	520 kg
Wheelbase	1562 mm
Tyres – front and rear	MT90S16
Wheels front and rear	3.00 x 16 cast-alloy
Top speed	156 km/h (with sidecar: approx. 125 km/h)
Fuel tank capacity	19.85 litres

Model FLH-80 1978 to 1979
Model FLH-80 Shrine 1979

Motor
Type	Alternator Shovelhead
Engine	Air cooled 45° ohv V-twin
Bore and stroke	3½ x 4¼ (3.5 x 4.25) in.
Bore and stroke	88.9 x 107.95 mm
Capacity	81.78 cu. in. (1340 cc)
Power output (SAE)	60(44)/4800 rpm
	64(47)/4600 DIN
Torque (Nm)	110/3700 rpm
Compression ratio	8.0
Carburettor	Keihin B78B (SAE)
	Keihin B78A (DIN)
Ignition	Electronic V-Fire
Electrical system	12-volt
Starter	Electric starter
Exhaust system	2 into 2, 2 cylindrical silencers

Transmission
Gearbox	4-speed (sidecar with reverse)
Overall reduction ratio	3.00 (1)
	1.82 (2)
	1.23 (3)
	1.00 (4)
	(R 2.66)
Secondary reduction ratio	2.32
Gearchange	Foot change
Final drive	Chain

Cycle parts
Frame	Twin-loop, with straight front downtubes
Front forks	Telescopic, hydraulically-damped
Rear suspension	Square-section swinging arm
Front brake	Hand-operated hydraulic disc
Rear brake	Foot-operated hydraulic disc

General data
Dry weight	320 kg (1978), 345 kg (1979)
Permissible total weight	492 kg (1978), 520 kg (1979)
Wheelbase	1562 mm
Tyres – front and rear	MT90S16
Wheels front and rear	3.00 x 16 cast-alloy
Top speed	160 km/h (with sidecar: approx. 130 km/h)
Fuel tank capacity	18.9 litres (1978) 19.85 litres (1979)

Model FLH-80 1980
Model FLH-80 Shrine 1980
Model FLH-80 Classic 1980
Model FLHS-80 1980

Motor
Type	Alternator Shovelhead
Engine	Air cooled 45° ohv V-twin
Bore and stroke	3½ x 4¼ (3.5 x 4.25) in.
Bore and stroke	88.9 x 107.95 mm
Capacity	81.78 cu. in. (1340 cc)
Power output (SAE)	60(44)/4800 rpm
	64(47)/4600 DIN
Torque (Nm)	110/3700 rpm
Compression ratio	8.0
Carburettor	Keihin B78D (SAE)
	Keihin B78A (DIN)
Ignition	Electronic V-Fire
Electrical system	12-volt
Starter	Electric starter
Exhaust system	2 into 2, 2 cylindrical silencers

Transmission
Gearbox	4-speed (sidecar with reverse)
Overall reduction ratio	3.00 (1)
	1.82 (2)
	1.23 (3)
	1.00 (4)
	(R 2.66)
Secondary reduction ratio	2.32
Gearchange	Foot change
Final drive	Chain

Cycle parts
Frame	Twin-loop, with straight front downtubes
Front forks	Telescopic, hydraulically-damped
Rear suspension	Square-section swinging arm
Front brake	Hand-operated hydraulic disc
Rear brake	Foot-operated hydraulic disc

General data
Dry weight	345 kg
Permissible total weight	520 kg
Wheelbase	1562 mm
Tyres – front and rear	MT90S16
Wheels front and rear	3.00 x 16 cast-alloy
Top speed	160 km/h (with sidecar: approx. 130 km/h)
Fuel tank capacity	19.85 litres

The FL80 police version (1978-80) with reduced compression of 7.25:1 had a power output of 57 hp, the FLH80 models (1979-84) are identical to the civilian versions, at least as far as engine and transmission are concerned.

Model FLH Series AA 1981 to 1982
Model FLH Classic Series AD 1981 to 1982
Model FLH Heritage Series AJ 1981 to 1982
Model FLH Shrine Series AC 1981 to 1982
Model FLHS Series AK 1981 to 1982

Motor
Type	Alternator Shovelhead
Engine	Air cooled 45° ohv V-twin
Bore and stroke	3½ x 4¼ (3.5 x 4.25) in.
Bore and stroke	88.9 x 107.95 mm
Capacity	81.78 cu. in. (1340 cc)
Power output (SAE)	65(48)/5400 rpm
	67(49)/6000 DIN
Torque (Nm)	97/3800 rpm
Compression ratio	7.4
Carburettor	Keihin B78E
Ignition	Electronic V-Fire II
Electrical system	12-volt
Starter	Electric starter
Exhaust system	2 into 2, 2 cylindrical silencers

Transmission
Gearbox	4-speed (sidecar with reverse)
Overall reduction ratio	3.00 (1)
	1.82 (2)
	1.23 (3)
	1.00 (4)
	(R 2.66)
Secondary reduction ratio	2.32 (1981), 2.14 (1982)
Gearchange	Foot change
Final drive	Chain

Cycle parts
Frame	Twin-loop, with straight front downtubes
Front forks	Telescopic, hydraulically-damped
Rear suspension	Square-section swinging arm
Front brake	Hand-operated hydraulic disc
Rear brake	Foot-operated hydraulic disc

General data
Dry weight	345 kg
Permissible total weight	535 kg
Wheelbase	1562 mm
Tyres – front and rear	MT90S16
Wheels front and rear	3.00 x 16 cast-alloy
Top speed	160 km/h (with sidecar: approx. 130 km/h)
Fuel tank capacity	19.85 litres (1981)
	19.2 litres (1982)

Model FLH Series AA 1983 to 1984
Model FLH Shrine Series AC (Chain) and **Series AL** (belt) 1983 to 1984
Model FLHS Series AK 1983 to 1984
Model FLHX Series AE and **Series AF** with sidecar

Motor
Type	Alternator Shovelhead
Engine	Air cooled 45° ohv V-twin
Bore and stroke	3½ x 4¼ (3.5 x 4.25) in.
Bore and stroke	88.9 x 107.95 mm
Capacity	81.78 cu. in. (1340 cc)
Power output (SAE)	65(48)/5400 rpm
	67(49)/6000 DIN
Torque (Nm)	97/3800 rpm
Compression ratio	7.4
Carburettor	Keihin B78E (1983)
	Keihin B88H (1984)
Ignition	Electronic V-Fire II
Electrical system	12-volt
Starter	Electric starter
Exhaust system	2 into 2, 2 cylindrical silencers

Transmission
Gearbox	4-speed (sidecar with reverse)
Overall reduction ratio	3.00 (1)
	1.82 (2)
	1.23 (3)
	1.00 (4)
	(R 2.66)
Secondary reduction ratio	2.14 (chain), 2.32 (belt)
Gearchange	Foot change
Final drive	Belt (Shrine: optional chain)

Cycle parts
Frame	Twin-loop, with straight front downtubes
Front forks	Telescopic, hydraulically-damped
Rear suspension	Square-section swinging arm
Front brake	Hand-operated hydraulic disc
Rear brake	Foot-operated hydraulic disc

General data
Dry weight	345 kg
Permissible total weight	535 kg
Wheelbase	1562 mm
Tyres – front and rear	MT90S16
Wheels front and rear	3.00 x 16 cast-alloy
Top speed	160 km/h (with sidecar: approx. 130 km/h)
Fuel tank capacity	19.2 litres

The Aermacchi singles
1961 to 1978

A brief history of the company

The roots of the Aermacchi company go back to before the First World War, when in 1912 one Giulio Macchi created the S. A. Nieuport Macchi with the aim of manufacturing powered aircraft just as was being done in the US, for example, by the Wright Brothers. At that time France was a leader in this field and Nieuport, like Blériot was one of the most important manufacturers. Macchi began by building Nieuports under licence.

A year later he was thus ready to commence production, albeit modestly, of the Nieuport-Macchi Parasol. A few years later the company's name was changed to Aeronautica Macchi, later abbreviated to Aermacchi. During the First World War the company naturally supplied aircraft to the Italian armed forces and by 1918 production amounted to at least well over 1,000 machines.

In 1921, the company's M-7 biplane, which was fitted with a 250-horsepower Isotta-Fraschini engine, won the world-famous Schneider Trophy air race for seaplanes.

After the 1922 seizure of power in Italy by the Fascists, Mussolini began to build up the armed forces. Part of this was provided by Aermacchi aircraft which were quite capable of competing directly with the Spitfires and Messerschmitts of the late Thirties. Their engines were mainly FIATs, but later Daimler-Benz Type DB-601A and DB-605 aircraft engines were used as Germany and Italy, the two Axis Powers, began to collaborate more closely in such technical matters.

Between 1918 and 1934 the company even built world record-breaking aircraft, the fastest of which was the MC-72 seaplane whose double contra-rotating propellers were powered by a modified FIAT AS-6 V-24 supercharged engine developing some 2,500 to 3,100 horsepower. This enabled it to reach 440 mph (709 km/h); the fastest speed ever attained by a piston-engined seaplane.

New beginning after the Second World War
After the hail of bombs of the Second World War Aermacchi was rebuilt, but since post-war Italy was not allowed to build aircraft for many years now concentrated, at least initially, more on the transport of people on the ground. However, in the Fifties aircraft production recommenced with the manufacture of trainers – the Types MB-326 and later the MB-339.

First of all the company attempted to market a small car, exactly what a war-shattered Italy so desperately needed. But nothing could be done in the face of the might of FIAT, who weren't letting the grass grow under their feet either, so Aermacchi turned to the two-wheeled market, which was showing phenomenal growth, thanks in no small way to Piaggio's (another former Italian aircraft manufacturer) Vespa scooters.

As early as 1946 production could begin of a three-wheeled small utility vehicle with a light truck body and the front part of a motorcycle; the Type MB-1. This category of vehicle, like the Vespa Ape and those of other motorcycle manufacturers, was always popular in Italy and the principle was still in use in the 1980s!

Even in the middle of the war work began on the prototype of an electrically-propelled motorcycle. In the long run, however, a 125 cc. (7.62 cu. in.) machine designed by Lino Tonti, a engineer with an excellent reputation, formed the basis of the progression to actual mass production. Various versions of this 125 continued to be built up to 1960; scooters and a short-lived twin-cylinder model completed Aermacchi's product line. Manufacture of competition machines started in 1957. The famous Ala d'Oro (Gold Wing) models continued at least until 1972, by then under the direction of Harley-Davidson.

Then in April 1960 H-D bought 50% of the Aermacchi company and instigated the separation of two- and three-wheeler production from aircraft construction, which up until then had been combined in a motley fashion in the same factory.

What H-D badly needed was a replacement for the future for its own small DKW-derived two-stroke singles and acquired with the Italian company the right to sell modified Aermacchis in the USA under another name. The first joint model of the venture was the Harley-Davidson Model C Sprint in the USA and the Aermacchi Ala Verde (Green Wing) in Italy. These 246 cc./15 cu. in. four-stroke-engined models were introduced for the first time in Italy in 1957 and were warmly welcomed. After four years of production they were thoroughly sorted and the component suppliers were familiar with them.

Alongside the standard Sprint was the somewhat hotter Model H Sprint H with more power and a sportier look (smaller, more streamlined fuel tank, shorter mudguards and Hi-Flo high-level exhaust system). These two four-strokes would last, with the larger 342 cc./21 cu. in. engines, until 1974.

The smallest two-stroke Aermacchis, the 47 cc./2.9 cu. in. Zeffiretto 48s, fell into the minibike category. Known as the Leggero (Italian for light) in the USA, they were aimed at those young people not yet old enough to drive their fathers' cars (which was usually allowed at the age of 16 in most States). They didn't cost much and were usually put away and forgotten as soon as their adolescent owners were finally old enough to drive cars. Accordingly, few survived; most being run into the ground in the 'Chicken Races' of their young and impetuous riders.

After two years of production a somewhat larger 65 cc. (4 cu. in.) engine appeared, while from 1972 the Shortster (similar to the Honda Monkey bike) joined the Leggero, the Shortster receiving a 90 cc./5.5 cu. in. engine the year after. After 1974 these minibikes were practically unsaleable; the 98 cc./6 cu. in.-engined enduro-styled Baja only being offered between 1970 and 1974.

For the 1968 model year H-D introduced a 125 cc./7.5 cu. in. two-stroke which, thanks to intensive development, remained in production for at least eleven years. Two-strokes of 175 cc./11 cu. in. and 250 cc./15 cu. in. were added to the range. For road-racing even two-strokes of 350 cc./21 cu. in. and 500 cc./30.5 cu. in. were built.

At the end of 1972 Harley-Davidson took over the remaining 50% of Aermacchi. From then on the motorcycles were no longer designated Aermacchi-Harley-Davidson but AMF-Harley-Davidson. In the meanwhile, at the end of 1978 the transatlantic motorcycle partnership was dissolved. The Italian branch was sold to Cagiva and the remaining stocks of motorcycles were flogged off with massive discounts on the US market during 1979.

Aermacchi models were available in H-D's normal export markets, usually with Harley tank badging.

Model description and technical information: the four-strokes

The first result of the American-Italian co-operation was the Ala Verde (Green Wing) model, which was easily adapted to the US market. The reference to wings was, of course, a reminder of the tradition of aircraft manufacture in Varese. This machine was renamed the Model C Sprint and possessed a 246 cc./15 cu. in. four-stroke overhead-valve engine which was air-cooled and produced some 18 horsepower.

Fitted with a modern spine-type frame which permitted a virtually horizontal forward-inclined cylinder, the Sprint was bang up to date and its telescopic front forks and swinging-arm rear suspension provided good ride and handling qualities matched by its reasonably-effective 7-inch (178 mm) drum brakes. Chrome plating was restricted to the handlebars, the wheel rims and

1961 Model Sprint C

spokes, the exhaust system and a few further, smaller items. The large tank was finished in a sporty two-tone bright red and white colour scheme. While the full imitation leather dualseat with a passenger strap promised much for the prospective owners' social lives, it was probably of lesser concern that the taillamp included lighting for the number plate. The four-speed gearbox and 6-volt electrical system were typical specification for a lightweight sporting motorcycle of the early 1960s.

Valuable publicity was gained for the little Aermacchi/Harleys when, in September 1964 a factory team took a Harley-Davidson streamliner powered by a 246 cc./15 cu. in. Sprint motor to the Bonneville Salt Flats and the machine, ridden by Harley-Davidson factory-supported racer Roger Reiman of Kewanee, Illinois, got to over 156 mph (251 km/h). Conditions were less than perfect however, so the team returned to Bonneville the following year and on October 21st 1965, powered by a (supposedly 'untouched out of the crate', according to H-D) 1966-specification 246 cc./ 15 cu. in. Sprint CR motor running on standard pump petrol, the streamliner was piloted to a record speed of 177.225 mph (285 km/h) by George Roeder (another factory-supported professional racer) of Monroeville, Ohio.

Model development

1962 model year
Apart from minor modifications such as a new design for the tank's two-tone paintwork, a waterproof contact breaker housing and improved advance/retard mechanism, stronger spokes, a more robust final-drive chain, the carburettor float chamber transferred to the other side to cure flooding problems and hinged footrests for the Sprint C, 1962 also saw the introduction of a new and more powerful model in the spring. Thanks to its slightly higher compression, larger intake valve and bigger-bore carburettor the Sprint H put out 21 horsepower. The two models could be distinguished easily as the Sprint H was styled along off-road lines; the beginnings of the 'street scrambler' look. The principal feature was the Hi-Flo high-level exhaust system curving up and over the gearbox on the right-hand side of the machine and well tucked in to protect the pillion passenger's legs from burns, but the mudguards were 'bobbed' (cut back), the fuel tank was smaller and more streamlined, higher handlebars and 18-inch wheels with trials-type tyres were fitted and the air cleaner was bigger and more effective for off-road excursions.

It has to be said, however, that the two little Harleys looked a bit feeble in comparison with the Honda Super Hawk of 1962 and their high prices were therefore not necessarily justified.

1963 model year
Again new paintwork and tank badges (the tank badge was changed to a new design based on H-D's bar-and-shield logo) for the third year of production. Both Sprints got higher handlebars with larger grips and new dualseats, to make the seating position conform more to US expectations (earlier Sprint Cs having imposed a riding position that required its American riders to crouch forwards in a European-style semi-racing stance, to which they were not accustomed).

In addition a more powerful generator and bigger battery were fitted, as well as a reinforced rear frame section. Further improvements were a stronger steering head bottom bearing, lightened valvegear, a new oil filter assembly and modified gearchange mechanism. For the first time the installation of a tachometer was planned (as option).

1964 model year
Variations on a theme: another new design for the tank's two-tone paintwork, but this time the badge remained untouched. Besides an improved oil pump, which provided a higher pressure, there were also changes to the clutch and valvegear. The new ignition switch was mounted on the left-hand side below the seat. The Sprint C still had telescopic front forks with chrome-plated oil seal holders and metal fork shrouds, while the Sprint H had rubber gaiters. For 1964 the latter's identity as an on/off-road model was reinforced by its being given the name Scrambler. With another carburettor, compression raised again and better-tuned exhaust system its power output went up to at least 25 (SAE) horsepower.

1965 model year
Apart from the annual change of tank paintwork the only change obvious to the eye was the Sprint H's exhaust system, which was now low-level, but somewhat shorter than that of the Sprint C. Both models got a crankcase breather which exited to atmosphere under the frame.

1966 model year
The tank badge changed to a new flattened-diamond design for 1966 onwards, and the tank paintwork was given a fresh motif. With their clumsy taillamps, the Sprints were beginning to look outmoded.

1967 model year
After six years of production a new engine had been designed for the four-stroke Sprints. The bore and stroke were slightly revised to make the new engine an oversquare 248 cc./15 cu. in. unit. The cylinder head and barrel were now both of cast aluminium alloy, the oversized finning making the engine appear larger than it actually was and making it easier to enlarge the engine in the future, as would actually happen in two years' time.

1964 Model Sprint H

1964 Model Sprint H

At the same time model names and designations were revised. The Sprint SS (Street Scrambler), as an off-road-capable road machine, took over from the Sprint C, but with increased ground clearance and an exhaust system that now had twin pipes and silencers, one on each side, while the Sprint H with its small tank and large rear tyre became more of a pure competition machine capable of some road use. For pure flat-track racing there was the Sprint CRS, which had been available since 1965 and which was devoid of all unnecessary parts like lighting and, in some events, brakes; obviously, this was not allowed on the roads.

1968 model year
On the engine side there was now a new manufacturing process. Before fitting the valves the light-alloy cylinder head was specially heat-treated in order to harden the valve seat; this was done probably because of customer complaints. Various new fixtures and fittings accompanied another new paint scheme for the tank.

1969 model year
The model range was now completely revised. The Sprint H was discontinued. A substantially-larger 342 cc./21 cu. in. engine gave a (not quite as substantial) power increase for the remaining model, which now did without a model name. The sole road version was called simply the Harley-Davidson SS-350 (again, a Street Scrambler), while the flat-track racing version became the ERS, a successor to the CRS. The ERS remained available for private competition enthusiasts up to 1972.

The new engine's extra capacity came from a slightly increased bore but a considerably extended stroke of 80 mm (3.15 in.), which meant a reversion to a long-stroke configuration.

The 350 Sprint still had its kickstart on the left-hand side and the four-speed gear lever on the right, with the Italian gearchange pattern of one up, three down. New Dell'Orto concentric-float carburettors and a strengthened dry clutch were fitted, as was a revised exhaust system and a peanut-style fuel tank.

1970 model year
No noteworthy alterations.

1971 model year
The former twin-model range was reinstated, an off-road-styled SX-350 being introduced alongside the now more road-only SS-350. By now there was less to distinguish them, the SS-350 having full electrical equipment, a low-level chrome-plated exhaust system with twin pipes and silencers and a full dualseat (the chrome-plated strip along the bottom of which, formerly kinked, was straightened this

1967 Model SS-250

year), while the SX-350 had abbreviated high-level mudguards, a high-level black-painted exhaust system running along the right-hand side of the machine, a shorter dualseat, braced handlebars and knobbly tyres to complete its off-road image.

1972 model year
The pure flat-track racing Model ERS was discontinued during the course of the year, as was (briefly) the SX-350. New AMF/Harley-Davidson tank graphics decorated all Harleys for this year.

1973 and 1974 model years
Although the idea of dropping the Sprints must already have been considered, AMF/H-D still didn't shrink from the expense of developing for 1973 a brand-new duplex-loop frame for the four-stroke engine, a five-speed gearbox, and finally an electric starter as well.

However, the four-stroke 350s were virtually unable to compete against the two-stroke competition, especially that from Japan, and their high price didn't help. Even the major new features, a last attempt to save the situation, were unable to rescue the sinking sales figures. Although sales increased steadily from 1968 to 1973 (1973 saw an increase to 6,568 Sprints sold compared with 6,300 the year before) they were nothing like the high point of the mid-1960s and Harley-Aermacchi's own two-strokes were already selling better and showing signs of greater potential. In their final year only 4,585 Sprints were sold.

1971 Aermacchi-Harley-Davidson Sprint SX-350

Model C Sprint 1961 to 1966

Motor
Engine	Air-cooled ohv single
Bore and stroke	66 x 72 mm (2.60 x 2.83 in.)
Capacity	246 cc (15.03 cu. in.)
Power output (SAE)	16(12)/6500 DIN 18(13)/6700 rpm
Compression ratio	8.5:1
Carburettor	Dell'Orto 24
Ignition	Battery-and-coil
Starter	Kickstarter
Electrical system	6-volt
Exhaust system	Long tubular silencer RHS

Transmission
Gearbox	4-speed
Overall reduction ratio	2.91 (1) 1.76 (2) 1.27 (3) 1.00 (4)
Gearchange	Foot change
Final drive	Chain

Cycle parts
Frame	Spine frame
Front forks	Hydraulically-damped telescopic
Rear suspension	Swinging-arm and hydraulic dampers
Front brake	Drum, via cable and handlebar lever
Rear brake	LHS foot-operated drum

General data
Dry weight	125 kg
Permissible total weight	280 kg
Wheelbase	1321 mm
Tyre front	3.00-17 (1961)
Tyre rear	3.00-17 (1961)
Wheels front	1.60 x 17 (1961)
Wheels rear	1.60 x 17 (1961) wire-spoked
Top speed	121 km/h
Fuel tank capacity	15.1 litres
Production figures	1961: Not available 1962: Not available 1963: 150 machines 1964: 230 machines 1965: 500 machines 1966: 600 machines
Catalogue price	$695

1974 Aermacchi-Harley-Davidson Sprint SX-350

Model H Sprint H 1962 to 1963

Motor

Engine	Air-cooled ohv single
Bore and stroke	66 x 72 mm (2.60 x 2.83 in.)
Capacity	246 cc (15.03 cu. in.)
Power output (SAE)	19.5(14)/7200 DIN 21(15)/7500 rpm
Compression ratio	9.2:1
Carburettor	Dell'Orto 27
Ignition	Battery-and-coil
Starter	Kickstarter
Electrical system	6-volt
Exhaust system	Tubular silencer high-level, RHS

Transmission

Gearbox	4-speed
Overall reduction ratio	2.91 (1) 1.76 (2) 1.27 (3) 1.00 (4)
Gearchange	Foot change
Final drive	Chain

Cycle parts

Frame	Spine frame
Front forks	Hydraulically-damped telescopic
Rear suspension	Swinging-arm and hydraulic dampers
Front brake	Drum, via cable and handlebar lever
Rear brake	LHS foot-operated drum

General data

Dry weight	125 kg
Permissible total weight	280 kg
Wheelbase	1321 mm
Tyre front	3.00-18
Tyre rear	3.50-18
Wheels front	1.60 x 18 wire-spoked
Wheels rear	1.85 x 18 wire-spoked
Top speed	130 km/h
Fuel tank capacity	10.0 litres
Production figures	1962: Not available 1963: 1416 machines
Catalogue price	$720

Model H Scrambler 1964 to 1966

Motor

Engine	Air-cooled ohv single
Bore and stroke	66 x 72 mm (2.60 x 2.83 in.)
Capacity	246 cc (15.03 cu. in.)
Power output (SAE)	22(16)/8400 DIN 25(18)/8700 rpm
Compression ratio	9.5:1
Carburettor	Dell'Orto 27
Ignition	Battery-and-coil
Starter	Kickstarter
Electrical system	6-volt
Exhaust system	Tubular silencer high-level, RHS (1964) Tubular silencer low-level RHS (1965-66)

Transmission

Gearbox	4-speed
Overall reduction ratio	2.91 (1) 1.76 (2) 1.27 (3) 1.00 (4)
Gearchange	Foot change
Final drive	Chain

Cycle parts

Frame	Spine frame
Front forks	Hydraulically-damped telescopic
Rear suspension	Swinging-arm and hydraulic dampers
Front brake	Drum, via cable and handlebar lever
Rear brake	LHS foot-operated drum

General data

Dry weight	125 kg
Permissible total weight	280 kg
Wheelbase	1321 mm
Tyre front	3.00-18
Tyre rear	3.50-18
Wheels front	1.60 x 18 wire-spoked
Wheels rear	1.85 x 18 wire-spoked
Top speed	135 km/h
Fuel tank capacity	10.0 litres
Production figures	1964: 1550 machines 1965: 2500 machines 1966: 4700 machines
Catalogue price	approx. $720

Model H Sprint H 1967 to 1968

Motor
Engine	Air-cooled ohv single
Bore and stroke	72 x 61 mm (2.83 x 2.40 in.)
Capacity	248 cc (15.16 cu. in.)
Power output (DIN)	19(14)/7000 DIN 21(15)/7250 rpm
Compression ratio	8.5:1
Carburettor	Dell'Orto 27
Ignition	Battery-and-coil
Starter	Kickstarter
Electrical system	6-volt
Exhaust system	Long tubular silencer RHS

Transmission
Gearbox	4-speed
Overall reduction ratio	2.91 (1) 1.76 (2) 1.27 (3) 1.00 (4)
Gearchange	Foot change
Final drive	Chain

Cycle parts
Frame	Spine frame
Front forks	Hydraulically-damped telescopic
Rear suspension	Swinging-arm and hydraulic dampers
Front brake	Drum, via cable and handlebar lever
Rear brake	LHS foot-operated drum

General data
Dry weight	125 kg
Permissible total weight	280 kg
Wheelbase	1321 mm
Tyre front	3.25-19
Tyre rear	3.50-18
Wheels front	1.85 x 19 wire-spoked
Wheels rear	1.85 x 18 wire-spoked
Top speed	125 km/h
Fuel tank capacity	9.8 litres
Production figures	1967: 2000 machines 1968: Not available
Price	approx. $750

Model Sprint 250 SS 1967 to 1968

Motor
Engine	Air-cooled ohv single
Bore and stroke	72 x 61 mm (2.83 x 2.40 in.)
Capacity	248 cc (15.16 cu. in.)
Power output (DIN)	19(14)/7000 DIN 21(15)/7250 rpm
Compression ratio	8.5:1
Carburettor	Dell'Orto 27
Ignition	Battery-and-coil
Electrical system	6-volt
Starter	Kickstarter
Exhaust system	Long tubular silencer RHS

Transmission
Gearbox	4-speed
Overall reduction ratio	2.91 (1) 1.76 (2) 1.27 (3) 1.00 (4)
Gearchange	Foot change
Final drive	Chain

Cycle parts
Frame	Spine frame
Front forks	Hydraulically-damped telescopic
Rear suspension	Swinging-arm and hydraulic dampers
Front brake	Drum, via cable and handlebar lever
Rear brake	LHS foot-operated drum

General data
Dry weight	125 kg
Permissible total weight	280 kg
Wheelbase	1321 mm
Tyre front	3.00-18
Tyre rear	3.50-18
Wheels front	1.60 x 18 wire-spoked
Wheels rear	1.85 x 18 wire-spoked
Top speed	130 km/h
Fuel tank capacity	19.1 litres
Production figures	1967: 7000 machines 1968: 4150 machines
Price	approx. $780

Model Sprint 350 SS 1969 to 1972

Motor
Type	6A
Engine	Air-cooled ohv single
Bore and stroke	73.8 x 80 mm (2.91 x 3.15 in.)
Capacity	342 cc (20.88 cu. in.)
Power output (DIN)	25(19)/7000 DIN
Compression ratio	9.0:1
Carburettor	Dell'Orto 27
Ignition	Battery-and-coil
Electrical system	6-volt
Starter	Kickstarter
Exhaust system	1 long tubular silencer RHS (1969-70) LHS (1971-72)

Transmission
Gearbox	4-speed
Overall reduction ratio	2.91 (1)
	1.76 (2)
	1.27 (3)
	1.00 (4)
Gearchange	Foot change RHS
Final drive	Chain

Cycle parts
Frame	Spine frame
Front forks	Hydraulically-damped telescopic
Rear suspension	Swinging-arm and hydraulic dampers
Front brake	Drum, via cable and handlebar lever
Rear brake	Drum, via rod and pedal

General data
Dry weight	150 kg
Permissible total weight	305 kg
Wheelbase	1365 mm (1971)
	1425 mm (1969)
Tyre front	3.25-19
Tyre rear	3.50-18
Wheels front	1.85 x 19 wire-spoked
Wheels rear	1.85 x 18 wire-spoked
Top speed	145 km/h
Fuel tank capacity	9.8 litres
Production figures	1969: 4575 machines
	1970: 4513 machines
	1971: 1500 machines
	1972: 3775 machines
Price	$795 (1969)
	$840 (1970)

Model Sprint 350 SX 1971 to 1972

Motor
Type	3C
Engine	Air-cooled ohv single
Bore and stroke	73.8 x 80 mm (2.91 x 3.15 in.)
Capacity	342 cc (20.88 cu. in.)
Power output (DIN)	25(19)/7000 DIN
Compression ratio	9.0:1
Carburettor	Dell'Orto 27
Ignition	Battery-and-coil ignition
Electrical system	6-volt
Starter	Kickstarter
Exhaust system	Tubular silencer high-level, RHS

Transmission
Gearbox	4-speed
Overall reduction ratio	2.91 (1)
	1.76 (2)
	1.27 (3)
	1.00 (4)
Gearchange	Foot change RHS
Final drive	Chain

Cycle parts
Frame	Spine frame
Front forks	Hydraulically-damped telescopic
Rear suspension	Swinging-arm and hydraulic dampers
Front brake	Drum, via cable and handlebar lever
Rear brake	Drum, via rod and pedal

General data
Dry weight	145 kg
Permissible total weight	305 kg
Wheelbase	1365 mm
Tyre front	3.50-19
Tyre rear	4.00-18
Wheels front	1.85 x 19 wire-spoked
Wheels rear	2.15 x 18 wire-spoked
Top speed	130 km/h
Fuel tank capacity	9.8 litres
Production figures	1971: 3920 machines
	1972: 2525 machines
Price	$870 (1971)

Model Sprint 350 SS 1973 to 1974

Motor
Engine	Air-cooled ohv single
Bore and stroke	73.8 x 80 mm (2.91 x 3.15 in.)
Capacity	342 cc (20.88 cu. in.)
Power output (DIN)	27(20)/7200 DIN
Compression ratio	9.0:1
Carburettor	Dell'Orto 30
Ignition	Battery-and-coil
Electrical system	6-volt
Starter	Electric starter
Exhaust system	1 long tubular silencer

Transmission
Gearbox	5-speed
Overall reduction ratio	2.78 (1)
	2.04 (2)
	1.53 (3)
	1.24 (4)
	1.00 (5)
Gearchange	Foot change
Final drive	Chain

Cycle parts
Frame	Twin-loop, full cradle
Front forks	Hydraulically-damped telescopic
Rear suspension	Swinging-arm and hydraulic dampers
Front brake	Drum, via cable and handlebar lever
Rear brake	Drum, via rod and pedal

General data
Dry weight	150 kg
Permissible total weight	305 kg
Wheelbase	1365 mm
Tyre front	3.25-19
Tyre rear	3.50-18
Wheels front	1.85 x 19 wire-spoked
Wheels rear	2.15 x 18 wire-spoked
Top speed	145 km/h
Fuel tank capacity	9.8 litres
Production figures	1973: 4137 machines
	1974: 2500 machines
Price	approx. $930

Model Sprint 350 SX 1973 to 1974

Motor
Engine	Air-cooled ohv single
Bore and stroke	73.8 x 80 mm (2.91 x 3.15 in.)
Capacity	342 cc (20.88 cu. in.)
Power output (DIN)	27(20)/7200 DIN
Compression ratio	9.0:1
Carburettor	Dell'Orto 30
Ignition	Battery-and-coil
Electrical system	6-volt
Starter	Electric starter
Exhaust system	Tubular silencer high-level, RHS

Transmission
Gearbox	5-speed
Overall reduction ratio	2.78 (1)
	2.04 (2)
	1.53 (3)
	1.24 (4)
	1.00 (5)
Gearchange	Foot change
Final drive	Chain

Cycle parts
Frame	Twin-loop, full cradle
Front forks	Hydraulically-damped telescopic
Rear suspension	Swinging-arm and hydraulic dampers
Front brake	Drum, via cable and handlebar lever
Rear brake	Drum, via rod and pedal

General data
Dry weight	145 kg
Permissible total weight	305 kg
Wheelbase	1365 mm
Tyre front	3.50-19
Tyre rear	4.00-18
Wheels front	1.85 x 19 wire-spoked
Wheels rear	2.15 x 18 wire-spoked
Top speed	130 km/h
Fuel tank capacity	9.8 litres
Production figures	1973: 2431 machines
	1974: 2085 machines
Price	approx. $980

Model description and technical information: the 50-, 65- and 90-cc. two-strokes

The period between 1965 and 1975 saw Harley-Davidson diversify their model range to a greater extent than any other time before or since. Beside the traditional V-twin models in their two capacity classes and in each case different levels of equipment, a gigantic gap yawned; the 246 cc./ 15 cu. in. Sprints were the next motorcycles down in capacity terms, closely followed by the aged two-strokes in the form of the Pacer and the Scat. But their 175 cc./11 cu. in. engines were due for one more year's sales with the Bobcat before they were discontinued, and while the old Topper scooter was still available in the mid-1960s, it too was hardly selling and about to be dropped.

New arrivals in the USA for 1965 were the 47 cc./ 2.9 cu. in. Leggero mopeds of Italian provenance, which were however regarded there only as toys for children and treated accordingly. Many did not survive, and after a relatively short honeymoon period they could only be sold to teenagers at massive discounts. Most will have breathed their last as transport on the beach or in short races between kids on their way to school; they were never suitable for longer distances.

In their first year, 1965, the M-50 Leggeros were available only with step-through frames to enable women to ride them in skirts and were finished in Holiday Red with Birch White tank panels. The small fuel tank mounted at the front took a full 1.3 Imp. gallons (6 litres). The Leggeros (Italian for light) turned out quite conventionally and, due to their tiny engine size, were enlarged to 65 cc. (4 cu. in.) two years later. The brochures promised a full 62% power increase as a result and the machines were sold both in the form of mopeds and as lightweight sporting motorcycles, depending on local legislation. With another capacity increase, these, the smallest of the Harley range, would carry on until the end of production in 1975.

All these engines were piston-ported, with an upright, but slightly forwards-inclined, cylinder. Over the eleven years of production three different capacities were used; 47 cc./2.9 cu. in., 65 cc./4 cu. in., and 90 cc./5.5 cu. in. All used the Dell'Orto carburettors which equipped all Aermacchi Harley-Davidsons and had state-of-the-art flywheel-magneto ignition and lighting systems. A three-speed hand-change gearbox and chain drive were equally up to the standard of that period.

Equally standard were telescopic front forks with twin rear suspension units, fitted to a spine-type (for lightness) frame, under which the engine was mounted. Only the Z-90 got a single-downtube full-

1965 Model M-50 in Holiday Red with Birch White tank panels

1967 Model M-50S

1967 Model M-65 in original condition

1974 Model X-90 Shortster

Further views of the X-90 – in blue ...

... and in red (tank badges not original)

loop frame to match that of the road-only SX-125. The wheels and tyres were 18 inches in diameter in the first year of production, then all models had 17-inch wheels until 1972, when the rear was given a 16-inch wheel with a fatter tyre. Later versions could even take two people.

Model development

1966 model year
The range was soon extended to include the sports Model M-50S with the same engine but its own frame and a European-style dualseat filling in the gap between the tank and the seat on the M-50. Both models now had 17-inch wheels and tyres. The M-50S's paint finish was basically red but with white panels on the sides of the front part of the tank. Apart from that, there were no technical changes. At the end of the model year it is alleged that some thousands of still-unsold Leggeros were simply thrown on to a waste dump...

1967 model year
It soon enough became clear that an engine of such tiny capacity was not going to sell well, so both bore and stroke were increased to enlarge the engine to 65 cc. (4 cu. in.), with the result that power increased to some 6.5 horsepower despite a reduction in compression ratio from 10.0:1 to 9.0:1. The white tank panel disappeared, giving way to a large Harley-Davidson tank badge – the new flattened-diamond design used for all Harleys from 1966 onwards – surrounded by white pinstriping. The three-speed gearbox remained, as did the 17-inch wheels and tyres. The sports model M-65S could now manage a top speed of just over 50 mph (80 km/h), finally fast enough for its customers, and from now on was available in shades other than the Italian national colour.

In 1966 over 16,000 M-50 Leggeros were imported. What happened to them all is unclear, but for 1967 sales targets were severely curtailed; only 5,267 more saleable M-65s found their way across the Atlantic.

1968 model year
In spite of marginal modifications, H-D imported 11,700 M-65s in this year of general recession; however, once again there were many unsold machines which this time would naturally have an effect on the sales figures over the years to come.

1969 model year
In this last year of the two M-65 models (the M-65 was discontinued at the end of the year) there was again a substantial reduction in the numbers of machines imported, as only 2,700 M-65s were sold in

1975 Model X-90 Shortster

1974 Model Z-90 Leggero

the US. Typical of the publicity of the period was this statement in the 1969 catalogue: 'Consumption is so amazingly low that walking will seem more expensive to you'. There were no changes worthy of the name.

1970 model year
After the standard M-65 had been discontinued, sales figures for the M-65S slowly increased from 2,000 to just under 4,000 machines sold each year of the succeeding three years until it too was discontinued.

1971 model year
The single surviving model, the M-65S Leggero, now received significantly wider tyres; 2.50-inch front and rear.

1972 model year
New AMF/Harley-Davidson tank graphics decorated all Harleys for this year. In its last year the M-65S got a smaller-diameter but wider rear tyre (3.00-16), which significantly improved roadholding. To the last there was only the normal three-speed gearbox. When all's said and done it was nevertheless amazing that a total of approximately 45,000 Leggeros could be unloaded on the market.

A use was also found for the M-65 engine unit in the MC-65 Shortster, a machine to compete with the little Honda Monkey or Dax machines. The first Harley minibike, with its small 10-inch wheels and off-road-style high-level exhaust system on the right-hand side, was an agile little machine, intended to be taken along in motorhomes or on boats.

1973 to 1975 model years
A last attempt to make the smallest Harleys interesting for buyers came with a further increase in capacity, this time to 90 cc./5.5 cu. in., the bigger engine being fitted to the Model Z-90 Leggero and to the Model X-90 Shortster. Both machines also had new 12-volt electrical systems.

The Z-90 Leggero had become, apparently at least, a genuine motorcycle, possessing a proper single-loop frame and a wider rear tyre. With its output increased to 8 horsepower Leggero riders could finally break the magic 60 mph (100 km/h) barrier. Completely new also was the four-speed gearbox with left-hand foot operation, designed for much higher speeds than that. Instead of the low-level and chrome-plated exhaust system on the right-hand side, the Z-90 now got a black-painted high-level system on the left-hand side which left the customers in no doubt as to the machine's intended sphere of use. The motorcycle was designed for a young generation of enthusiasts who wanted a road machine that could be used off-road – a trailbike, in fact. While weight-saving was essential for off-road machines, however, it was perhaps a step too far for a road motorcycle to have the diameter of its brake drums reduced by 0.6 inch (15 mm) compared with the previous Model M-65s.

The Model X-90 Shortster appeared on the market as the successor to the 1972 Model MC-65. It combined the previous year's minibike cycle parts with the same 90 cc./5.5 cu. in. engine used in the Z-90 Leggero; obviously with the exception that its small wheels and tyres needed a different overall gear ratio. Its maximum speed, nevertheless, was now 53 mph (85 km/h).

Both the X-90 and Z-90 continued to sell quite well until sales dropped off during the 1975 model year and both were discontinued at the end of that year.

The range of Harley Davidson models up to 100 cc.:

Leggero
50	1965-1966
50 Sport	1966
65	1967-1969
65 Sport	1967-1972
Z-90	1973-1975

Shortster
MC-65	1972
X-90	1973-1975

Model M-50 1965 to 1966

Motor
Engine	Single-cylinder air-cooled piston-ported 2-stroke
Bore and stroke	38.8 x 40 mm (1.53 x 1.58 in.)
Capacity	47 cc (2.89 cu. in.)
Power output (SAE)	4(3)/7000 rpm
Compression ratio	10.0:1
Carburettor	Dell'Orto
Ignition	Flywheel magneto
Electrical system	6-volt
Starter	Kickstarter
Exhaust system	Long tubular silencer RHS

Transmission
Gearbox	3-speed
Reduction ratios	3.08 (1) (1966)
	1.78 (2) (1966)
	1.17 (3) (1966)
Secondary reduction ratio	2.08 (1966)
Gearchange	Hand change on LHS
Final drive	Chain

Cycle parts
Frame	Spine frame
Front forks	Hydraulically-damped telescopic
Rear suspension	Swinging-arm with spring units
Front brake	Drum, via cable and handlebar lever
Rear brake	Drum, via rod and pedal

General data
Dry weight	85 kg
Permissible total weight	245 kg
Wheelbase	1120 mm
Tyre front	2.00-18 (1965), 2.00-17 (1966)
Tyre rear	2.00-18 (1965), 2.00-17 (1966)
Wheels front	1.35 x 18 or 1.35 x17 wire-spoked
Wheels rear	1.35 x 18 or 1.35 x17 wire-spoked
Top speed	50 km/h
Fuel tank capacity	6.0 litres
Production figures	1965: 9000 machines
	1966: 5700 machines
Price	$225

Model M-50 Sport 1966

Motor
Engine	Single-cylinder air-cooled piston-ported 2-stroke
Bore and stroke	38.8 x 40 mm (1.53 x 1.58 in.)
Capacity	47 cc (2.89 cu. in.)
Power output (SAE)	4(3)/7000 rpm
Compression ratio	10.0:1
Carburettor	Dell'Orto
Ignition	Flywheel magneto
Electrical system	6-volt
Starter	Kickstarter
Exhaust system	Long tubular silencer RHS

Transmission
Gearbox	3-speed
Reduction ratios	3.08 (1)
	1.78 (2)
	1.17 (3)
Secondary reduction ratio	2.08
Gearchange	Hand change on LHS
Final drive	Chain

Cycle parts
Frame	Spine frame
Front forks	Hydraulically-damped telescopic
Rear suspension	Swinging-arm with spring units
Front brake	Drum, via cable and handlebar lever
Rear brake	Drum, via rod and pedal

General data
Dry weight	85 kg
Permissible total weight	245 kg
Wheelbase	1120 mm
Tyre front	2.00-17
Tyre rear	2.00-17
Wheels front	1.35 x 17 wire-spoked
Wheels rear	1.35 x 17 wire-spoked
Top speed	60 km/h
Fuel tank capacity	9.5 litres
Production figures	1966: 10500 machines
Price	$275

The figures given are total production figures, only a few of which were sold in the US.

Model M-65 1967 to 1969

Motor
Type	-
Engine	Single-cylinder air-cooled piston-ported 2-stroke
Bore and stroke	44 x 42 mm (1.73 x 1.65 in.)
Capacity	64 cc (3.9 cu. in.)
Power output (SAE)	6.5(5)/7800 rpm
Compression ratio	9.0:1
Carburettor	Dell'Orto ME 18BS
Ignition	Flywheel magneto
Electrical system	6-volt
Starter	Kickstarter
Exhaust system	Long tubular silencer RHS

Transmission
Gearbox	3-speed
Reduction ratios	3.08 (1)
	1.78 (2)
	1.17 (3)
Secondary reduction ratio	2.08
Gearchange	Hand change on LHS
Final drive	Chain

Cycle parts
Frame	Spine frame
Front forks	Hydraulically-damped telescopic
Rear suspension	Swinging-arm with spring units
Front brake	Drum, via cable and handlebar lever
Rear brake	Drum, via rod and pedal

General data
Dry weight	90 kg
Permissible total weight	245 kg
Wheelbase	1138 mm
Tyre front	2.00-17
Tyre rear	2.00-17
Wheels front	1.35 x 17 wire-spoked
Wheels rear	1.35 x 17 wire-spoked
Top speed	75 km/h
Fuel tank capacity	6.0 litres
Production figures	1967: 2000 machines
	1968: 1200 machines
	1969: 950 machines
Price	$235

Model M-65 Sport 1967 to 1972

Motor
Type	8A
Engine	Single-cylinder air-cooled piston-ported 2-stroke
Bore and stroke	44 x 42 mm (1.73 x 1.65 in.)
Capacity	64 cc (3.9 cu. in.)
Power output (SAE)	6.5(5)/7800 rpm
Compression ratio	9.0:1 (1967-71), 8.5:1 (1972)
Carburettor	Dell'Orto ME 18BS
Ignition	Flywheel magneto
Electrical system	6-volt
Starter	Kickstarter
Exhaust system	Long tubular silencer RHS

Transmission
Gearbox	3-speed
Reduction ratios	3.08 (1)
	1.78 (2)
	1.17 (3)
Secondary reduction ratio	2.08
Gearchange	Hand change on LHS
Final drive	Chain

Cycle parts
Frame	Spine frame
Front forks	Hydraulically-damped telescopic
Rear suspension	Swinging-arm with spring units
Front brake	125 mm drum, via cable and handlebar lever
Rear brake	125 mm drum, via rod and pedal

General data
Dry weight	90 kg
Permissible total weight	245 kg
Wheelbase	1138 mm
Tyre front	2.00-17 (1967-70)
	2.50-17 (1971-72)
Tyre rear	2.00-17 (1967-70)
	2.50-17 (1971)
	3.00-16 (1972)
Wheels front and rear	1.35 x 17 wire-spoked
	1972, wheels rear
	1.85 x 16 wire-spoked
Top speed	82 km/h
Fuel tank capacity	9.5 litres
Production figures	1967: 3267 machines
	1968: 10500 machines
	1969: 1750 machines
	1970: 2080 machines
	1971: 3100 machines
	1972: 3708 machines
Price	$265

The figures given are total production figures, only a few of which were sold in the US.

Model Z-90 1973 to 1975

Motor
Type	3D
Engine	Single-cylinder air-cooled piston-ported 2-stroke
Bore and stroke	48 x 50 mm (1.89 x 1.97 in.)
Capacity	90 cc (5.52 cu. in.)
Power output (SAE)	8(6)/7500 rpm
Compression ratio	9.2:1
Carburettor	Dell'Orto
Ignition	Flywheel magneto
Electrical system	12-volt
Starter	Kickstarter
Exhaust system	1 long high-level silencer LHS

Transmission
Gearbox	4-speed
Reduction ratios	2.50 (1)
	1.40 (2)
	0.92 (3)
	0.72 (4)
Secondary reduction ratio	2.50
Gearchange	Foot change LHS
Final drive	Chain

Cycle parts
Frame	Spine frame with single downtube
Front forks	Hydraulically-damped telescopic
Rear suspension	Swinging-arm with spring units
Front brake	110 mm drum, via cable and handlebar lever
Rear brake	110 mm drum, via rod and pedal

General data
Dry weight	90 kg
Permissible total weight	250 kg
Wheelbase	1181 mm
Tyre front	2.50-17
Tyre rear	3.25-16
Wheels front	1.35 x 17 wire-spoked
Wheels rear	1.85 x 16 wire-spoked
Top speed	95 km/h
Fuel tank capacity	9.1 litres
Production figures	1973: 8250 machines
	1974: 7168 machines
	1975: 2652 machines
Price	Not available

Model MC-65 1972

Motor
Type	5C
Engine	Single-cylinder air-cooled piston-ported 2-stroke
Bore and stroke	44 x 42 mm (1.73 x 1.65 in.)
Capacity	64 cc (3.9 cu. in.)
Power output (SAE)	6.5(5)/7800 rpm
Compression ratio	8.5:1
Carburettor	Dell'Orto
Ignition	Flywheel magneto
Electrical system	6-volt
Starter	Kickstarter
Exhaust system	1 long high-level silencer LHS

Transmission
Gearbox	3-speed
Reduction ratios	3.08 (1)
	1.78 (2)
	1.17 (3)
Secondary reduction ratio	2.17
Gearchange	Hand change on LHS
Final drive	Chain

Cycle parts
Frame	Spine frame
Front forks	Hydraulically-damped telescopic
Rear suspension	Swinging-arm with spring units
Front brake	Drum, via cable and handlebar lever
Rear brake	Drum, via rod and pedal

General data
Dry weight	90 kg
Permissible total weight	175 kg
Wheelbase	1120 mm
Tyre front	3.00-10
Tyre rear	3.00-10
Wheels front	1.60 x 10 steel disc
Wheels rear	1.60 x 10 steel disc
Top speed	80 km/h
Fuel tank capacity	5.3 litres
Production figures	8000 machines
Price	Not available

The figures given are total production figures, only a few of which were sold in the US.

Model X-90 1973 to 1975

Motor

Type	2D
Engine	Single-cylinder air-cooled piston-ported 2-stroke
Bore and stroke	48 x 50 mm (1.89 x 1.97 in.)
Capacity	90 cc (5.52 cu. in.)
Power output (SAE)	8(6)/7500 rpm
Compression ratio	9.2:1
Carburettor	Dell'Orto
Ignition	Flywheel magneto
Electrical system	12-volt
Starter	Kickstarter
Exhaust system	1 long high-level silencer LHS

Transmission

Gearbox	4-speed
Reduction ratios	2.50 (1)
	1.40 (2)
	0.92 (3)
	0.72 (4)
Secondary reduction ratio	3.58
Gearchange	Foot change LHS
Final drive	Chain

Cycle parts

Frame	Spine frame
Front forks	Hydraulically-damped telescopic
Rear suspension	Swinging-arm with spring units
Front brake	110 mm drum, via cable and handlebar lever
Rear brake	110 mm drum, via rod and pedal

General data

Dry weight	95 kg
Permissible total weight	175 kg
Wheelbase	1035 mm
Tyre front	3.00-10
Tyre rear	3.00-10
Wheels front	1.60 x 10 steel disc
Wheels rear	1.60 x 10 steel disc
Top speed	95 km/h
Fuel tank capacity	5.3 litres
Production figures	1973: 8250 machines
	1974: 7019 machines
	1975: 1586 machines
Price	Not available

The figures given are total production figures, only a few of which were sold in the US.

Model description and technical information: the 100 cc. two-strokes

Due to the differing regulations concerning desert-racing and special off-road disciplines such as enduros in America, Harley-Davidson were obliged to react and produced a machine that was only in production for four years; the Baja 100. Originally conceived as a pure off-road sports machine (the Model MSR-100) with no on-road capability at all, the Baja was imported into the USA with the model designation 8B and sold well. With its very high-compression (11.5:1) engine it put out a considerable 13 horsepower at 8,000 rpm, which since it only weighed 198 lb (90 kg), offered explosive road performance.

So at West Juneau Avenue in Milwaukee, a chance was seen to market this machine for more civilized use. This succeeded reasonably well with the on/off-road version MSX-100L, or rather MSR-100L (model designation 6C), available starting from 1971, L standing for Lights. Only slightly more heavy and scarcely any less powerful than its competition-orientated forerunner, this was an off-road bike for new up-and-coming talents.

The machine had a two-stroke engine with square bore and stroke whose technical characteristics corresponded in principle to those of the M-65 models. It was equipped with a relatively large 0.95-inch (24 mm) Dell'Orto carburettor and was the first Harley two-stroke with a five-speed gearbox. It had a large, comfortable seat – a legacy of its desert-racing background, rather than an attempt to make it civilized for about-town use – and an expensive high-level exhaust system painted matt black and provided with a chrome-plated metal wire grille to protect the rider from burns.

On top of that was a front mudguard wrapped closely around the front wheel, which would be replaced in 1972 by a mudguard mounted on the underside of the steering head. The little 98 cc./6 cu. in. engine delivered over eleven horsepower at 8,000 rpm. Top gear being extremely high, the rider could choose between different drive and driven sprockets to arrive at the correct gearing required. The MSR models were available until the end of 1973.

Then in 1974 the Model SR-100 Baja, a successor machine, but one that was more closely tailored to off-road motorcycling, arrived in the USA. The engine's specifications were identical, except that it now featured a separate oil-injection lubrication system; before then, two-strokes had had to be refuelled with petrol and oil pre-mixed in precisely the correct proportions. The cylinder head was redesigned, with much deeper cooling fins, and a hard chrome-plated cylinder bore was used. Road-legal knobbly tyres were fitted as standard (previously, these had only been fitted to pure off-road motorcycles), as was a matt black-painted high-level exhaust system. Once again, in spite of the appearance of its large seat, the SR-100 was a single-seater. In spite of these changes, the model designation of 6C was retained.

All Baja models used a frame based on the sharply-curved spine frame of the two-stroke 125s, but with an additional engine front mounting fitted by the men in Varese.

1972 Model Baja 100: A pure off-roader for desert racing

Model MSR-100L 1971 to 1973

Motor
Type	6C
Engine	Single-cylinder air-cooled piston-ported 2-stroke
Bore and stroke	50 x 50 mm (1.97 x 1.97 in.)
Capacity	98 cc (5.99 cu. in.)
Power output (DIN)	11.5(8.5)/8000 DIN
Compression ratio	9.5:1
Carburettor	Dell'Orto 24 mm
Ignition	Battery-and-coil
Electrical system	12-volt
Starter	Kickstarter
Exhaust system	1 long silencer, high-level, LHS

Transmission
Gearbox	5-speed
Reduction ratios	2.06 (1)
	1.33 (2)
	0.92 (3)
	0.75 (4)
	0.58 (5)
Secondary reduction ratio	Gearbox sprocket 12, 13, 14 or 15 teeth
	Rear sprocket 42 or 72 teeth
Gearchange	Foot change
Final drive	Chain

Cycle parts
Frame	Spine frame
Front forks	Hydraulically-damped telescopic
Rear suspension	Swinging-arm with spring units
Front brake	Drum, via cable and handlebar lever
Rear brake	Drum, via rod and pedal

General data
Dry weight	100 kg
Permissible total weight	180 kg
Wheelbase	1321 mm
Tyre front	3.00-21
Tyre rear	3.50-18
Wheels front	1.85 x 21 wire-spoked
Wheels rear	2.15 x 18 wire-spoked
Top speed	Dependent on gearing
Fuel tank capacity	9.5 litres
Production figures	1971: 1200 machines
	1972: 1200 machines
	1973: 986 machines
Price	Not available

Model SR-100 1974

Motor
Type	6C
Engine	Single-cylinder air-cooled piston-ported 2-stroke
Bore and stroke	50 x 50 mm (1.97 x 1.97 in.)
Capacity	98 cc (5.99 cu. in.)
Power output (DIN)	11.5(8.5)/8000 DIN
Compression ratio	9.5:1
Carburettor	Dell'Orto 24 mm
Ignition	Battery-and-coil
Electrical system	12-volt
Starter	Kickstarter
Exhaust system	1 long silencer, high-level, LHS

Transmission
Gearbox	5-speed
Reduction ratios	2.06 (1)
	1.33 (2)
	0.92 (3)
	0.75 (4)
	0.58 (5)
Secondary reduction ratio	Gearbox sprocket 12 teeth
	Rear sprocket 42 or 72 teeth
Gearchange	Foot change
Final drive	Chain

Cycle parts
Frame	Spine frame
Front forks	Hydraulically-damped telescopic
Rear suspension	Swinging-arm with spring units
Front brake	Drum, via cable and handlebar lever
Rear brake	Drum, via rod and pedal

General data
Dry weight	100 kg
Permissible total weight	180 kg
Wheelbase	1321 mm
Tyre front	3.00-21
Tyre rear	3.50-18
Wheels front	1.85 x 21 wire-spoked
Wheels rear	2.15 x 18 wire-spoked
Top speed	Dependent on gearing
Fuel tank capacity	9.5 litres
Production figures	1396 machines
Price	Not available

Model description and technical information: the 125 cc. two-strokes

By and large ten quite successful years of sales were granted by the, groundbreaking by Harley standards, 125 motorcycle that first saw the light of the American motorcycle world in 1968 as the Rapido. Aermacchi at this time could already look back on 17 years of successful production of 125 cc./7.5 cu. in. motorcycles, as it was in 1951 that the first two-stroke Aermacchi of this piston displacement appeared on the Italian market. With the Rapido, the gap between the mopeds and the Sprints, gaping for so long, could finally be closed.

Obviously descended from the legendary 125 cc./7.5 cu. in. Ala d'Oro (Gold Wing) of 1967, the model designated simply M or Model M-125 represented what was technically feasible at that time. An output of 10 horsepower and a decent foot change four-speed gearbox were the icing on the cake of a project made easier by the changeover from the small Leggeros. The oversquare engine configuration meant that it had some reserves of power in hand.

The Rapido's spine frame, like that of the Leggeros, was at the time as good as anything else from Europe, or from Japan. Compared with the usual type of frame, this was clearly lighter and made the motorcycle look more up to date. While the Aermacchi wasn't as fast as it looked, a top speed of 68 mph (110 km/h) was nevertheless completely respectable. A Ceriani telescopic front fork took care of the front end, while the two rear suspension units were fully enclosed. Together with the chrome-plated low-level exhaust system and the long, well-padded seat with its little European-style racing hump at the rear, the whole machine looked very sporty indeed.

Model development

1969 model year

Already in its second model year the modest M-125 became the comprehensively revised ML-125. A strengthened frame gave greater torsional rigidity, and the somewhat smaller tank permitted the fitting of a longer seat. In fact this would be the last year for the road version, the machines remaining in stock being sold off during 1970.

In addition, and eventually as replacement, an on/off-road version appeared in the form of the MLS-125 which remained on sale until 1972. The MLS was quite well suited for road use as the front mudguard was mounted close to the tyre as was usual for most Street Scrambler models – by now beginning to be called trailbikes – in the USA. The raised chrome-plated exhaust system curved upwards over the engine/gearbox unit on the left-hand side and then ran horizontally to the rear, its prettily-punched and also chrome-plated heat shield looking like that of a much larger-capacity machine.

1970 and 1971 model years

The road version having been discontinued, the MLS-125 Scrambler continued to sell quite well all by itself, aided by the new 18-inch wheels and wider

1968 Model M-125 Rapido

tyres which gave improved sales appeal as well as better grip, and by the new and slightly smaller tank which improved the looks of the machine. The chrome-plating of the exhaust system gave way to matt black paint with a protective metal wire grille.

1972 model year
In the final year of the original MLS-125 Scrambler, 19-inch wheels and tyres were fitted. New AMF/Harley-Davidson tank graphics decorated all Harleys for this year.

1973 model year
A completely new model, at least according to the designation, TX-125, was the follow-up machine to the MLS-125 but was sold only for a single model year. Strangely enough the model designation (7A) was not changed, nor even when in the following year the penultimate Rapido appeared in the form of the SX-125. The TX-125 retained a spine frame, but now with twin downtubes and appeared much more civilized than the previous model. It had a separate oil-injection lubrication system, a five-speed gearbox and two rear sprockets offering a choice of reduction ratios for on and off-road motorcycling. Its battery-and-coil ignition is equally worthy of mention, as was the 12-volt electrical system. The TX-125 had a 19-inch front wheel and an 18-inch rear.

1974 model year
As replacement for the TX-125 that was available for the one year only, the SX-125 appeared, with a completely redesigned all aluminium-alloy engine with a hard chrome-plated cylinder bore. It thus represented the ultimate development of the 125 cc. Aermacchi engine.

1975 model year
The successor to the SX-125, the SXT-125, can be identified by its disc front brake. Colours available were Sunburst Blue or Burgundy.

1976 and 1977 model years
The last stage of development was represented by the SS-125 road machine. Sold alongside the SXT-125 trailbike, it was manufactured until 1977 and sold up to 1978; even in Germany. The frame of both was a new full-loop design with twin downtubes, similar to that of the 175s, the engine's oil being kept in a reservoir in the frame top tube. A spring-loaded automatic final drive chain tensioner was fitted to both.

The SS-125 should have opened up sales to new groups of buyers as the sales of the trailbikes were beginning to suffer thanks to the ever-stronger Japanese competition. However, the attempt went wrong from the start, probably thanks to misguided attempts to cut costs. Unlike the SXT-125's disc front brake, the road machine had to make do with a drum and, unlike the 175, it did not have a tachometer as standard. It did, however, have rear suspension units with three-way adjustable spring preload.

1974 Model SX-125

1969 Model ML 125 Rapido Scrambler

The range of Rapido models:

1st series
M-125	1968-1969
ML-125 Road model	1969-1970 (1,000 manufactured)
ML-125 Scrambler	1969-1970 (3,275 manufactured)
MLS-125 Scrambler	1970-1972

2nd series
TX-125	1973

3rd series
SX-125	1974-1975
SXT-125	1975-1977
SS-125	1976-1977

Model M-125 1968 to 1969

Motor
Type	-
Engine	Single-cylinder air-cooled piston-ported 2-stroke
Bore and stroke	56 x 50 mm (2.21 x 1.97 in.)
Capacity	123 cc (7.52 cu. in.)
Power output (DIN)	12.5(9)/7200 DIN
Compression ratio	7.65:1
Carburettor	Dell'Orto
Ignition	Flywheel magneto
Electrical system	6-volt
Starter	Kickstarter
Exhaust system	1 long silencer LHS

Transmission
Gearbox	3-speed
Reduction ratios	Not available
Secondary reduction ratio	Not available
Gearchange	Foot change
Final drive	Chain

Cycle parts
Frame	Spine frame
Front forks	Hydraulically-damped telescopic
Rear suspension	Swinging-arm with spring units
Front brake	Drum, via cable and handlebar lever
Rear brake	Drum, via rod and pedal

General data
Dry weight	75 kg
Permissible total weight	240 kg
Wheelbase	-
Tyre front	2.50-17
Tyre rear	3.00-17
Wheels front	1.50 x 17 wire-spoked
Wheels rear	1.85 x 17 wire-spoked
Top speed	110 km/h
Fuel tank capacity	9.5 litres
Production figures	1968: 5000 machines 1969: 1000 machines
Price	$395

Model ML-125 Scrambler 1969

Motor
Type	-
Engine	Single-cylinder air-cooled piston-ported 2-stroke
Bore and stroke	56 x 50 mm (2.21 x 1.97 in.)
Capacity	123 cc (7.52 cu. in.)
Power output (DIN)	12.5(9)/7200 DIN
Compression ratio	7.65:1
Carburettor	Dell'Orto
Ignition	Flywheel magneto
Electrical system	6-volt
Starter	Kickstarter
Exhaust system	1 long silencer, high-level, RHS

Transmission
Gearbox	3-speed
Reduction ratios	Not available
Secondary reduction ratio	Gearbox sprocket 12, 13, 14 or 15 teeth Rear sprocket 50 or 62 teeth
Gearchange	Foot change
Final drive	Chain

Cycle parts
Frame	Spine frame
Front forks	Hydraulically-damped telescopic
Rear suspension	Swinging-arm with spring units
Front brake	Drum, via cable and handlebar lever
Rear brake	Drum, via rod and pedal

General data
Dry weight	70 kg
Permissible total weight	240 kg
Wheelbase	-
Tyre front	2.50-17
Tyre rear	3.00-17
Wheels front	1.50 x 17 wire-spoked
Wheels rear	1.85 x 17 wire-spoked
Top speed	Dependent on gearing
Fuel tank capacity	9.5 litres
Production figures	1969: 3275 machines
Price	$395

The figures given are total production figures, only a few of which were sold in the US.

Model MLS-125 1970 to 1972

Motor
Type	7A
Engine	Single-cylinder air-cooled piston-ported 2-stroke
Bore and stroke	56 x 50 mm (2.21 x 1.97 in.)
Capacity	123 cc (7.52 cu. in.)
Power output (DIN)	12.5(9)/7200 DIN
Compression ratio	7.65:1
Carburettor	Dell'Orto
Ignition	Battery-and-coil
Electrical system	6-volt
Starter	Kickstarter
Exhaust system	1 long silencer, high-level, LHS

Transmission
Gearbox	4-speed
Reduction ratios	2.50 (1)
	1.40 (2)
	0.92 (3)
	0.72 (4)
Secondary reduction ratio	Gearbox sprocket 12, 13, 14 or 15 teeth
	Rear sprocket 50 or 62 teeth
Gearchange	Foot change
Final drive	Chain

Cycle parts
Frame	Spine frame
Front forks	Telescopic
Rear suspension	Swinging-arm
Front brake	Drum, via cable and handlebar lever
Rear brake	Drum, via rod and pedal

General data
Dry weight	95 kg
Permissible total weight	260 kg
Wheelbase	1242 mm
Tyre front	3.00-18 (1970-71)
	3.00-19 (1972)
Tyre rear	3.50-18 (1970-71)
	3.50-19 (1972)
Wheels front	1.60 x 18 wire-spoked
	1.60 x 19 wire-spoked
Wheels rear	1.85 x 18 wire-spoked
	1.85 x 19 wire-spoked
Top speed	approx. 120 km/h
Fuel tank capacity	9.1 litres
Production figures	1970: 4059 machines
	1971: 5200 machines
	1972: 6000 machines
Price	Not available

Model TX-125 1973

Motor
Type	7A
Engine	Single-cylinder air-cooled piston-ported 2-stroke
Bore and stroke	56 x 50 mm (2.21 x 1.97 in.)
Capacity	123 cc (7.52 cu. in.)
Power output (DIN)	11(8)/7200 DIN
Compression ratio	8.7:1
Carburettor	Dell'Orto 27 mm
Ignition	Battery-and-coil
Electrical system	12-volt
Starter	Kickstarter
Exhaust system	1 long silencer, high-level, LHS

Transmission
Gearbox	5-speed
Reduction ratios	2.06 (1)
	1.33 (2)
	0.92 (3)
	0.75 later 0.79 (4)
	0.58 later 0.75 (5)
Secondary reduction ratio	4.75 (Road)
	5.92 (Off-road)
Gearchange	Foot change
Final drive	Chain

Cycle parts
Frame	Single downtube full-cradle
Front forks	Hydraulically-damped telescopic
Rear suspension	Swinging-arm with spring units
Front brake	130 mm drum, via cable and handlebar lever
Rear brake	130 mm drum, via rod and pedal

General data
Dry weight	105 kg
Permissible total weight	265 kg
Wheelbase	1245 mm
Tyre front	3.00-19
Tyre rear	3.50-18
Wheels front	1.60 x 19 wire-spoked
Wheels rear	1.85 x 18 wire-spoked
Top speed	112 km/h (Road version)
Fuel tank capacity	9.1 litres
Production figures	9225 machines
Price	Not available

Model SX-125 1974 to 1975

Motor
Type	7A
Engine	Single-cylinder air-cooled piston-ported 2-stroke
Bore and stroke	56 x 50 mm (2.21 x 1.97 in.)
Capacity	123 cc (7.52 cu. in.)
Power output (DIN)	13.5(10)/7800 DIN
Compression ratio	10.7:1
Carburettor	Dell'Orto 27 mm
Ignition	Battery-and-coil
Electrical system	12-volt
Starter	Kickstarter
Exhaust system	1 long silencer, high-level, LHS

Transmission
Gearbox	5-speed
Reduction ratios	2.06 (1)
	1.33 (2)
	0.92 (3)
	0.75
	0.58
Secondary reduction ratio	4.75 (Road), 5.90 (Off-road)
Gearchange	Foot change
Final drive	Chain

Cycle parts
Frame	Single downtube full-cradle
Front forks	Hydraulically-damped telescopic
Rear suspension	Swinging-arm with spring units
Front brake	130 mm drum, via cable and handlebar lever
Rear brake	130 mm drum, via rod and pedal

General data
Dry weight	105 kg
Permissible total weight	265 kg
Wheelbase	1354 mm
Tyre front	3.00-19
Tyre rear	3.50-18
Wheels front	1.60 x 19 wire-spoked
Wheels rear	1.85 x 18 wire-spoked
Top speed	121 km/h (Road version)
Fuel tank capacity	9.1 litres
Production figures	4000 machines

Model SXT-125 1975 to 1977

Motor
Type	3F
Engine	Single-cylinder air-cooled piston-ported 2-stroke
Bore and stroke	56 x 50 mm (2.21 x 1.97 in.)
Capacity	123 cc (7.52 cu. in.)
Power output (DIN)	13(10)/7000 DIN
Compression ratio	10.8:1
Carburettor	Dell'Orto VHB 27 AD
Ignition	Battery-and-coil
Electrical system	12-volt
Starter	Kickstarter
Exhaust system	1 long silencer, high-level, LHS

Transmission
Gearbox	5-speed
Reduction ratios	2.08 (1)
	1.31 (2)
	0.95 (3)
	0.76 (4)
	0.61 (5)
Secondary reduction ratio	4.36
Gearchange	Foot change
Final drive	Chain

Cycle parts
Frame	Single downtube full-cradle
Front forks	Hydraulically-damped telescopic
Rear suspension	Swinging-arm
Front brake	135 mm drum, via cable and handlebar lever
Rear brake	135 mm drum, via rod and pedal

General data
Dry weight	110 kg
Permissible total weight	275 kg
Wheelbase	1354 mm
Tyre front	3.00-19
Tyre rear	3.50-18
Wheels front	1.60 x 19 wire-spoked
Wheels rear	1.85 x 18 wire-spoked
Top speed	104 km/h
Fuel tank capacity	10.5 litres
Production figures	1975: 2500 machines
	1976: 6056 machines
	1977: 48 machines

Model SS-125 1976 to 1977

Motor
Type	6F
Engine	Single-cylinder air-cooled piston-ported 2-stroke
Bore and stroke	56 x 50 mm (2.21 x 1.97 in.)
Capacity	123 cc (7.52 cu. in.)
Power output (DIN)	13(10)/7000 DIN
Compression ratio	10.8:1
Carburettor	Dell'Orto VHB 27 AD
Ignition	Battery-and-coil
Electrical system	12-volt
Starter	Kickstarter
Exhaust system	1 long silencer, high-level, LHS

Transmission
Gearbox	5-speed
Reduction ratios	2.08 (1) 1.31 (2) 0.95 (3) 0.76 (4) 0.61 (5)
Secondary reduction ratio	4.21
Gearchange	Foot change
Final drive	Chain

Cycle parts
Frame	Full cradle, with single downtube
Front forks	Hydraulically-damped telescopic
Rear suspension	Swinging-arm
Front brake	135 mm drum, via cable and handlebar lever
Rear brake	135 mm drum, via rod and pedal

General data
Dry weight	110 kg
Permissible total weight	275 kg
Wheelbase	1354 mm
Tyre front	3.00-19
Tyre rear	3.50-18
Wheels front	1.60 x 19 wire-spoked
Wheels rear	1.85 x 18 wire-spoked
Top speed	104 km/h
Fuel tank capacity	10.5 litres
Production figures	1976: 1560 machines 1977: 488 machines

Model description and technical information: the 175 cc. two-strokes

Once it had become clear that there was little mileage left in the 125s, engine capacity was increased to produce two model variations on the theme of a 174 cc./11 cu. in. unit, which with its 17 horsepower, should be able to compete once more with the foreign competition. Both bore and stroke were increased to make the bigger engine only slightly oversquare (making for good torque output and reduced piston speeds), but the bigger machine otherwise followed closely the construction of the 125s. As with the SX-125 of that year, the all aluminium-alloy engine had a hard chrome-plated cylinder bore, but further goodies followed. For example a direct ignition system, which made it possible to remove the lights and battery for competitions and was of the new contact breakerless Capacitor Discharge Ignition (CDI) type. A resettable trip odometer was also fitted.

The 175s were fitted with the same full-loop frame with twin downtubes as the last 125s. Further details were the quickly-detachable rear wheel, rear suspension units with five-way adjustable spring preload and the larger (compared with the 125) 2.9 Imp. gallon (13 litre) fuel tank. Naturally the engine had a separate oil-injection lubrication system; the oil being kept in a reservoir in the frame top tube. Equally naturally, a 12-volt electrical system was fitted, and a five-speed gearbox.

The SX-175 trailbike, with tyres of 3.00-19 front and (initially) 3.25-18 rear, appeared in 1974 and remained on sale for three years, while the road-only SS-175, with 3.25-19 front and 4.00-18 rear tyres, was only available for 1976 and 1977. With their 69 mph (111 km/h) top speed, the 175s were scarcely any faster than the 65 mph (104 km/h) of the later 125s, but their ignition systems were more modern and more reliable than those of the 125s.

Model development

1975 model year
The rear tyre width was increased to 3.50-18 for better grip.

1976 model year
The road-only SS-175 was identical to the trailbike except for wheel and tyre sizes, styling of seat, tank and lights, and its low-level chrome-plated exhaust on the right-hand side of the machine. The trailbike's rear tyre width was increased again to 4.00-18.

1977 model year
For their last model year there were no significant changes.

Model SS-175 1976 to 1977

Motor
Type	4F
Engine	Single-cylinder air-cooled piston-ported 2-stroke
Bore and stroke	61 x 59.6 mm (2.40 x 2.35 in.)
Capacity	174 cc (10.63 cu. in.)
Power output (DIN)	17(13)/6750 DIN
Compression ratio	10.67:1
Carburettor	Dell'Orto VHB 27 AD
Ignition	Battery-and-coil CDI
Electrical system	12-volt
Starter	Kickstarter
Exhaust system	1 long silencer, low-level, LHS

Transmission
Gearbox	5-speed
Reduction ratios	2.54 (1) 1.76 (2) 1.29 (3) 1.00 (4) 0.81 (5)
Secondary reduction ratio	3.57
Gearchange	Foot change
Final drive	Chain

Cycle parts
Frame	Twin-loop, with oil reservoir in top tube
Front forks	Hydraulically-damped telescopic
Rear suspension	Swinging-arm
Front brake	135 mm drum, via cable and handlebar lever
Rear brake	135 mm drum, via rod and pedal

General data
Dry weight	125 kg
Permissible total weight	295 kg
Wheelbase	1419 mm
Tyre front	3.25-19
Tyre rear	4.00-18
Wheels front	1.85 x 19 wire-spoked
Wheels rear	2.15 x 18 wire-spoked
Top speed	111 km/h
Fuel tank capacity	10.5 litres
Production figures	1976: 1461 machines 1977: 110 machines

Model SX-175 1974 to 1976

Motor
Type	5D
Engine	Single-cylinder air-cooled piston-ported 2-stroke
Bore and stroke	61 x 59.6 mm (2.40 x 2.35 in.)
Capacity	174 cc (10.63 cu. in.)
Power output (DIN)	17(13)/6750 DIN
Compression ratio	10.67:1
Carburettor	Dell'Orto VHB 27 AD
Ignition	Battery-and-coil CDI
Electrical system	12-volt
Starter	Kickstarter
Exhaust system	1 long silencer, high-level, LHS

Transmission
Gearbox	5-speed
Reduction ratios	2.54 (1) 1.76 (2) 1.29 (3) 1.00 (4) 0.81 (5)
Secondary reduction ratio	3.57
Gearchange	Foot change
Final drive	Chain

Cycle parts
Frame	Twin-loop, with oil reservoir in top tube
Front forks	Hydraulically-damped telescopic
Rear suspension	Swinging-arm
Front brake	135 mm drum, via cable and handlebar lever
Rear brake	135 mm drum, via rod and pedal

General data
Dry weight	120 kg
Permissible total weight	295 kg
Wheelbase	1366 mm
Tyre front	3.00-19
Tyre rear	3.25-18, from 1975: 3.50-18 from 1976: 4.00-18
Wheels front	1.60 x 18 wire-spoked from 1976: 1.85 x 18 wire-spoked
Wheels rear	1.85 x 18 wire-spoked from 1976: 2.15 x 18 wire-spoked
Top speed	111 km/h
Fuel tank capacity	10.5 litres
Production figures	1974: 3612 machines 1975: 8500 machines

Model description and technical information: the 250 cc. two-strokes

For fourteen years the good old four-stroke Aermacchi had survived on the US market, but now it had no more tricks left to fight off the ever-stiffer competition from the increasing numbers of Japanese and other European rivals; even off-road, it was no longer up to the mark. In spite of intensive development time had passed it by and the two-strokes now ruled the roost. However good a motorcycle it was, in 1974 it had to give way to the new 250 trailbike that was now coming across the Atlantic. Between 1974 and 1978 over 14,000 two-stroke 250s were sold.

There were two versions, the road-only SS-250 and the on/off-road SX-250 trailbike. Technically, nothing changed in the four years for which the 250s were on sale.

Both were based on exactly the same mechanical components. The frame and cycle parts remained as far as possible identical to those of the 175s; rear suspension units with five-way adjustable spring preload were fitted. The road-only model was naturally fitted with road-pattern tyres, 19-inch front, while the trailbike had a 21-inch front with semi-knobbly tyres; both had 4.00-18 rear tyres and the usual drum brakes, slightly larger in diameter at 5.5 inches (140 mm) than those on the later 125s and 175s. Power output was of the order of 20 horsepower (according to the catalogue the SX-250 gave some 15% more), with a compression ratio of 10.3:1 and ignition was by thyristor-controlled Capacitor Discharge Ignition (CDI). Using the same stroke as the 175s, but with a drastically bored-out cylinder, the engine's stroke-to-bore ratio was an excellent 0,83:1 but this was pushing the limits of the unit's development potential.

The road-only SS-250 was identical to the trailbike except for wheel and tyre sizes, styling of seat, tank and lights, and its low-level chrome-plated exhaust on the right-hand side of the machine. The trailbike featured a high-level matt black-painted exhaust system that curved upwards over the engine/gearbox unit on the left-hand side and then ran horizontally to the rear, with a protective chrome-plated metal wire grille. The SS-250 had a top speed of about 75 mph (121 km/h), while the SX-250 could manage 71 mph (114 km/h); basically not bad, but not good enough either to compete effectively with the faster and cheaper Japanese machinery.

Model development

1976 and 1977 model years
Harley-Davidson made a front disc brake available as an option for the road-only SS-250, the bigger Harleys having been fitted with disc brakes since 1973; models with these brakes were called the SST-250. However, such things were now commonplace and no longer had the sales appeal that they had aroused a few years before. The SX-250 was discontinued at the end of the year.

1978 model year
The SS-250 and SST-250 were thus the last machines from the Aermacchi programme which were imported into the USA; the SX-250 not being on sale in this model year. In June 1978 H-D sold its Italian branch to the Italian firm of Cagiva, which still exists today. The remaining stocks of motorcycles were flogged off at massive discounts on the US market during 1979, as small-displacement two-strokes by then were simply not in demand. Harley-Davidson was thus left once more, for the first time since 1947, with only its large-capacity V-twins.

Production figures

Model	1975	1976	1977	1978
SS-250/SST-250	3000	1416	1416	479
SX-250	11000	3125	558	-

The figures given are total production figures, only a few of which were sold in the US.

1975 Model SS 175 or 250

Model SS-250 1975 to 1978
Model SST-250 1976 to 1978

Motor
Type	SS-250: 9E; SST-250: 7F
Engine	Single-cylinder air-cooled piston-ported 2-stroke
Bore and stroke	72 x 59.6 mm (2.84 x 2.35 in.)
Capacity	243 cc (14.81 cu. in.)
Power output (DIN)	20(15)/7000 DIN
Compression ratio	10.3:1
Carburettor	Dell'Orto PHB 32 AD
Ignition	Flywheel magneto CDI
Electrical system	12-volt
Starter	Kickstarter
Exhaust system	1 long silencer LHS, low-level

Transmission
Gearbox	5-speed
Reduction ratios	2.54 (1)
	1.76 (2)
	1.29 (3)
	1.00 (4)
	0.81 (5)
Secondary reduction ratio	3.57
Gearchange	Foot change
Final drive	Chain

Cycle parts
Frame	Twin-loop, full cradle
Front forks	Hydraulically-damped telescopic
Rear suspension	Swinging-arm
Front brake	SS-250: 140 mm drum, via cable and handlebar lever
	SST-250: Hand-operated hydraulic disc 227 mm diameter
Rear brake	140 mm drum, via rod and pedal

General data
Dry weight	125 kg
Permissible total weight	295 kg
Wheelbase	1419 mm
Tyre front	3.25-19
Tyre rear	4.00-18
Wheels front	1.85 x 19 wire-spoked
Wheels rear	2.15 x 18 wire-spoked
Top speed	121 km/h
Fuel tank capacity	10.5 litres
Price	**SS-250:** 1975: $1.130

Model SX-250 1975 to 1977

Motor
Type	6D
Engine	Single-cylinder air-cooled piston-ported 2-stroke
Bore and stroke	72 x 59.6 mm (2.84 x 2.35 in.)
Capacity	243 cc (14.81 cu. in.)
Power output (DIN)	20(15)/7000 DIN
Compression ratio	10.3:1
Carburettor	Dell'Orto PHB 32 AD
Ignition	Flywheel magneto CDI
Electrical system	12-volt
Starter	Kickstarter
Exhaust system	1 long silencer LHS, high-level

Transmission
Gearbox	5-speed
Reduction ratios	2.54 (1)
	1.76 (2)
	1.29 (3)
	1.00 (4)
	0.81 (5)
Secondary reduction ratio	3.57
Gearchange	Foot change
Final drive	Chain

Cycle parts
Frame	Twin-loop, full cradle
Front forks	Hydraulically-damped telescopic
Rear suspension	Swinging-arm
Front brake	140 mm drum, via cable and handlebar lever
Rear brake	140 mm drum, via rod and pedal

General data
Dry weight	130 kg
Permissible total weight	295 kg
Wheelbase	1366 mm
Tyre front	3.00-21
Tyre rear	4.00-18
Wheels front	1.60 x 21 wire-spoked
Wheels rear	2.15 x 18 wire-spoked
Top speed	114 km/h
Fuel tank capacity	10.5 litres
Price	1974: $1.142
	1975: $1.142

Sportster.
The XL models
1957 to 1985

Model description and technical information

As successor to the relatively short-lived Model K range, Harley-Davidson had high hopes for their new small V-twin with its snappy name of Sportster – cast loud and clear into the primary chaincase. The sales figures of the Model Ks – which never had proper names – were always shaded by those of their bigger brethren and had not grown as expected in the face of the competition from imported British motorcycles. So in the middle of the 1950s, H-D decided to develop a smaller V-twin engine of 54 cu. in. (883 cc.) capacity with overhead valves. As with the Models KH and KHK, most round up the Sportster's capacity to class it as a Fifty-Five.

The Sportster that arrived on the market in the autumn of 1956 was thus state of the art. In that first year only the standard Model XL was available at a price tag of just over 1100 dollars. It was thus only 64 greenbacks cheaper than the basic Model FL – but one must remember that 'base' models just didn't exist as far as Harley was concerned; every new machine had to be ordered with one of the Groups of accessories – and 170 dollars more expensive than the Model KH (or, what is more to the point, a bargain 100 bucks more than the super-sports Model KHK) the much less ponderous motorcycle that preceded it. With a quite low (by Harley standards) kerb weight of 496 lb (225 kg) the Sportster got off the mark quickly enough and went on to a top speed of over 99 mph (160 km/h), fast enough to keep up with the British sports models.

To underline its high speed potential, an H-D 100 emblem was fixed to the tank and to the optional panniers (when fitted).

The fact that engine and (four-speed) gearbox were of unit construction linked the new model firmly to the old Model Ks, but particularly to their racing derivatives which were sold from 1958 to 1966 in parallel to the Sportster and carried on winning all the most prestigious US races, road and flat-track, up to 1969. Even the shorter Linkert

Model range

Fifty-Fours (883 cc.) 1957 to 1971
Model XL	Sportster		(1957-1959)
Model XLC	Sportster Competition		(1958)
Model XLH	Sportster H	3A *	(1958-1971)
Model XLCH	Sportster Competition H	4A *	(1958-1971)

* Model designations, 1970-on

Sixty-Ones (997 cc.) 1972 to 1985
Model XLH	Sportster H	3A [1]	(1972-1985)
Model XLCH	Sportster CH	4A	(1972-1979)
Model XLH Liberty	Sportster H	3A	(1976)
Model XLCH Liberty	Sportster CH	4A	(1976)
Model XLH Anniversary Edition	Sportster H	3A	(1978)
Model XLT	Sportster Touring	2G	(1977-1978)
Model XLCR	Café Racer	7F	(1977-1978)
Model XLS	Roadster	4E [2]	(1979-1985)
Model XR-1000	XR-1000	CD	(1983-1984)
Model XLX	XLX-61	CC	(1983-1985)

[1] CA, from 1981-on
[2] CB, from 1981-on

1957 Model XL: Birch White with Black tank panels and big CycleRay headlamp

1958 Model XL in Calypso Red with Birch White tank panels

1959 Model XLH: Hi-Fi Turquoise with Birch White tank panels, nacelle headlamp and valanced mudguards

carburettor with the new larger air cleaner contributed to the impression of compactness. The slim frame was naturally developed from that of the Models K and KHK that had been introduced in 1956 and proved so well-suited to its new powerplant that it was even retained, in slightly lightened form, for the Sixty-One Sportster of 1972-on.

The cylinder heads and cylinders, on the other hand, in contrast to those of the preceding models, were once more of cast iron, the heads resembling those of the Shovelhead engine which would power the Harley Big Twins from 1966-on; the combustion chambers, however were hemi-spherical for the first time.

The spark plugs, as with all their kind of that period, could be dismantled for cleaning. The proven 6-volt electrical system had a new generator and the Jubilee Trumpet horn was fitted. The traditional Harley pushbike pedal-shaped kickstart was beginning to look a little old fashioned but would have to wait until 1963 before being replaced by something more modern.

18-inch wheels were standard equipment and the customer could specify a solo seat or twinseat for every new XL. As with the Model Ks, control layout was derived from flat track racing; handlebar lever-operated clutch, right foot gearchange and left foot rear brake. The tank held 3.6 Imp. gallons (16.5 litres) and the badge for this year was a round Lucite disc; H-D's first use of plastic for badges. Standard colour schemes this year were: Pepper Red with Black tank panels and Pepper Red mudguards, Black with Pepper Red tank panels and Black mudguards, Skyline Blue with Birch White tank panels and Blue mudguards, Birch White with Black tank panels and Birch White mudguards, Police Silver, Metallic Midnight Blue (tank and mudguards), with Birch White tank panels. At no extra cost were: Birch White, solid or black panels and matching mudguards, Metallic Midnight Blue (tank and mudguards), with Birch White tank panels.

Model development

1958 model year
For the new model year, outwardly, not much changed particularly. However, three new models were added to the range. Additional to the standard XL was a hotter model with higher compression and more power; Harley-Davidson supplied both models in competition versions. The range thus looked as follows:

XL   Basic model with standard compression ratio.
XLH Basic model with higher compression ratio (H stands for Hot, or High Compression).
XLC Competition model with standard compression ratio (C stands for Competition).

1959 Model XLCH: Special C-model badge on Model 165/Hummer tank, silenced high-level exhaust and Eyebrow headlamp

XLCH Competition model with higher compression ratio (CH stands for Competition Hot or Competition/High Compression).

Note that while these are the usual interpretations applied to the XL-series model designations, they were never confirmed by the factory in 1958 and have been hotly disputed by Harley experts and historians ever since. Some believe that the C stands for California, on the basis that it was Californian dealers who collectively placed sufficient orders for H-D to go ahead with the competition versions; others deny this.

Just why anyone would want a competition model with low compression, at, presumably, a price very similar to that of the hotter versions, remains a mystery right up to the present day. Since probably no-one actually ordered one of the XLC models it is unlikely that any were sold or made and the option was discontinued for 1959.

The competition models differed from the standard versions by having no lights or speedometer, straight-through exhaust pipes without silencers, a smaller (1.8 Imp. gallons/8.3 litres) tank – derived from the Model 165 and Hummer range, via the KR models – and a solo seat. Also the rear mudguard was simply cut off above the taillamp location. Both were fitted with magneto ignition and shorter gearing. Furthermore the competition models got their own Checkered Flag tank sticker, embedded into a V-emblem. The XLCH in particular got in addition even-larger and cleaner intake ports and bigger valves, aluminium tappets and pistons with domed crowns; as a result, it was rumoured to be capable of 130 mph (210 km/h).

Two-tone paint schemes were standard for this year, in the following combinations: Calypso Red/Birch White, Black/Birch White, Skyline Blue/Birch White and Sabre Gray Metallic/Birch white. The first colour was applied to the tank top and mudguards, while the second was applied to the tank bottom. Any solid standard colour could be ordered at no extra cost for those who wanted more choice.

1959 model year
The last model year for the standard Sportster brought small changes like a Nylon primary chain tensioner. The XL and XLH got headlamps set in chrome-plated pressed-steel nacelles for a more modern appearance and valanced front mudguards. A two-into-two exhaust system and a version of the Buddy Seat (a big seat for two slim people) were available as extra-cost options for all models, while a trip odometer and new Arrow-Flite (red ball with chrome arrow) tank badges (for the XL and XLH only – the XLCH kept its special tank badging) were likewise amongst the improvements worth mentioning. The XLC had already disappeared during the 1958 model year. That year's colour schemes were: Calypso Red, Black, Skyline Blue, Hi-Fi Red or Hi-Fi Turquoise tank and

1963 Model XLH in Tango Red, but with staggered 'shorty duals' exhaust system

1964 Model XLH: Hi-Fi Blue with Birch White tank panels

mudguards with Birch White tank panels. Any solid standard colour without tank panels was available at no extra cost.

The XLH and XLCH got hotter intake and exhaust cams. The XLCH could now be registered for the road, but then lost its off-road straight-through exhaust pipes and got instead a high-level system terminating in a single silencer on the right-hand side of the motorcycle; it also got full lighting, albeit minimal, with the Eyebrow headlamp cover that is still a Sportster trademark today, and a 19-inch front wheel.

1960 model year
After the deletion of the standard XL only the XLH and the XLCH remained in the catalogue. Harley-Davidson maintained this policy of two Sportster models until 1977, when for the first time additional models were once more introduced. The rear suspension units had revised and carefully-matched damping characteristics and standard colours available were Black or Skyline Blue with Birch White tank panels, while Hi-Fi Red, Hi-Fi Green or Hi-Fi Blue with Birch White tank panels were available at extra cost; again, any solid standard colour without tank panels was available at no extra cost.

1961 model year
The only new features for this year were stronger seat springs and new (oval) tank badges – for both XLH and XLCH. Standard colours available were unchanged apart from the substitution of Pepper Red for Skyline Blue.

1962 model year
For the XLH, the headlamp nacelle/front fork upper cover was now made of polished aluminium alloy rather than chrome-plated pressed-steel. The oval tank badges introduced the previous year were slightly larger. Standard colours available were again unchanged apart from the substitution of Tango Red for Pepper Red and Hi-Fi Purple for Hi-Fi Green.

1963 model year
Third gear ratio was revised on both models, the kickstart pedal was redesigned and the tank badges were changed to the new design based on H-D's bar-and-shield logo, and fitted to all Harleys of this year. Standard colours available were again unchanged apart from the addition of Horizon Metallic Blue. The XLCH also got a rubber-mounted headlamp and an improved magneto with an ignition switch, and the headlamp dipswitch was now mounted directly beside the throttle twistgrip. It also got another feature that was to become a famous Sportster trademark; the staggered 'shorty duals' exhaust system, with each cylinder's exhaust gases venting through a separate chrome-plated downpipe and silencer, both on the right-hand side of the machine.

1964 model year
Both models were fitted with better front brakes with full-width cast aluminium-alloy brake drums. Also the tappet guides were now of aluminium. Standard colours available were Black or Fiesta Red with Birch

1964 Model XLH

1965 Model XLH, but with staggered 'shorty duals' exhaust system

1968 Model XLH

1968 Model XLH, much-modified; the primary case is a KR flat-tracker look-alike part – a popular accessory for Sportsters

1969 Model XLH, with redesigned headlamp nacelle

1971 Model XLH with boat tail-shaped fibre-glass seat unit

White tank panels, while Hi-Fi Red and Hi-Fi Blue with Birch White tank panels were available at extra cost.

1965 model year
The first big change. With the Electra-Glide, the Sportsters were both fitted with a 12-volt electrical system, but the XLH's starter motor took a little while longer. Two 6-volt batteries were connected in series to become a 12-volt unit (there wasn't room in a Sportster frame for a battery the size of the Electra-Glide's), the generator was upgraded to suit the demands of the new system and a new High Fidelity horn was fitted. For the XLH, an automatic ignition advance/retard mechanism was fitted in the timer unit (contact breaker housing), the fuel tank was now smaller, holding 3.1 Imp. gallons (14 litres) and the rear suspension units now had adjustable spring preload.

Both models got racing-style ball-ended handlebar control levers. Standard colours were Black or Holiday Red with Birch White tank panels, while Hi-Fi Red or Hi-Fi Blue with Birch White tank panels were available at extra cost.

1966 model year
The engines featured redesigned intake ports and bigger valves, which with the new Tillotson diaphragm carburettors (which had to be protected from the heat of the engine by a heat insulator mounted between the intake manifold and cylinder heads) and the soon-to-be-famous new P cams produced a 15% power increase. The air cleaner was fitted with a round cover and so was christened the Ham Can by owners. Tank badges? New again – naturally . . . a new flattened-diamond design for all Harleys from 1966 onwards.

1967 model year
The Sportster XLH's electric starter finally arrived, as an option, powered by a new 32 Ah battery. The XLCH still only had a kickstart. On both models, the rear suspension units were revised.

1968 model year
For the XLH the kickstart was discontinued. Anyone who desperately wanted such a thing had to choose an XLCH. Both models received a new front fork with longer spring travel and revised damping, the primary drive cover was redesigned and the XLCH's Peanut tank was available as an option on the XLH.

1969 model year
The new standard exhaust system arrived with a crossover balance tube between the two downpipes. In addition a new taillamp and, as every year, new colour schemes.

1970 model year
The XLCH's magneto ignition was replaced by the XLH's battery-and-coil ignition system. The XLH got the Eyebrow headlamp of the XLCH. As an option a

Various …

… Sportsters …

... of between ...

... 1967 and 1971

fibre-glass boat tail-shaped seat unit and rear mudguard was made available.

1971 model year
The new wet clutch now had a single and much-heavier spring. The contact breaker points and condenser moved behind an inspection plate set in the timing cover, the previous timer unit (contact breaker housing) now redundant and deleted. Thus the ignition timing was derived much more directly from the camshaft, with less risk of inaccuracy due to the backlash inherent in a long gear train.

1972 model year
The big news for this year was the introduction of the bored-out 60.85 cu. in./997 cc. engine, making the Sportster Harley's first genuine Sixty-One, and producing a power increase from 58 to 61 horsepower and a flatter torque curve, a performance assisted by the new Bendix-Zenith carburettor. The seat was thinner with less padding and the oil tank was relocated closer to the chain. New AMF/Harley-Davidson tank graphics decorated all Harleys for this year.

1973 model year
For the first time in the history of the Sportster and one year after the Electra-Glide, a front disc brake was fitted. In addition, and again for the first time, H-D started fitting all their V-twins with front forks made by the Japanese manufacturer, Kayaba; these were not only cost-effective, but their quality was of a high standard as well.

1974 model year
The only technical innovation worth mentioning was the self-closing throttle twistgrip with a return spring – a novelty for Harley-Davidson.

1975 model year
The rear brake pedal was changed over to the right-hand side and the gearchange lever moved to the left-hand side as a result of Federal legislation. Front forks were now by Showa. A particularly sweeping S-shaped graphic decorated the tank for this year.

1976 model year
While the two-model policy remained in force, two special models appeared this year, inspired by the celebrations of the 200th anniversary of the American Declaration of Independence; the XLH and XLCH Liberty Editions with a special metal-flake black paint finish and a huge Liberty Edition graphic featuring the American Eagle taking up most of the top surface of the tank.

1977 model year
The XLH and XLCH were joined this year by two further additions to the Sportster range; neither lasted as long as their elder sisters. The Sportster Touring, or Model XLT, was fitted for its intended purpose with windscreen and panniers. Unfortunately for H-D, not too many people wanted to go touring on a Sportster and the XLT was discontinued in 1978 through lack of sales.

The Sportster Café Racer, or Model XLCR, was a complete departure for H-D. Another creation of Willie G. Davidson himself, it was sporty in the European idiom, rather than reflecting the US preference for flat track racing. It had its own frame, developed from that of the standard models', special black-painted cast aluminium-alloy wheels with polished rims and spoke edges, a siamesed exhaust system, twin front disc brakes, a small handlebar fairing and a tank and seat (with a fibre-glass tail section) that were styled as one unit to fit the image conveyed by the name. An all-black paint finish rounded off the whole. Obviously, it was a bit too radical for American tastes to do well and was sold only in 1977 and 1978.

All models received the Japanese Keihin carburettors that are still common today.

1978 model year
For 1978 H-D treated the Sportsters to twin front disc brakes, as they did the Big Twins. The XLCR already had twin front discs, but it, like the XLT, had not sold at all well and was about to be discontinued.

The Models XLH and XLCH were changed over to the new electronic ignition for more accurate and maintenance-free ignition timing and better combustion, received a siamesed exhaust system, the frame of the XLCR and single rear disc brakes finally replaced the old drums. Cast aluminium-alloy wheels, replacing the previous wire-spoked wheels, were available as optional extras.

In time for Harley-Davidson's 75th birthday, the Company brought out the XLH Anniversary Edition, with gold-painted cast aluminium-alloy wheels and the appropriate emblems as well as a special paint finish in Midnight Black with gold trim.

1979 model year
For 1979 came a new model in the shape of the Model XLS; the Sportster range now stood at three models. The XLS, which would later be christened the Roadster, was inspired by the success of the FX models, particularly the Low Rider. With its extended front fork, 16-inch rear wheel, siamesed exhaust, drag bars on 3.5-inch (90 mm) risers, sissy bar with leather pouch, highway pegs, two-piece seat and chromed rear sprocket, the XLS stood ready to awaken or fulfil the Peter Fonda fantasies in any owner.

The rear disc brake's master cylinder needing to be fitted precisely where the kickstart shaft emerged from the engine/gearbox side cover, the kickstart was now deleted, whether as standard equipment or as a special-order item. This was also the last year of the Model XLCH.

1980 model year
Nothing new apart from the Hugger option, which, with its thinner seat padding and shortened rear

1975 Model XLH

1976 Model XLH

suspension units was conceived to cater for shorter riders by lowering the seat height. Obviously, there were new tank graphics for all models . . .

The Sportster range was back to two models; the XLH (Sportster) and the XLS (Roadster).

1981 model year
In the year of the buy-back from AMF both models were available with different fuel tanks, and with wire-spoked or cast aluminium-alloy wheels; the rear wheels could be of 16-inch diameter or 18-inch, as the customer wished. The XLH received shorter front forks, the XLS got new Buckhorn handlebars and the new Shorty exhaust system.

1982 model year
This year saw a new lighter frame with battery (XLS only) and oil tank (both) relocated again, and a bigger rear brake disc. Thicker cylinder head gaskets reduced compression ratios to 8.0:1 so that the engines could use the lead-free fuel that was about to be introduced,

1983 model year
The XLH came up with a new seat; its tank capacity was now 2.7 Imp. gallons (12.5 litres). The exhaust system was less restrictive to allow the engine to breathe more freely. Both models now had a vacuum-controlled ignition advance/retard.

The XLS inherited the tank of the FXRS and with it the corresponding instrument panel. The speedometer, however, remained in the traditional location for Sportsters; mounted on the front fork top yoke, between the handlebar mounting clamps.

Two further models appeared. The XLX was conceived as an economy model; it didn't even warrant its own sales designation. All components normally chrome-plated, even the exhaust, were largely painted in matt black. The solo seat was suitable only for riders on their own or for those with very understanding and intimate partners. The Peanut tank was good enough only for a very restricted range and even the tachometer was missing.

The XR-1000 was something completely different: Admittedly fitted with the same frame as the XLX, the XR could offer for road use racing-derived cast aluminium-alloy crossflow heads with intake ports on the right and exhaust on the left, polished and bored by Jerry Branch, XR pistons giving a 9.0:1 compression ratio, light aluminium pushrods, valve clearances adjusted by eccentric rocker shafts, iron cylinders (shortened half an inch to make room for the Branch heads), vacuum-advance V-Fire III electronic ignition, twin 1.4 inch (36 mm) Dell'Orto carburettors (incidentally, the only time the Italian company supplied equipment for a 'genuine' Harley), oiled-felt air filters, separate high-level exhausts on the left-hand side, satin-textured cases, 9-spoke cast aluminium-alloy wheels, triple disc brakes. The result: 71 horsepower @ 5,600 rpm and 115 mph (185 km/h).

Some idea of the problems facing H-D at this time when exporting motorcycles from the US can be gained from the fact that in 1983 they had to reduce the Sportster's rated power output so that models producing 50 horsepower were ready for German customers to satisfy the requirements of their insurance companies.

1984 model year
For this year all Sportsters, like the Super Glide models, changed to a single, much larger-diameter and drilled front brake disc. The feeling of 'brake now – stop later' experienced with Harley brakes in the past was largely obviated by these discs, which were much larger and gave much better braking performance than the previous twin discs. In addition there was a new electric starter and a modified diaphragm-spring clutch for the standard models.

The XR-1000 racer also got improved brakes with two larger front brake discs. Its only other new feature was an optional paint scheme in Harley-Davidson's racing colours of Orange and Black. At the end of the 1984 model year they were discontinued.

1985 model year
The only models remaining were the economy XLX, the Standard XLH and the Roadster XLS; there were no technical modifications. The XR-1000 was also listed in the 1985 catalogue, but only to sell off unsold 1983 and 1984 models, since according to Harley-Davidson no '85 frame number exists. For private owners the trade offered tuning kits, with which the XR-1000 could be pumped up to 95 horsepower (70 KW).

Various …

… views of the …

... legendary XR-1000

The XLCR, better known ...

... as the Café Racer

Various ...

...1000 cc. ...

... or Sixty-One ...

... Sportsters

1980 Model XLH

1982 Model XLS

1983 Model XLX

The following Accessory Groups were available for these models:

Group	1957	1958	1959	1960	1961	1962	1963	1964	1965	1966	1967	1968	1969
Deluxe #SP1	x	x	x	x	x	x	x	x					
Deluxe #SP1A			x	x	x								
Deluxe #SP1B			x		x								
Deluxe #SP2		x	x	x	x	x	x	x					
Deluxe #SP2A				x	x	x	x	x					
Deluxe #SP2B					x								
Deluxe #SP3		x	x	x									
Deluxe #SP3H						x	x	x					
Deluxe #SPH1									x	x			
Deluxe #SPH2									x	x			
Deluxe #SPH3									x	x			
Deluxe #SPH4										x			
Deluxe #SPH5										x			
Deluxe #SPH6										x			
Deluxe #SPH7										x			
Deluxe #SPCH1									x	x			
Deluxe #SPCH2									x	x			
Deluxe #SPCH3										x			
Deluxe #SPCH4										x			
Deluxe (H+CH)											x	x	x
Highway Cruiser (H)											x	x	x

Catalogue prices ($ US)

Model	1957	1958	1959	1960	1961	1962	1963	1964	1965	1966	1967	1968	1969
XL	1103	1155	1175	-	-	-	-	-	-	-	-	-	-
XLC	-	N./Av.	-	-	-	-	-	-	-	-	-	-	-
XLH	-	N./Av.	1200	1225	1250	1250	1270	1295	1396	1415	1650	1765	1765
XLCH	-	N./Av.	1285	1310	1335	1335	1355	1360	1400	1411	1600	1698	1698

Production figures

Model	1957	1958	1959	1960	1961	1962	1963	1964	1965	1966	1967	1968
XL	1983	579	42	-	-	-	-	-	-	-	-	-
XLC	-	?	-	-	-	-	-	-	-	-	-	-
XLH	-	711	947	-	-	-	432	810	955	900	2000	1975
XLCH	-	239	1059	-	-	-	1001	1950	2815	3900	2500	4900
alle X *	-	-	-	2765	2014	1998	-	-	-	-	-	-

* Unfortunately more accurate figures aren't available.

Production figures

Model	1969	1970	1971
XLH	2700	3033	3950
XLCH	5100	5527	6825

Production figures

Model	1972	1973	1974	1975	1976	1977	1978	1979	1980	1981
XLH	7500	9875	13295	13515	12844	12742	11271*	6525	11841	8442
XLCH	10650	10825	10535	5895	5238	4074	2758	141	-	-
XLCR	-	-	-	-	-	1923	1201	9	-	-
XLS	-	-	-	-	-	-	2	5123	2926	1660
XLT	-	-	-	-	-	1099	6	-	-	-

* Plus 2323 XLH Anniversary Edition models (on the occasion of Harley-Davidson's 75th birthday).

Production figures

Model	1980	1981	1982	1983	1984	1985
XLH	11841	8442	5015	2230	4442	4074
XLS	2926	1660	1261	1616	1135	615
XLX	-	-	-	4892	4284	1824
XR-1000	-	-	-	1018	759	-

Model XL 1957 to 1959

Motor
Engine	Air cooled 45° ohv V-twin
Bore and stroke	3 x 3¹³⁄₁₆ (3 x 3.81) in.
Bore and stroke	76.2 x 96.84 mm
Capacity	53.90 cu. in. (883 cc)
Power output (SAE)	42(31)/5500 rpm
Compression ratio	7.5:1
Carburettor	Linkert M-Type
Ignition	Battery-and-coil
Electrical system	6-volt
Starter	Kickstarter
Exhaust system	2 into 1, 1R/H silencer optional 2 into 2, 2 R/H silencers

Transmission
Gearbox	4-speed
Reduction ratios	2.52 (1)
	1.83 (2)
	1.38 (3)
	1.00 (4)
Secondary reduction ratio	2.43 (51:21 teeth)
Gearchange	Foot change
Final drive	Chain

Cycle parts
Frame	Twin-loop frame
Front forks	Telescopic, hydraulically-damped
Rear suspension	Swinging-arm
Front brake	Drum, via cable and handlebar lever
Rear brake	Drum, via rod and pedal

General data
Dry weight	225 kg
Permissible total weight	400 kg
Wheelbase	1435 mm
Tyres – front and rear	3.50S18
Wheels front and rear	2.15 x 18 wire-spoked
Top speed	162 km/h
Fuel tank capacity	16.5 litres

Model XLC 1958

Motor
Engine	Air cooled 45° ohv V-twin
Bore and stroke	3 x 3¹³⁄₁₆ (3 x 3.81) in.
Bore and stroke	76.2 x 96.84 mm
Capacity	53.90 cu. in. (883 cc)
Power output (SAE)	approx. 50(37)/6000 rpm
Compression ratio	7.5:1
Carburettor	Linkert M-Type
Ignition	Magneto
Electrical system	6-volt
Starter	Kickstarter
Exhaust system	2 into 2, Dragpipes None silencer

Transmission
Gearbox	4-speed
Reduction ratios	2.52 (1)
	1.83 (2)
	1.38 (3)
	1.00 (4)
Secondary reduction ratio	2.55 (51:20 teeth)
Gearchange	Foot change
Final drive	Chain

Cycle parts
Frame	Twin-loop frame
Front forks	Telescopic, hydraulically-damped
Rear suspension	Swinging-arm
Front brake	Drum, via cable and handlebar lever
Rear brake	Drum, via rod and pedal

General data
Dry weight	195 kg
Permissible total weight	400 kg
Wheelbase	1448 mm
Tyres – front and rear	3.50H18
Wheels front and rear	2.15 x 18 wire-spoked
Top speed	approx. 180 km/h
Fuel tank capacity	8.3 litres

Model XLH 1958 to 1965
Model XLCH 1958

Motor
Engine	Air cooled 45° ohv V-twin
Bore and stroke	3 x 3¹³⁄₁₆ (3 x 3.81) in.
Bore and stroke	76.2 x 96.84 mm
Capacity	53.90 cu. in. (883 cc)
Power output (SAE)	55(40)/6300 rpm
Compression ratio	9.0:1
Carburettor	Linkert M-Type
Ignition	XLH: Battery-and-coil
	XLCH: Magneto
Electrical system	6-volt (1958-64)
	12-volt (1965)
Starter	Kickstarter
Exhaust system	XLH: 2 into 1, 1R/H silencer optional 2 into 2, 2 R/H silencers
	XLCH: 2 into 1, 1 silencer RHS high-level

Transmission
Gearbox	4-speed
Reduction ratios	2.52 (1)
	1.83 (2)
	1.38 (3) 1.323, XLH '63-on
	1.00 (4)
Secondary reduction ratio	2.43 (51:21 teeth) XLH
	2.55 (51:20 teeth) XLCH
Gearchange	Foot change
Final drive	Chain

Cycle parts
Frame	Twin-loop frame
Front forks	Telescopic, hydraulically-damped
Rear suspension	Swinging-arm
Front brake	Drum, via cable and handlebar lever
Rear brake	Drum, via rod and pedal

General data
Dry weight	230 kg
Permissible total weight	400 kg
Wheelbase	1435 mm (XLH)
	1448 mm (XLCH)
Tyres – front and rear	3.50S18
Wheels – front and rear	2.15 x 18 wire-spoked
Top speed	170 km/h
Fuel tank capacity	16.5 litres (to 1964)
	14 litres (1965)
	8.3 litres (XLCH)

Model XLCH 1959

Motor
Engine	Air cooled 45° ohv V-twin
Bore and stroke	3 x 3¹³⁄₁₆ (3 x 3.81) in.
Bore and stroke	76.2 x 96.84 mm
Capacity	53.90 cu. in. (883 cc)
Power output (SAE)	approx 65(48)/6500 rpm
Compression ratio	9.0:1
Carburettor	Linkert M-Type
Ignition	Magneto
Electrical system	6-volt
Starter	Kickstarter
Exhaust system	2 into 2, Dragpipes None silencer

Transmission
Gearbox	4-speed
Reduction ratios	2.52 (1)
	1.83 (2)
	1.38 (3)
	1.00 (4)
Secondary reduction ratio	2.55 (51:20 teeth)
Gearchange	Foot change
Final drive	Chain

Cycle parts
Frame	Twin-loop frame
Front forks	Telescopic, hydraulically-damped
Rear suspension	Swinging-arm
Front brake	Drum, via cable and handlebar lever
Rear brake	Drum, via rod and pedal

General data
Dry weight	195 kg
Permissible total weight	400 kg
Wheelbase	1448 mm
Tyre front	3.25V19 optional 3.50V19
Tyre rear	4.00V18
Wheels front	2.15 x 19 wire-spoked
Wheels rear	2.15 x 18 wire-spoked
Top speed	approx. 210 km/h
Fuel tank capacity	8.3 litres

Model XLCH 1960 to 1965

Motor
Engine	Air cooled 45° ohv V-twin
Bore and stroke	3 x 3$\frac{13}{16}$ (3 x 3.81) in.
Bore and stroke	76.2 x 96.84 mm
Capacity	53.90 cu. in. (883 cc)
Power output (SAE)	55(40)/6300 rpm
Compression ratio	9.0:1
Carburettor	Linkert M-Type
Ignition	Magneto ignition
Electrical system	6-volt (1960-64)
	12-volt (1965)
Starter	Kickstarter
Exhaust system	2 into 1, 1 silencer RHS high-level
	from 1963: 2 into 2, 2 silencers RHS low

Transmission
Gearbox	4-speed
Reduction ratios	2.52 (1)
	1.83 (2)
	1.38 (3) 1.323, '63-on
	1.00 (4)
Secondary reduction ratio	2.55 (51:20 teeth)
Gearchange	Foot change
Final drive	Chain

Cycle parts
Frame	Twin-loop frame
Front forks	Telescopic, hydraulically-damped
Rear suspension	Swinging-arm
Front brake	Drum, via cable and handlebar lever
Rear brake	Drum, via rod and pedal

General data
Dry weight	200 kg
Permissible total weight	400 kg
Wheelbase	1448 mm
Tyre front	3.25S19 optional 3.50S19
Tyre rear	4.00S18
Wheels front	2.15 x 19 wire-spoked
Wheels rear	2.15 x 18 wire-spoked
Top speed	170 km/h
Fuel tank capacity	8.3 litres

Model XLH 1966 to 1971

Motor
Engine	Air cooled 45° ohv V-twin
Bore and stroke	3 x 3$\frac{13}{16}$ (3 x 3.81) in.
Bore and stroke	76.2 x 96.84 mm
Capacity	53.90 cu. in. (883 cc)
Power output (SAE)	58(43)/6800 rpm
Compression ratio	9.0:1
Carburettor	Tillotson 27162-66C
Ignition	Battery-and-coil
Electrical system	12-volt
Starter	1966: Kickstarter
	from 1967: optional Electric starter
Exhaust system	2 into 1, 1 R/H silencer

Transmission
Gearbox	4-speed
Reduction ratios	2.52 (1)
	1.83 (2)
	1.38 (3)
	1.00 (4)
Secondary reduction ratio	1966-69: 2.43 (51:21 teeth) optional
	1967-71: 2.55 (51:20 teeth)
Gearchange	Foot change
Final drive	Chain

Cycle parts
Frame	Twin-loop frame
Front forks	Telescopic, hydraulically-damped
Rear suspension	Swinging-arm
Front brake	203 mm drum, via cable and handlebar lever
Rear brake	203 mm drum, via rod and pedal

General data
Dry weight	240 kg
Permissible total weight	400 kg
Wheelbase	1966: 1435 mm
	1967-71: 1486 mm
Tyre front	3.25S19 optional 3.50S19 (1966-69)
	3.75S19 (1970-71)
Tyre rear	4.00S18 (1966-69
	4.25S18 (1970-71)
Wheels front	2.15 x 19 wire-spoked
Wheels rear	2.15 x 18 wire-spoked
Top speed	175 km/h
Fuel tank capacity	14 litres

Model XLCH 1966 to 1971

Motor

Engine	Air cooled 45° ohv V-twin
Bore and stroke	3 x 3¹³⁄₁₆ (3 x 3.81) in.
Bore and stroke	76.2 x 96.84 mm
Capacity	53.90 cu. in. (883 cc)
Power output (SAE)	58(43)/6800 rpm
Compression ratio	9.0:1
Carburettor	Tillotson 27162-66C
Ignition	Magneto (1966-69)
	Battery-and-coil (1970-71)
Electrical system	12-volt
Starter	Kickstarter
Exhaust system	2 into 1, 1 R/H silencer

Transmission

Gearbox	4-speed
Reduction ratios	2.52 (1)
	1.83 (2)
	1.38 (3)
	1.00 (4)
Secondary reduction ratio	1966-69: 2.55 (51:20 teeth) optional
	1967-69: 2.43 (51:21 teeth)
	1970-71: 2.68 (51:19 teeth)
Gearchange	Foot change
Final drive	Chain

Cycle parts

Frame	Twin-loop frame
Front forks	Telescopic, hydraulically-damped
Rear suspension	Swinging-arm
Front brake	203 mm drum, via cable and handlebar lever
Rear brake	203 mm drum, via rod and pedal

General data

Dry weight	205 kg
Permissible total weight	400 kg
Wheelbase	1448 mm
Tyre front	3.25H19 optional 3.50H19 (1966-69)
	3.75H19 (1970-71)
Tyre rear	4.00H18 (1966-69)
	4.25H18 (1970-71)
Wheels front	2.15 x 19 wire-spoked
Wheels rear	2.15 x 18 wire-spoked
Top speed	196 km/h
Fuel tank capacity	8.5 litres

Model XLH 1972
Model XLCH 1972

Motor

Engine	Air cooled 45° ohv V-twin
Bore and stroke	3³⁄₁₆ x 3¹³⁄₁₆ (3.19 x 3.81) in.
Bore and stroke	80.96 x 96.84 mm
Capacity	60.85 cu. in. (997 cc)
Power output (SAE)	61(45)/6200 rpm
Torque (Nm)	43/(58)/4000 rpm
Compression ratio	9.0:1
Carburettor	Bendix-Zenith 16P12
Ignition	Battery-and-coil
Electrical system	12-volt
Starter	XLH: Electric starter
	XLCH: Kickstarter
Exhaust system	2 into 1 (Siam), 1 silencer RHS
	2 into 2, 2 Standard-silencers RHS

Transmission

Gearbox	4-speed
Reduction ratios	2.52 (1)
	1.83 (2)
	1.38 (3)
	1.00 (4)
Secondary reduction ratio	2.55 (51:20 teeth)
Gearchange	Foot change
Final drive	Chain

Cycle parts

Frame	Twin-loop frame
Front forks	Telescopic, hydraulically-damped
Rear suspension	Swinging-arm
Front brake	203 mm drum, via cable and handlebar lever
Rear brake	203 mm drum, via rod and pedal

General data

Dry weight	238 kg (XLH)
	215 kg (XLCH)
Permissible total weight	408 kg
Wheelbase	1486 mm
Tyre front	3.75S19
Tyre rear	4.25S18
Wheels front	2.15 x 19 wire-spoked optional
	2.50 x 19 wire-spoked
Wheels rear	2.50 x 18 wire-spoked
Top speed	165 km/h
Fuel tank capacity	8.5 litres optional approx. 15 litres

Model XLH 1973 to 1974
Model XLCH 1973 to 1974

Motor

Engine	Air cooled 45° ohv V-twin
Bore and stroke	3³⁄₁₆ x 3¹³⁄₁₆ (3.19 x 3.81) in.
Bore and stroke	80.96 x 96.84 mm
Capacity	60.85 cu. in. (997 cc)
Power output (SAE)	61(45)/6200 rpm
Torque (Nm)	43/(58)/4000 rpm
Compression ratio	9.0:1
Carburettor	Bendix-Zenith 16P12
Ignition	Battery-and-coil
Electrical system	12-volt
Starter	XLH: Electric starter
	XLCH: Kickstarter
Exhaust system	2 into 1, 1 silencer RHS
	2 into 2, 2 standard-silencers RHS

Transmission

Gearbox	4-speed
Reduction ratios	2.52 (1)
	1.83 (2)
	1.38 (3)
	1.00 (4)
Secondary reduction ratio	2.55 (51:20 teeth)
Gearchange	Foot change
Final drive	Chain

Cycle parts

Frame	Twin-loop frame
Front forks	Telescopic, hydraulically-damped
Rear suspension	Swinging-arm
Front brake	292 mm drum, via cable and handlebar lever
Rear brake	203 mm drum, via rod and pedal

General data

Dry weight	238 kg (XLH)
	215 kg (XLCH)
Permissible total weight	408 kg
Wheelbase	1486 mm
Tyre front	3.75S19
Tyre rear	4.25S18
Wheels front	2.15 x 19 wire-spoked optional
	2.50 x 19 wire-spoked
Wheels rear	2.50 x 18 wire-spoked
Top speed	165 km/h
Fuel tank capacity	8.5 litres optional approx. 15 litres

Model XLH 1975 to 1976
Model XLCH 1975 to 1976

Motor

Engine	Air cooled 45° ohv V-twin
Bore and stroke	3³⁄₁₆ x 3¹³⁄₁₆ (3.19 x 3.81) in.
Bore and stroke	80.96 x 96.84 mm
Capacity	60.85 cu. in. (997 cc)
Power output (SAE)	61(45)/6200 rpm
	57(42)/6000 DIN
Torque (Nm)	43/(58)/4000 rpm
Compression ratio	9.0:1
Carburettor	Bendix-Zenith 16P12
Ignition	Battery-and-coil
Electrical system	12-volt
Starter	XLH: Electric starter
	XLCH: Kickstarter
Exhaust system	2 into 1, 1 silencer RHS
	2 into 2, 2 standard-silencers RHS

Transmission

Gearbox	4-speed
Reduction ratios	2.52 (1)
	1.83 (2)
	1.38 (3)
	1.00 (4)
Secondary reduction ratio	2.43 (51:21 teeth)
Gearchange	Foot change
Final drive	Chain

Cycle parts

Frame	Twin-loop frame
Front forks	Telescopic, hydraulically-damped
Rear suspension	Swinging-arm
Front brake	Hydraulically-operated 292 mm disc
Rear brake	203 mm drum, via rod and pedal

General data

Dry weight	238 kg (XLH)
	215 kg (XLCH)
Permissible total weight	408 kg
Wheelbase	1486 mm
Tyre front	3.75S19 optional
	MM90S19
Tyre rear	4.25S18
Wheels front	2.15 x 19 wire-spoked optional
	2.50 x 19 wire-spoked
Wheels rear	2.50 x 18 wire-spoked
Top speed	165 km/h
Fuel tank capacity	8.5 litres optional approx. 15 litres

Model XLH 1977 to 1978
Model XLCH 1977 to 1978

Motor
Engine	Air cooled 45° ohv V-twin
Bore and stroke	3³⁄₁₆ x 3¹³⁄₁₆ (3.19 x 3.81) in.
Bore and stroke	80.96 x 96.84 mm
Capacity	60.85 cu. in. (997 cc)
Power output (SAE)	61(45)/6200 rpm
	57(42)/6000 DIN
Torque (Nm)	43/(58)/4000 rpm
Compression ratio	9.0:1
Carburettor	Keihin 27153-77
Ignition	1977: Battery-and-coil
	1978: Electronic, contactless
Electrical system	12-volt
Starter	XLH: Electric starter
	XLCH: Kickstarter
Exhaust system	2 into 1, R/H silencer, optional 2 into 2

Transmission
Gearbox	4-speed
Reduction ratios	2.52 (1)
	1.83 (2)
	1.38 (3)
	1.00 (4)
Secondary reduction ratio	2.43 (51:21 teeth)
Gearchange	Foot change
Final drive	Chain

Cycle parts
Frame	Twin-loop frame
Front forks	Telescopic, hydraulically-damped
Rear suspension	Swinging-arm
Front brake:	
1977	Hydraulically-operated 292 mm disc
1978	Twin 254 mm discs
Rear brake	
1977	203 mm drum, via rod and pedal
1978	Hydraulically-operated 292 mm disc

General data
Dry weight	238 kg (XLH)
	215 kg (XLCH)
Permissible total weight	408 kg
Wheelbase	1486 mm
Tyre front	3.75S19 optional MM90S19
Tyre rear	4.25S18 optional MN90S18
Wheels front	2.50 x 19 wire-spoked
Wheels rear	2.50 x 18 wire-spoked
Top speed	165 km/h
Fuel tank capacity	8.5 litres optional approx. 15 litres

Model XLT 1977 to 1978

Motor
Engine	Air cooled 45° ohv V-twin
Bore and stroke	3³⁄₁₆ x 3¹³⁄₁₆ (3.19 x 3.81) in.
Bore and stroke	80.96 x 96.84 mm
Capacity	60.85 cu. in. (997 cc)
Power output (SAE)	61(45)/6200 rpm
	57(42)/6000 DIN
Torque (Nm)	43/(58)/4000 rpm
Compression ratio	9.0:1
Carburettor	Keihin 27153-77
Ignition	1977: Battery-and-coil
	1978: Electronic, contactless
Electrical system	12-volt
Starter	Electric starter
Exhaust system	2 into 1, R/H silencer, optional 2 into 2

Transmission
Gearbox	4-speed
Reduction ratios	2.52 (1)
	1.83 (2)
	1.38 (3)
	1.00 (4)
Secondary reduction ratio	2.43 (51:21 teeth)
Gearchange	Foot change
Final drive	Chain

Cycle parts
Frame	Twin-loop frame
Front forks	Telescopic, hydraulically-damped
Rear suspension	Swinging-arm
Front brake	Hydraulically-operated 292 mm disc
Rear brake	
1977	203 mm drum, via rod and pedal
1978	Hydraulically-operated 292 mm disc

General data
Dry weight	245 kg
Permissible total weight	408 kg
Wheelbase	1486 mm
Tyre front	3.75S19
Tyre rear	4.25S18
Wheels front	2.50 x 19 wire-spoked
Wheels rear	2.50 x 18 wire-spoked
Top speed	160 km/h
Fuel tank capacity	approx. 13.6 litres

Model XLCR 1977 to 1978

Motor
Engine	Air cooled 45° ohv V-twin
Bore and stroke	3³⁄₁₆ x 3¹³⁄₁₆ (3.19 x 3.81) in.
Bore and stroke	80.96 x 96.84 mm
Capacity	60.85 cu. in. (997 cc)
Power output (DIN)	61(45)/6200 rpm
	57(42)/6000 DIN
Torque (Nm)	43/(58)/4000 rpm
Compression ratio	9.0:1
Carburettor	Keihin 27153-77
Ignition	1977: Battery-and-coil
	1978: Electronic, contactless
Electrical system	12-volt
Starter	Electric starter
Exhaust system	2 into 2, 1 silencer

Transmission
Gearbox	4-speed
Reduction ratios	2.52 (1)
	1.83 (2)
	1.38 (3)
	1.00 (4)
Secondary reduction ratio	2.43 (51:21 teeth)
Gearchange	Foot change
Final drive	Chain

Cycle parts
Frame	Twin-loop frame
Front forks	Telescopic, hydraulically-damped
Rear suspension	Swinging-arm
Front brake	Hand-operated twin 254 mm discs
Rear brake	Foot-operated 292 mm hydraulic disc

General data
Dry weight	235 kg
Permissible total weight	408 kg
Wheelbase	1486 mm
Tyre front	3.75S19
Tyre rear	4.25S18
Wheels front	2.50 x 19 wire-spoked
Wheels rear	2.50 x 18 wire-spoked
Top speed	170 km/h
Fuel tank capacity	15.1 litres

Model XLH 1979 to 1981
Model XLCH 1979

Motor
Engine	Air cooled 45° ohv V-twin
Bore and stroke	3³⁄₁₆ x 3¹³⁄₁₆ (3.19 x 3.81) in.
Bore and stroke	80.96 x 96.84 mm
Capacity	60.85 cu. in. (997 cc)
Power output (DIN)	60(44)/6200 DIN (1979-80)
	56(41)/5800 DIN (1981)
Torque (Nm)	77.5/4150 DIN (1979-80)
	74/4400 DIN (1981)
Compression ratio	9.0:1
Carburettor	Keihin B83A (1979)
	Keihin B83D (1980-81)
Ignition	Electronic, contactless
Electrical system	12-volt
Starter	XLH: Electric starter
	XLCH: Kickstarter
Exhaust system	2 into 2, 1 silencer RHS/LHS (1979-80)
	2 into 2, 2 RHS silencers (1981)

Transmission
Gearbox	4-speed
Reduction ratios	2.52 (1)
	1.83 (2)
	1.38 (3)
	1.00 (4)
Secondary reduction ratio	2.24
Gearchange	Foot change
Final drive	Chain

Cycle parts
Frame	Twin-loop frame
Front forks	Telescopic, hydraulically-damped
Rear suspension	Swinging-arm
Front brake	Hand-operated twin 254 mm discs
Rear brake	Foot-operated 292 mm hydraulic disc

General data
Dry weight	236 kg (XLH), 233 kg (XLCH)
Permissible total weight	408 kg
Wheelbase	1486 mm
Tyre front	MJ90S19
Tyre rear	MN90S18 (1979-80)
	MT90S16 (1981)
Wheels front	2.50 x 19 wire-spoked optional 2.15 x 19 cast-alloy
Wheels rear	1979-80: 2.50 x 18 wire-spoked optional 2.50 x 18 cast-alloy 1981: 3.00 x 16 wire-spoked optional 3.00 x 16 cast-alloy
Top speed	165 km/h
Fuel tank capacity	8.5 litres

Model XLS 1979 to 1981

Motor
Engine	Air cooled 45° ohv V-twin
Bore and stroke	3³⁄₁₆ x 3¹³⁄₁₆ (3.19 x 3.81) in.
Bore and stroke	80.96 x 96.84 mm
Capacity	60.85 cu. in. (997 cc)
Power output (SAE)	60(44)/6200 DIN (1979-80)
	56(41)/5800 DIN (1981)
Torque (Nm)	77.5/4150 DIN (1979-80)
	74/4400 DIN (1981)
Compression ratio	9.0:1
Carburettor	Keihin B83A (1979)
	Keihin B83D (1980-81)
Ignition	Electronic, contactless
Electrical system	12-volt
Starter	Electric starter
Exhaust system	2 into 2, 1 silencer RHS/LHS (1979-80)
	2 into 2, 2 R/H silencers (1981)

Transmission
Gearbox	4-speed
Reduction ratios	2.52 (1)
	1.83 (2)
	1.38 (3)
	1.00 (4)
Secondary reduction ratio	2.24
Gearchange	Foot change
Final drive	Chain

Cycle parts
Frame	Twin-loop frame
Front forks	Telescopic, hydraulically-damped
Rear suspension	Swinging-arm
Front brake	Hand-operated twin 254 mm discs
Rear brake	Foot-operated 292 mm hydraulic disc

General data
Dry weight	239 kg
Permissible total weight	408 kg
Wheelbase	1514 mm
Tyre front	MJ90S19
Tyre rear	MT90S16
Wheels front	2.50 x 19 wire-spoked optional 2.15 x 19 cast-alloy
Wheels rear	3.00 x 16 wire-spoked optional 3.00 x 16 cast-alloy
Top speed	165 km/h
Fuel tank capacity	13.6 litres

Model XLH 1982

Motor
Engine	Air cooled 45° ohv V-twin
Bore and stroke	3³⁄₁₆ x 3¹³⁄₁₆ (3.19 x 3.81) in.
Bore and stroke	80.96 x 96.84 mm
Capacity	60.85 cu. in. (997 cc)
Power output (SAE)	56(41)/5800 DIN
Torque (Nm)	74/4400 DIN
Compression ratio	9.0:1
Carburettor	Keihin B83D
Ignition	Electronic, contactless
Electrical system	12-volt
Starter	Electric starter
Exhaust system	2 into 2, 2 RHS silencers

Transmission
Gearbox	4-speed
Reduction ratios	2.52 (1)
	1.83 (2)
	1.38 (3)
	1.00 (4)
Secondary reduction ratio	2.24
Gearchange	Foot change
Final drive	Chain

Cycle parts
Frame	Twin-loop frame
Front forks	Telescopic, hydraulically-damped
Rear suspension	Swinging-arm
Front brake	Hand-operated twin 254 mm discs
Rear brake	Foot-operated 292 mm hydraulic disc

General data
Dry weight	236 kg
Permissible total weight	408 kg
Wheelbase	1524 mm
Tyre front	MJ90S19
Tyre rear	MT90S16
Wheels front	2.50 x 19 wire-spoked optional 2.15 x 19 cast-alloy
Wheels rear	3.00 x 16 wire-spoked optional 3.00 x 16 cast-alloy
Top speed	168 km/h
Fuel tank capacity	11.2 litres

Model XLS 1982

Motor
Engine	Air cooled 45° ohv V-twin
Bore and stroke	3³⁄₁₆ x 3¹³⁄₁₆ (3.19 x 3.81) in.
Bore and stroke	80.96 x 96.84 mm
Capacity	60.85 cu. in. (997 cc)
Power output (DIN)	56(41)/5800 DIN
Torque (Nm)	74/4400 DIN
Compression ratio	9.0:1
Carburettor	Keihin B83D
Ignition	Electronic, contactless
Electrical system	12-volt
Starter	Electric starter
Exhaust system	2 into 2, 2 R/H silencers

Transmission
Gearbox	4-speed
Reduction ratios	2.52 (1)
	1.83 (2)
	1.38 (3)
	1.00 (4)
Secondary reduction ratio	2.24
Gearchange	Foot change
Final drive	Chain

Cycle parts
Frame	Twin-loop frame
Front forks	Telescopic, hydraulically-damped
Rear suspension	Swinging-arm
Front brake	Hand-operated twin 254 mm discs
Rear brake	Foot-operated 292 mm hydraulic disc

General data
Dry weight	240 kg
Permissible total weight	408 kg
Wheelbase	1543 mm
Tyre front	MJ90S19
Tyre rear	MT90S16
Wheels front	2.50 x 19 wire-spoked optional 2.15 x 19 cast-alloy
Wheels rear	3.00 x 16 wire-spoked optional 3.00 x 16 cast-alloy
Top speed	168 km/h
Fuel tank capacity	12.9 litres

Model XLH 1983 to 1985
Model XLS 1983 to 1985
Model XLX 1983 to 1985

Motor
Engine	Air cooled 45° ohv V-twin
Bore and stroke	3³⁄₁₆ x 3¹³⁄₁₆ (3.19 x 3.81) in.
Bore and stroke	80.96 x 96.84 mm
Capacity	60.85 cu. in. (997 cc)
Power output (DIN)	52(38)/6000 DIN optional 50(37)/6000 DIN
Torque (Nm)	72/4400 DIN (52 PS) 70/4000 DIN (50 PS)
Compression ratio	9.0:1
Carburettor	Keihin B83D
Ignition	Electronic, contactless
Electrical system	12-volt
Starter	Electric starter
Exhaust system	2 into 2, 2 R/H silencers

Transmission
Gearbox	4-speed
Reduction ratios	2.52 (1)
	1.83 (2)
	1.38 (3)
	1.00 (4)
Secondary reduction ratio	2.24
Gearchange	Foot change
Final drive	Chain

Cycle parts
Frame	Twin-loop frame
Front forks	Telescopic, hydraulically-damped
Rear suspension	Swinging-arm
Front brake	Hand-operated twin 254 mm discs from 1984 1 drilled 292 mm hydraulic disc
Rear brake	Foot-operated 292 mm hydraulic disc

General data
Dry weight	236 kg (XLH), 235 kg (XLX), 240 kg (XLS)
Permissible total weight	408 kg
Wheelbase	1524 mm (XLH and XLX), 1543 mm (XLS)
Tyre front	MJ90S19
Tyre rear	MT90S16
Wheels front	2.50 x 19 wire-spoked optional 2.15 x 19 cast-alloy
Wheels rear	3.00 x 16 wire-spoked optional 3.00 x 16 cast-alloy
Top speed	162 km/h
Fuel tank capacity	11.2 litres (XLH) 15.0 litres (XLS) 8.3 litres (XLX)

Model XR-1000 1983 to 1984

Motor

Engine	Air cooled 45° ohv V-twin
Bore and stroke	3$\frac{3}{16}$ x 3$\frac{13}{16}$ (3.19 x 3.81) in.
Bore and stroke	80.96 x 96.84 mm
Capacity	60.85 cu. in. (997 cc)
Power output (SAE)	70(51)/6200 rpm
Torque (Nm)	Not available
Compression ratio	9.0:1
Carburettor	2 Dell'Orto PHF 36
Ignition	Electronic, contactless
Electrical system	12-volt
Starter	Electric starter
Exhaust system	2 into 2, 2 silencers high-level, LHS

Transmission

Gearbox	4-speed
Reduction ratios	2.52 (1)
	1.83 (2)
	1.38 (3)
	1.00 (4)
Secondary reduction ratio	2.24
Gearchange	Foot change
Final drive	Chain

Cycle parts

Frame	Twin-loop frame
Front forks	Telescopic, hydraulically-damped
Rear suspension	Swinging-arm
Front brake	1 drilled 292 mm hydraulic disc
Rear brake	Foot-operated 292 mm hydraulic disc

General data

Dry weight	240 kg
Permissible total weight	408 kg
Wheelbase	1543 mm
Tyre front	100/90V19 optional MM90V19
Tyre rear	130/90V16 optional MT90V16
Wheels front	2.15 x 19 cast-alloy
Wheels rear	3.00 x 16 cast-alloy
Top speed	185 km/h
Fuel tank capacity	9.5 litres

Super Glide.
The FX models
1971 to 1986

Model description and technical information

The policy of maintaining two model ranges in their core market, the big V-twin, kept Harley-Davidson going from 1953-onwards, in addition to any income from the sales, sometimes substantial, of single-cylinder motorcycles of German or Italian origin.

When at the beginning of the 1960s the first Japanese motorcycles timidly began to conquer the reawakening US motorcycle market, the first reaction in Milwaukee was rather dismissive as they simply did not understand the innovative strength behind the Hondas, and later the Suzukis and so on, nor did they comprehend the depth of the flood of machines that was about to engulf them.

Up until 1965 Harley's sales had fluctuated around a level of 10,000 to 15,000 motorcycles annually, when the developing boom in two-wheeler sales dragged H-D along with it so that, without any really new models, Harley suddenly found itself selling, for the first time since the post-War boom of 1948-9, more than 20,000 motorcycles; in 1966 this rose to over 36,000 machines. However, between then and 1970 this so very positive development was not followed through and so sales slumped to no more than 29,000 per year, while the Japanese inexorably continued their triumphant advance.

The answer was obvious: a new and attractive model of motorcycle was badly needed! This need combined with the fact that even the Japanese, on the one hand pioneers of the motorcycle, became, for the youth of the 1960s on the other hand, the very essence of the Establishment that these young people were revolting against in the first place, which manifested itself throughout the western industrialized nations in the form of the occasionally extremely radical outbreaks of student unrest. Right about this time came the movie 'Easy Rider', today revered as a cult film. With Peter Fonda in the leading role, it had apehanger handlebars, loud exhausts, wild colours and stripped frames; from that day on these have become objects of desire for many Harley riders who, still today, love to slip into the role of Peter Fonda and ride off to give free rein to their dreams.

Model range

Seventy-Fours (74 cu. in./1207 cc.) or '1200' models, 1971 to 1980

Model	Name	Feature	Years
Model FX	Super Glide	Kickstart only	(1971-78)
Model FX Liberty	Super Glide	Kickstart only	(1976)
Model FXE	Super Glide	Electric starter	(1974-80)
Model FXE Liberty	Super Glide	Electric starter	(1976)
Model FXS	Low Rider	First factory custom	(1977-80)
Model FXEF	Fat Bob	Shorter mudguards	(1979-80)

Liberty Edition (on the 200th anniversary of the American Declaration of Independence).

Eighties (82 cu. in./1340 cc.) or '80' models 1978 to 1986

Model	Name	Feature	Years
Model FXE	Super Glide	Electric starter	(1980-84)
Model FXEF	Fat Bob	Shorter mudguards	(1979-85)
Model FXS	Low Rider	Factory custom	(1979-82)
Model FXWG	Wide Glide	Wide front forks	(1980-86)
Model FXB	Sturgis	Belt drive	(1980-82)
Model FXSB	Low Rider	Belt drive	(1983-85)
Model FXDG	Disc Glide	Disc rear wheel	(1983)

No FX model was approved for sidecar use.

1971: The first Model FX Super Glide, complete with boat tail-shaped fibre-glass seat unit

After the film had made clear to the Harley management just what sort of machines were wanted, they lost no time in rushing out the sort of thing that would start the tills ringing once more.

The H-D recipe for a successful motorcycle of this sort was very simple and based on what the customizers themselves were doing: they took existing components and mixed them together with a touch of magic provided by Willie G. Davidson to conjure up an innovative and fresh model that would become the basis for a whole new range. As big was beautiful, the Electra-Glide supplied the frame, engine and gearbox, while the Sportster provided the forks, front wheel and fittings for that essential stripped and lean front end look. A tank patterned on that of the Electra-Glide's, but smaller and slimmer with a Stars and Stripes 'Number One' logo, was mated with the Sportster's fibre-glass boat tail-shaped seat unit and rear mudguard. High Buckhorn handlebars and a 2-into-1 exhaust system rounded off the package, which was finished in a shimmering white paint finish with red and blue highlights called 'Sparkling America'. Forward-mounted footrests, forerunners of Highway Pegs, were available as optional extras for a relaxed and cool-looking laid-back riding position. The combination of parts from the FL and XL model ranges naturally led to the new model being designated FX; but to the world at large it was the Super Glide.

Engine (the customary Shovelhead), transmission and frame (well-suited to the new chopper configuration, in spite of having rear suspension) were therefore well-known and familiar; it was just the combination of the Electra-Glide frame with a lighter-weight front end and a 19-inch wheel (a size that hadn't been used on a Big Twin since the last Model Vs of 1936).

Another familiar echo from the past was the return of the kickstart, an essential feature for prospective FX owners, since Peter Fonda had so forcefully shown that only a Real Man could kick a Big Harley into life.

One feature was not carried over to the new range; it was never designed for sidecars. But then, 'Easy Rider' in sidecar outfits . . .? Wallace and Gromit perhaps, but not Peter Fonda, Dennis Hopper and Jack Nicholson!

If the seat-tail unit was perhaps not the most successful element of the Super Glide, it was at least a departure from the usual Harley norms and drew the eye backwards along the smoothly-flowing fibre-glass surfaces to the rounded rear end with a circular stop/taillamp embedded to the rear of the stepped-level dualseat.

Although its sales were additional to those of the FLs and XLs, and thus a welcome contribution to AMF/H-D's profits, the Super Glide was hardly a success in its first year; exactly 4,700 were sold in 1971, compared to 6,675 Electra-Glides and 10,775 Sportsters.

Model development

1972 model year
The new model's sales were slowly improving; the sales statistics showed 6,500 Super Glides sold,

317

1971 Model FX Super Glide

1971 Model FX Super Glide in 'Sparkling America' finish, complete with No. 1 Stars and Stripes logo

1973 Model FX Super Glide: Sprint tank

1972 Model FX Super Glide

against 9,700 Electra-Glides and 18,150 Sportsters.

The fibre-glass seat-tail unit was not a universal success, so after a year it was relegated to the options list for those who still wanted it and a conventional, Sportster-like rear mudguard was fitted as standard, with a thin, low, banana seat that hugged the frame at the front and curved up over the rear mudguard in true custom style. Beyond that there was a new oil pump, and new AMF/Harley-Davidson tank graphics decorated all Harleys for this year.

1973 model year

This was the last year that the Super Glide range consisted only of the kickstart-equipped Model FX. The Electra-Glide's front and rear disc brakes were fitted, which improved stopping power considerably.

The ignition switch moved to the left, under the tank, and the rear brake master cylinder was mounted above the rider's right-hand footrest. The Sprint Scramblers' small and streamlined teardrop tank, containing 3 Imp. gallons (13.6 litres), was fitted to improve the long, low and lean custom look, and firmer springs were used. In addition, H-D started fitting all their V-twins with front forks made by the Japanese manufacturer, Kayaba; these were not only cost-effective, but their quality was of a high standard as well.

1974 model year
The demand for an electric starter had now become loud enough that the men in West Juneau Avenue could no longer ignore it, and the Model FXE (E for Electric start) was created. The FXE naturally used the Electra-Glide's starter and drive components, but got away with the Sportster's smaller and lighter battery; evidently it was a popular move, as combined sales of FX and FXE now exceeded those of the FLs for the first time. Other technical changes were the fitting of a self-closing throttle twistgrip with a return spring – a novelty for Harley-Davidson – and a new gearchange lever.

1975 model year
The only technical innovation worth mentioning was a change in the first gear reduction ratio, by which Harley-Davidson addressed the frequent complaint that getting the Super Glide off the line wasn't easy. This was the year in which Harley-Davidson achieved its best-ever sales record, with over 75,000 machines sold.

1976 model year
Sales were now beginning to take off – the FX and FXE combined now outsold the Sportster for the first time and from here on, the Glides were to be Harley's best-selling V-twin. The much more reliable Keihin carburettors were finally fitted to the Super Glides. In time for the celebrations of the 200th anniversary of the American Declaration of Independence, Harley-Davidson introduced Liberty Editions of the FX and FXE with a special metal-flake black paint finish and a huge Liberty Edition graphic featuring the American Eagle taking up most of the top surface of the tank.

1977 model year
The formerly quite meagre Super Glide range was extended by the introduction of the Model FXS Low Rider in the spring of 1977 (i.e., half-way through the 1977 model year, which had begun, as all model years do, in the autumn of the previous year).
With this machine H-D introduced to the Super Glide range innovations like the now-famous Showa front forks, cast aluminium-alloy wheels and twin-disc front brakes, which were gradually extended to the other two models.
The Low Rider featured a slightly longer wheelbase thanks to a raked-out steering head angle and front forks extended by approximately 2 inches (50 mm), shorter rear suspension units, lower seat height and flatter handlebars for a lower riding position (hence the name), a 2-into-1 exhaust system exiting through a massive megaphone-shaped slash-cut silencer, cast aluminium-alloy wheels painted black but with polished rims and spoke edges, white-lettered tyres and a special metallic grey paint finish with orange graphics. The Fatbob twin fuel tanks and instrument panel reappeared, the latter holding a speedometer and tachometer. Although the passenger did not sit quite as low as the rider, the FXS was a genuine two-seater.

The first of the factory customs, the Low Rider introduced into mass production many features which had long been used by the custom-builders such as forward-mounted footrests, or Highway Pegs, and near-straight 'drag bars' mounted on rearwards-curving risers – all these details were lapped up by the fans and the FXS and its derivatives went on to become the best-sellers of the Super Glide range until 1981, and were only beaten then when the FXWG Wide Glide took over the top slot.

The FXS arrived with the normal Shovelhead engine which had, however, been given a far-from-normal finish; black wrinkle-finish paint covered the crankcases, cylinders and even the cylinder heads, but the primary chaincase inspection covers, the timing cover and gearbox end cover and the rocker boxes were very highly polished. Electric start and kickstart were fitted. The ever-larger rectangular air cleaner had a black panel with the label '1200' to identify the engine's capacity, the Seventy-Fours abandoning their roots in American standard units and beginning the change to the decimal system.

1978 model year
The FX was offered with a kickstart for the last time – for the time being; it would become available again with the introduction of the FXWG.

While Harley-Davidson brought out Anniversary Editions of the FLH and XLH models in time for the Company's 75th birthday, strangely, nothing similar was done for the FX models.

Apart from the larger air cleaners and more restrictive exhaust silencers now mandated by new legislation, Harley also introduced Motorola V-Fire electronic ignition for more accurate and maintenance-free ignition timing and better combustion. The valves and valve guides were strengthened in the course of the model year and the camshaft was revised. FXS-pattern cast aluminium-alloy wheels were available as optional extras for the other Super Glides.

In the meanwhile, the Japanese were becoming interested in the sales potential of Harley's new factory customs – especially after the introduction of the Low Rider – and started converting their own existing machines, just as Harley had done. However, they didn't have the magic ingredient – Willie G. Davidson. Some of the early Honda, Kawasaki, Suzuki and Yamaha 'Special', or 'Limited

1976 Model FXE Super Glide

1977 Model FXS Low Rider: First of the factory customs

1980 Model FXWG Wide Glide

1980 Model FXB Sturgis

1978 Model FX Super Glide

1979 Model FXS Low Rider

1980 Model FXE-80 Super Glide

1981 Model FXS Low Rider

Edition' or simply 'Ltd' models were truly abysmal motorcycles and no doubt made Harley laugh for a while. But the Japanese have a habit of catching on quickly . . .

1979 model year
The range was extended again in the new model year by the addition of the two Model FXEF Fat Bobs with either the 74 cu. in./1207 cc. engine or the new larger-capacity 82 cu. in. (1340 cc.) unit introduced on the Electra-Glide the previous year.

The bigger engine was also made available with the Model FXS Low Rider. In an uncharacteristic fit of modesty, when identifying the two engines by marking the air cleaner housing panels, Harley rounded-down the capacity of the bigger models' engines to make them '80's. While we're on the subject of names, 'Fat Bob' stems from the post-Second World War period when riders were stripping surplus weight from their Big Twins to make them faster against the pesky British competition. Part of the process was the cutting-down, or 'bobbing' of items like mudguards, but nothing could be done with the standard big fat twin petrol tanks, so they were left alone – to produce 'fat bobs'. So the two Fat Bobs proudly displayed their twin tanks with an instrument panel like that of the Electra-Glide's, but holding a speedometer and tachometer, as with the FXS Low Rider. They were fitted with Buckhorn handlebars and, naturally, cut-off 'Fat Bob' mudguards, the rear one of which allowed the number plate to be mounted under the taillamp for the first time, instead of above it.

1980 model year

H-D took a big step towards maintenance-free operation with the introduction of the new Model FXB Sturgis, based on a standard FXS but with primary and final drive by toothed belt developed specifically for that application, instead of by chain. The German company Glas had for the first time used the toothed belt in a camshaft drive at the beginning of the 1960s, but Harley-Davidson, having seen the success of aftermarket conversions for belt drives, started research in 1977 and worked closely with engineers from the Gates Rubber Company, a huge supplier of drivebelts of all types to the automotive industries, to develop a Kevlar-reinforced polyurethane toothed belt that could take the abuse implied by motorcycle final drive layouts.

The FXB's name came from a rally held in August of every year near Sturgis, in the Black Hills of South Dakota. Tens of thousands of riders, the vast majority of whom are on Harleys, attend a festival of motorcycle events including motocross racing, a ½ mile race, hill climbs, road tours and short track racing. For the majority, however, the rally is a place for cruising, drinking, seeing and being seen and it has become a gigantic Harley party, one of the most important social events in the Harley enthusiast's calendar and the embodiment of Harley culture; the nearest thing in European terms is possibly the Isle of Man TT. The use of this name for a new model is to evoke all that is nearest and dearest to the hearts of Harley fans and, once the FXB and the FXSB were discontinued, H-D used it again to great effect in the Dyna Glide series.

To accommodate the drivebelt, which was substantially wider than the equivalent chain, the swinging arm had to be widened. The front forks were extended by approximately 2 inches (50 mm), as on the Low Rider, a sissy bar was fitted and cast aluminium-alloy spoked wheels.

Another new model introduced this year was the Model FXWG Wide Glide. With its wide-spaced front fork legs (hence the name), the narrow 21-inch front wheel and kickstart, it was a machine entirely after the hearts of the Harley enthusiasts, especially when they saw its cut-off 'Fat Bob' mudguards, sissy bar (naturally) and the Fatbob twin tank (here in full Electra-Glide sized 4.2 Imp gallon/18.9 litre capacity) with its black paintwork with orange flames applied as standard by the factory! Further details were the small headlamp and the wire-spoked wheels, now a rarity when everything else was getting cast aluminium-alloy wheels, and the Sportster-styled staggered 'shorty duals' chrome-plated exhaust system. The Harley enthusiasts loved it and it became the bestselling model in the FX range in its first year of production.

Both the Models FXB and FXWG were Eighties, coming with the larger-capacity engine from the outset, and the standard Model FXE was now also available with the bigger powerplant. The Motorola V-Fire electronic ignition now had electronic advance and retard.

1981 model year

The ever-tightening grip of legislation curbing vehicle emissions and pollution is shown by the need to adopt an even more sophisticated 2nd-generation electronic ignition system; V-Fire II. So that the engines could use the lead-free fuel that was about to be introduced, their compression ratios had to be reduced to 7.4:1; strangely, this does not seem to be reflected in a corresponding drop in the Motor Company's published power outputs for its biggest engines. The valve guides were extended and better valve stem seals fitted. The rear brake caliper was now supplied by Girling, the manufacturer of braking systems and components.

As with the Electra-Glides, the Seventy-Fours, aided no doubt in this case by the fact that the Eighties weren't significantly more expensive, were discontinued this year.

1982 model year

The valves and their springs were again improved on all big Harleys. The Sturgis was provided with gold-painted cast aluminium-alloy spoked wheels. With the introduction of the successor FXR models of the five-speed rubber-mounted engine/transmission Super Glide II range, sales of the FX models, as was only to be expected, decreased rapidly. From a total of well over 22,000 FX models sold in 1981, less than 8,000 machines were sold in 1982. Only the Low Rider and Sturgis, of the original FX range, were still selling in reasonable numbers.

1983 model year

In spite of their imminent obsolescence, two new models were added. On the one hand was the Model FXSB, which combined the FXS and FXB but took the Low Rider name; the Sturgis name took a break for a few years. The FXSB kept the primary and secondary belt drive and was fitted with the Wide Glide's 21-inch front wheel.

The other was the Model FXDG Disc Glide, which featured for the first time a cast aluminium-alloy rear wheel in the form of a solid disc. It had toothed-belt final drive and a matt black-painted Sportster-styled staggered 'shorty duals' exhaust system and was available for this one year only.

1984 model year
The whole raison d'être of the FX series was brought into question even more sharply by the launch of the Softail models only two years after the introduction of the more manoeuvrable FXR range. Nevertheless, the FX models still succeeded in achieving substantial sales figures up to 1985. The range still consisted of the Model FXE Super Glide (discontinued at the end of the year) and the Fat Bob Model FXEF, both with chain final drive, and the Models FXSB Low Rider and Model FXWG Wide Glide, both with full or final-only belt drive.

1985 model year
The standard Model FXE Super Glide having been discontinued, the Fat Bob FXEF finally also got the benefit of the toothed-belt final drive. All models were fitted with the new Evolution engine, which was fitted also to the other big Harleys. The engine's compression ratio was increased to a more effective (compared with the old Shovelhead) 8.5:1. In addition the Super Glides were also fitted with the wet multi-plate clutch of the Electra-Glides. All except the Wide Glide were discontinued at the end of the year.

1986 model year
In their final year only the Model FXWG Wide Glide remained of the original FX range; all the rest had been discontinued. The Wide Glide's swan-song was a sales total of some 1,200 machines. At the end of the model year, the last Harley with a kickstart disappeared.

1983 Model FXE Super Glide

1985 Model FXSB Low Rider

1984 Model FXDG Disc Glide

1985 Model FXEF Fat Bob

1984 Model FXWG Wide Glide

1986 Model FXWG with Evolution engine

Production figures

Model	1971	1972	1973	1974	1975	1976	1977	1978	1979	1980
FX-1200	4700	6500	7625	3034	3060	3857	2049	1774	-	-
FXE-1200	-	-	-	6199	9350	13838	9400	8314	3117	3169
FXEF-1200	-	-	-	-	-	-	-	-	4678	*
FXS-1200	-	-	-	-	-	-	3742	9787	3827	3
FXB-80	-	-	-	-	-	-	-	-	-	1470
FXE-80	-	-	-	-	-	-	-	-	-	**
FXEF-80	-	-	-	-	-	-	-	-	5264	4773
FXS-80	-	-	-	-	-	-	-	-	9433	5922
FXWG-80	-	-	-	-	-	-	-	-	-	6085
Total	4700	6500	7625	9233	12410	17695	15191	19875	26319	18253

* Included in the FXE models figure.
** Included in the FXEF-80 models figure.

Production figures

Model	1981	1982	1983	1984	1985	1986
FXE	3085	1617	1215	667	-	-
FXEF	3691	*	*	1440	2324	-
FXS	7223	1816	-	-	-	-
FXB	3543	1833	-	-	-	-
FXSB	-	-	3277	2877	2359	-
FXWG	5166	2348	2873	2227	4171	1199
FXDG	-	-	810	-	-	-
Total	22708	7614	8175	7211	8854	1199

* Included in the FXE models figure.

Model FX 1971 to 1972 Series 2C

Motor
Type	Alternator Shovelhead
Engine	Air cooled 45° ohv V-twin
Bore and stroke	3⁷⁄₁₆ x 3³¹⁄₃₂ (3.44 x 3.97) in.
Bore and stroke	87.31 x 100.81 mm
Capacity	73.66 cu. in. (1207 cc)
Power output (DIN)	62(46)/5400 rpm
	58(43)/5150 DIN
Torque (Nm)	95/3900 rpm
	88/3500 DIN
Compression ratio	8.0:1
Carburettor	Bendix-Zenith
Ignition	Battery-and-coil
Electrical system	12-volt
Starter	Kickstarter
Exhaust system	2 into 1, 1 cylindrical silencer; optional 2 into 2, 2 cylindrical silencers

Transmission
Gearbox	4-speed
Reduction ratios	3.00 (1)
	1.82 (2)
	1.23 (3)
	1.00 (4)
Secondary reduction ratio	2.22
Gearchange	Foot change
Final drive	Chain

Cycle parts
Frame	Twin-loop, with straight front downtubes
Front forks	Telescopic, hydraulically-damped
Rear suspension	Square-section swinging arm
Brakes	Cable front/ Hydraulic rear
Front brake	Drum, via cable and handlebar lever
Rear brake	Hydraulically-operated drum, via pedal

General data
Dry weight	275 kg
Permissible total weight	440 kg
Wheelbase	1594 mm
Tyre front	3.75S19
Tyre rear	5.10S16
Wheels front	2.50 x 19 wire-spoked
Wheels rear	3.00 x 16 wire-spoked
Top speed	156 km/h
Fuel tank capacity	13.6 litres

Model FX 1973 to 1973 Series 2C
Model FXE 1974 Series 9D

Motor
Type	Alternator Shovelhead
Engine	Air cooled 45° ohv V-twin
Bore and stroke	3⁷⁄₁₆ x 3³¹⁄₃₂ (3.44 x 3.97) in.
Bore and stroke	87.31 x 100.81 mm
Capacity	73.66 cu. in. (1207 cc)
Power output (DIN)	62(46)/5400 rpm
	58(43)/5150 DIN
Torque (Nm)	95/3900 rpm
	88/3500 DIN
Compression ratio	8.0:1
Carburettor	Bendix-Zenith
Ignition	Battery-and-coil
Electrical system	12-volt
Starter	Kickstarter (FX)
	Electric starter (FXE)
Exhaust system	2 into 1, 1 cylindrical silencer; optional 2 into 2, 2 cylindrical silencers

Transmission
Gearbox	4-speed
Reduction ratios	3.00 (1)
	1.82 (2)
	1.23 (3)
	1.00 (4)
Secondary reduction ratio	2.22
Gearchange	Foot change
Final drive	Chain

Cycle parts
Frame	Twin-loop, with straight front downtubes
Front forks	Telescopic, hydraulically-damped
Rear suspension	Square-section swinging arm
Brakes	Hydraulic disc brakes
Disc diameter	292 mm
Front brake	Hand-operated disc
Rear brake	Foot-operated disc

General data
Dry weight	275 kg
Permissible total weight	440 kg
Wheelbase	1600 mm
Tyre front	3.75S19
Tyre rear	5.10S16
Wheels front	2.50 x 19 wire-spoked
Wheels rear	3.00 x 16 wire-spoked
Top speed	156 km/h
Fuel tank capacity	13.6 litres

Model FX 1975 Series 2C
Model FXE 1975 Series 9D

Motor
Type	Alternator Shovelhead
Engine	Air cooled 45° ohv V-twin
Bore and stroke	3⁷/₁₆ x 3³¹/₃₂ (3.44 x 3.97) in.
Bore and stroke	87.31 x 100.81 mm
Capacity	73.66 cu. in. (1207 cc)
Power output (DIN)	62(46)/5400 rpm
	58(43)/5150 DIN
Torque (Nm)	95/3900 rpm
	88/3500 DIN
Compression ratio	8.0:1
Carburettor	Bendix-Zenith
Ignition	Battery-and-coil
Electrical system	12-volt
Starter	Kickstarter (FX)
	Electric starter (FXE)
Exhaust system	2 into 1, 1 cylindrical silencer; optional 2 into 2, 2 cylindrical silencers

Transmission
Gearbox	4-speed
Reduction ratios	2.45 (1)
	1.82 (2)
	1.23 (3)
	1.00 (4)
Secondary reduction ratio	2.22
Gearchange	Foot change
Final drive	Chain

Cycle parts
Frame	Twin-loop, with straight front downtubes
Front forks	Telescopic, hydraulically-damped
Rear suspension	Square-section swinging arm
Brakes	Hydraulic disc brakes
Disc diameter	292 mm
Front brake	Hand-operated disc
Rear brake	Foot-operated disc

General data
Dry weight	275 kg
Permissible total weight	440 kg
Wheelbase	1600 mm
Tyre front	3.75S19
Tyre rear	5.10S16
Wheels front	2.50 x 19 wire-spoked
Wheels rear	3.00 x 16 wire-spoked
Top speed	156 km/h
Fuel tank capacity	13.6 litres

Model FX 1976 Series 2C
Model FXE 1976 Series 9D

Motor
Type	Alternator Shovelhead
Engine	Air cooled 45° ohv V-twin
Bore and stroke	3⁷/₁₆ x 3³¹/₃₂ (3.44 x 3.97) in.
Bore and stroke	87.31 x 100.81 mm
Capacity	73.66 cu. in. (1207 cc)
Power output (DIN)	62(46)/5400 rpm
	58(43)/5150 DIN
Torque (Nm)	95/3900 rpm
	88/3500 DIN
Compression ratio	8.0:1
Carburettor	Keihin
Ignition	Battery-and-coil
Electrical system	12-volt
Starter	Kickstarter (FX)
	Electric starter (FXE)
Exhaust system	2 into 2, 2 cylindrical silencers

Transmission
Gearbox	4-speed
Reduction ratios	2.45 (1)
	1.82 (2)
	1.23 (3)
	1.00 (4)
Secondary reduction ratio	2.22
Gearchange	Foot change
Final drive	Chain

Cycle parts
Frame	Twin-loop, with straight front downtubes
Front forks	Telescopic, hydraulically-damped
Rear suspension	Square-section swinging arm
Brakes	Hydraulic disc brakes
Disc diameter	292 mm
Front brake	Hand-operated disc
Rear brake	Foot-operated disc

General data
Dry weight	275 kg
Permissible total weight	440 kg
Wheelbase	1600 mm
Tyre front	3.75S19
Tyre rear	5.10S16
Wheels front	2.50 x 19 wire-spoked
Wheels rear	3.00 x 16 wire-spoked
Top speed	156 km/h
Fuel tank capacity	13.6 litres

Model FX 1977 Series 2C
Model FXE 1977 Series 9D

Motor

Type	Alternator Shovelhead
Engine	Air cooled 45° ohv V-twin
Bore and stroke	$3^{7}/_{16}$ x $3^{31}/_{32}$ (3.44 x 3.97) in.
Bore and stroke	87.31 x 100.81 mm
Capacity	73.66 cu. in. (1207 cc)
Power output (DIN)	60(44)/5200 rpm
	58(43)/5150 DIN
Torque (Nm)	95/3900 rpm
	88/3500 DIN
Compression ratio	8.0:1
Carburettor	Keihin
Ignition	Battery-and-coil
Electrical system	12-volt
Starter	Kickstarter (FX)
	Electric starter (FXE)
Exhaust system	2 into 1, 1 cylindrical silencer RHS; optional 2 into 2, 2 R/H silencers

Transmission

Gearbox	4-speed
Reduction ratios	2.45 (1)
	1.82 (2)
	1.23 (3)
	1.00 (4)
Secondary reduction ratio	2.22
Gearchange	Foot change
Final drive	Chain

Cycle parts

Frame	Twin-loop, with straight front downtubes
Front forks	Telescopic, hydraulically-damped
Rear suspension	Square-section swinging arm
Brakes	Hydraulic disc brakes
Disc diameter	292 mm
Front brake	Hand-operated disc
Rear brake	Foot-operated disc

General data

Dry weight	275 kg
Permissible total weight	440 kg
Wheelbase	1600 mm
Tyre front	3.75S19
Tyre rear	5.10S16
Wheels front	2.50 x 19 wire-spoked
Wheels rear	3.00 x 16 wire- or cast-alloy
Top speed	156 km/h
Fuel tank capacity	13.6 litres

Model FXS 1977½ Series 2F

Motor

Type	Alternator Shovelhead
Engine	Air cooled 45° ohv V-twin
Bore and stroke	$3^{7}/_{16}$ x $3^{31}/_{32}$ (3.44 x 3.97) in.
Bore and stroke	87.31 x 100.81 mm
Capacity	73.66 cu. in. (1207 cc)
Power output (DIN)	60(44)/5200 rpm
	58(43)/5150 DIN
Torque (Nm)	95/3900 rpm
	88/3500 DIN
Compression ratio	8.0:1
Carburettor	Keihin
Ignition	Battery-and-coil
Electrical system	12-volt
Starter	Kickstarter
Exhaust system	2 into 1, 1 cylindrical silencer RHS

Transmission

Gearbox	4-speed
Reduction ratios	2.45 (1)
	1.82 (2)
	1.23 (3)
	1.00 (4)
Secondary reduction ratio	2.22
Gearchange	Foot change
Final drive	Chain

Cycle parts

Frame	Twin-loop, with straight front downtubes
Front forks	Telescopic, hydraulically-damped
Rear suspension	Square-section swinging arm
Brakes	Hydraulic disc brakes
Disc diameter	254 mm front, 292 mm rear
Front brake	Hand-operated discs
Rear brake	Foot-operated disc

General data

Dry weight	275 kg
Permissible total weight	440 kg
Wheelbase	1613 mm
Tyre front	3.75S19
Tyre rear	5.10S16
Wheels front	2.15 x 19 cast-alloy
Wheels rear	3.00 x 16 cast-alloy
Top speed	156 km/h
Fuel tank capacity	13.6 litres

Model FX 1978 Series 2C
Model FXE 1978 Series 9D

Motor
Type	Alternator Shovelhead
Engine	Air cooled 45° ohv V-twin
Bore and stroke	3⁷⁄₁₆ x 3³¹⁄₃₂ (3.44 x 3.97) in.
Bore and stroke	87.31 x 100.81 mm
Capacity	73.66 cu. in. (1207 cc)
Power output (DIN)	60(44)/5200 rpm
	58(43)/5150 DIN
Torque (Nm)	95/3900 rpm
	88/3500 DIN
Compression ratio	8.0:1
Carburettor	Keihin B81A 38 mm
Ignition	Electronic, contactless
Electrical system	12-volt
Starter	Kickstarter (FX)
	Electric starter (FXE)
Exhaust system	2 into 1, 1 cylindrical silencer RHS; optional 2 into 2, 2 R/H silencers

Transmission
Gearbox	4-speed
Reduction ratios	2.45 (1)
	1.82 (2)
	1.23 (3)
	1.00 (4)
Secondary reduction ratio	2.22
Gearchange	Foot change
Final drive	Chain

Cycle parts
Frame	Twin-loop, with straight front downtubes
Front forks	Telescopic, hydraulically-damped
Rear suspension	Square-section swinging arm
Brakes	Hydraulic disc brakes
Disc diameter	292 mm
Front brake	Hand-operated disc
Rear brake	Foot-operated disc

General data
Dry weight	275 kg
Permissible total weight	440 kg
Wheelbase	1600 mm
Tyre front	MM90S16, later MJ90S16
Tyre rear	MT90S16
Wheels front	2.50 x 19 wire-spoked
Wheels rear	3.00 x 16 wire- or cast-alloy
Top speed	156 km/h
Fuel tank capacity	12.1 litres

Model FXS 1978 Series 2F

Motor
Type	Alternator Shovelhead
Engine	Air cooled 45° ohv V-twin
Bore and stroke	3⁷⁄₁₆ x 3³¹⁄₃₂ (3.44 x 3.97) in.
Bore and stroke	87.31 x 100.81 mm
Capacity	73.66 cu. in. (1207 cc)
Power output (DIN)	60(44)/5200 rpm
	58(43)/5150 DIN
Torque (Nm)	95/3900 rpm
	88/3500 DIN
Compression ratio	8.0:1
Carburettor	Keihin B81A 38 mm
Ignition	Electronic, contactless
Electrical system	12-volt
Starter	Kickstarter
Exhaust system	2 into 1, 1 cylindrical silencer RHS

Transmission
Gearbox	4-speed
Reduction ratios	2.45 (1)
	1.82 (2)
	1.23 (3)
	1.00 (4)
Secondary reduction ratio	2.22
Gearchange	Foot change
Final drive	Chain

Cycle parts
Frame	Twin-loop, with straight front downtubes
Front forks	Telescopic, hydraulically-damped
Rear suspension	Square-section swinging arm
Brakes	Hydraulic disc brakes
Disc diameter	254 mm front, 292 mm rear
Front brake	Hand-operated disc
Rear brake	Foot-operated disc

General data
Dry weight	275 kg
Permissible total weight	440 kg
Wheelbase	1613 mm
Tyre front	MJ90S19
Tyre rear	MT90S16
Wheels front	2.15 x 19 cast-alloy
Wheels rear	3.00 x 16 cast-alloy
Top speed	156 km/h
Fuel tank capacity	13.2 litres

Model FXE 1979 Series 9D
Model FXEF 1979 Series 5E

Motor
Type	Alternator Shovelhead
Engine	Air cooled 45° ohv V-twin
Bore and stroke	$3\frac{7}{16}$ x $3^{31}/_{32}$ (3.44 x 3.97) in.
Bore and stroke	87.31 x 100.81 mm
Capacity	73.66 cu. in. (1207 cc)
Power output (DIN)	60(44)/5200 rpm
	58(43)/5300 DIN
Torque (Nm)	95/3900 rpm
	88/3500 DIN
Compression ratio	8.0:1
Carburettor	Keihin B81A 38 mm
Ignition	Electronic, contactless
Electrical system	12-volt
Starter	Electric starter
Exhaust system	2 into 1, 1 cylindrical silencer RHS; optional 2 into 2, 2 R/H silencers

Transmission
Gearbox	4-speed
Reduction ratios	2.45 (1)
	1.82 (2)
	1.23 (3)
	1.00 (4)
Secondary reduction ratio	2.22
Gearchange	Foot change
Final drive	Chain

Cycle parts
Frame	Twin-loop, with straight front downtubes
Front forks	Telescopic, hydraulically-damped
Rear suspension	Square-section swinging arm
Brakes	Hydraulic disc brakes
Disc diameter	254 mm front, 292 mm rear
Front brake	Hand-operated disc
Rear brake	Foot-operated disc

General data
Dry weight	290 kg
Permissible total weight	492 kg
Wheelbase	1600 mm
Tyre front	MJ90S19
Tyre rear	MT90S16
Wheels front	2.50 x 19 wire-spoked
Wheels rear	3.00 x 16 wire- or cast-alloy
Top speed	156 km/h
Fuel tank capacity	12.1 litres (FXE)
	13.2 litres (FXEF)

Model FXS 1979 Series 2F

Motor
Type	Alternator Shovelhead
Engine	Air cooled 45° ohv V-twin
Bore and stroke	$3\frac{7}{16}$ x $3^{31}/_{32}$ (3.44 x 3.97) in.
Bore and stroke	87.31 x 100.81 mm
Capacity	73.66 cu. in. (1207 cc)
Power output (DIN)	60(44)/5200 rpm
	58(43)/5150 DIN
Torque (Nm)	95/3900 rpm
	88/3500 DIN
Compression ratio	8.0:1
Carburettor	Keihin B81A 38 mm
Ignition	Electronic, contactless
Electrical system	12-volt
Starter	Kickstarter
Exhaust system	2 into 1, 1 cylindrical silencer RHS

Transmission
Gearbox	4-speed
Reduction ratios	2.45 (1)
	1.82 (2)
	1.23 (3)
	1.00 (4)
Secondary reduction ratio	2.22
Gearchange	Foot change
Final drive	Chain

Cycle parts
Frame	Twin-loop, with straight front downtubes
Front forks	Telescopic, hydraulically-damped
Rear suspension	Square-section swinging arm
Brakes	Hydraulic disc brakes
Disc diameter	254 mm front, 292 mm rear
Front brake	Hand-operated disc
Rear brake	Foot-operated disc

General data
Dry weight	290 kg
Permissible total weight	492 kg
Wheelbase	1613 mm
Tyre front	MJ90S19
Tyre rear	MT90S16
Wheels front	2.15 x 19 cast-alloy
Wheels rear	3.00 x 16 cast-alloy
Top speed	156 km/h
Fuel tank capacity	13.2 litres

Model FXEF 1979 Series 6E

Motor
Type	Alternator Shovelhead
Engine	Air cooled 45° ohv V-twin
Bore and stroke	3½ x 4¼ (3.5 x 4.25) in.
Bore and stroke	88.9 x 107.95 mm
Capacity	81.78 cu. in. (1340 cc)
Power output (DIN)	60(44)/4800 rpm
	64(47)/4600 DIN
Torque (Nm)	91/3900 rpm
Compression ratio	8.0:1
Carburettor	Keihin B78B 38 mm (SAE)
	Keihin B78A 38 mm (DIN)
Ignition	Electronic, contactless
Electrical system	12-volt
Starter	Electric starter
Exhaust system	2 into 1, 1 cylindrical silencer RHS; optional 2 into 2, 2 R/H silencers

Transmission
Gearbox	4-speed
Reduction ratios	2.45 (1)
	1.82 (2)
	1.23 (3)
	1.00 (4)
Secondary reduction ratio	2.22
Gearchange	Foot change
Final drive	Chain

Cycle parts
Frame	Twin-loop, with straight front downtubes
Front forks	Telescopic, hydraulically-damped
Rear suspension	Square-section swinging arm
Brakes	Hydraulic disc brakes
Disc diameter	254 mm front, 292 mm rear
Front brake	Hand-operated disc
Rear brake	Foot-operated disc

General data
Dry weight	290 kg
Permissible total weight	492 kg
Wheelbase	1600 mm
Tyre front	MJ90S19
Tyre rear	MT90S16
Wheels front	2.50 x 19 wire-spoked
Wheels rear	3.00 x 16 wire- or cast-alloy
Top speed	156 km/h
Fuel tank capacity	13.2 litres

Model FXS-80 1979 Series 7G

Motor
Type	Alternator Shovelhead
Engine	Air cooled 45° ohv V-twin
Bore and stroke	3½ x 4¼ (3.5 x 4.25) in.
Bore and stroke	88.9 x 107.95 mm
Capacity	81.78 cu. in. (1340 cc)
Power output (DIN)	60(44)/4800 rpm
	64(47)/4600 DIN
Torque (Nm)	91/3900 rpm
Compression ratio	8.0:1
Carburettor	Keihin B78B 38 mm (SAE)
	Keihin B78A 38 mm (DIN)
Ignition	Electronic, contactless
Electrical system	12-volt
Starter	Kickstarter
Exhaust system	2 into 1, 1 cylindrical silencer RHS

Transmission
Gearbox	4-speed
Reduction ratios	2.45 (1)
	1.82 (2)
	1.23 (3)
	1.00 (4)
Secondary reduction ratio	2.22
Gearchange	Foot change
Final drive	Chain

Cycle parts
Frame	Twin-loop, with straight front downtubes
Front forks	Telescopic, hydraulically-damped
Rear suspension	Square-section swinging arm
Brakes	Hydraulic disc brakes
Disc diameter	254 mm front, 292 mm rear
Front brake	Hand-operated disc
Rear brake	Foot-operated disc

General data
Dry weight	290 kg
Permissible total weight	492 kg
Wheelbase	1613 mm
Tyre front	MJ90S19
Tyre rear	MT90S16
Wheels front	2.15 x 19 cast-alloy
Wheels rear	3.00 x 16 cast-alloy
Top speed	155 km/h
Fuel tank capacity	13.2 litres

Model FXE 1980 Series 9D
Model FXEF 1980 Series 5E

Motor
Type	Alternator Shovelhead
Engine	Air cooled 45° ohv V-twin
Bore and stroke	3⁷⁄₁₆ x 3³¹⁄₃₂ (3.44 x 3.97) in.
Bore and stroke	87.31 x 100.81 mm
Capacity	73.66 cu. in. (1207 cc)
Power output (DIN)	60(44)/5200 rpm
	58(43)/5300 DIN
Torque (Nm)	95/3900 rpm
	93/3500 DIN
Compression ratio	8.0:1
Carburettor	Keihin B81D 38 mm (SAE)
Ignition	Electronic, contactless
Electrical system	12-volt
Starter	Electric starter
Exhaust system	2 into 1, 1 cylindrical silencer RHS; optional 2 into 2, 2 R/H silencers

Transmission
Gearbox	4-speed
Reduction ratios	2.45 (1)
	1.82 (2)
	1.23 (3)
	1.00 (4)
Secondary reduction ratio	2.22
Gearchange	Foot change
Final drive	Chain

Cycle parts
Frame	Twin-loop, with straight front downtubes
Front forks	Telescopic, hydraulically-damped
Rear suspension	Square-section swinging arm
Brakes	Hydraulic disc brakes
Disc diameter	254 mm front, 292 mm rear
Front brake	Hand-operated disc
Rear brake	Foot-operated disc

General data
Dry weight	290 kg
Permissible total weight	492 kg
Wheelbase	1600 mm
Tyre front	MJ90S19
Tyre rear	MT90S16
Wheels front	2.50 x 19 wire-spoked
Wheels rear	3.00 x 16 wire- or cast-alloy
Top speed	156 km/h
Fuel tank capacity	12.1 litres (FXE)
	13.2 litres (FXEF)

Model FXS 1980 Series 2F

Motor
Type	Alternator Shovelhead
Engine	Air cooled 45° ohv V-twin
Bore and stroke	3⁷⁄₁₆ x 3³¹⁄₃₂ (3.44 x 3.97) in.
Bore and stroke	87.31 x 100.81 mm
Capacity	73.66 cu. in. (1207 cc)
Power output (DIN)	60(44)/5200 rpm
	58(43)/5300 DIN
Torque (Nm)	95/3900 rpm
	93/3500 DIN
Compression ratio	8.0:1
Carburettor	Keihin B81D 38 mm (SAE)
Ignition	Electronic, contactless
Electrical system	12-volt
Starter	Kickstarter
Exhaust system	2 into 1, 1 cylindrical silencer RHS

Transmission
Gearbox	4-speed
Reduction ratios	2.45 (1)
	1.82 (2)
	1.23 (3)
	1.00 (4)
Secondary reduction ratio	2.22
Gearchange	Foot change
Final drive	Chain

Cycle parts
Frame	Twin-loop, with straight front downtubes
Front forks	Telescopic, hydraulically-damped
Rear suspension	Square-section swinging arm
Brakes	Hydraulic disc brakes
Disc diameter	254 mm front, 292 mm rear
Front brake	Hand-operated disc
Rear brake	Foot-operated disc

General data
Dry weight	290 kg
Permissible total weight	492 kg
Wheelbase	1613 mm
Tyre front	MJ90S19
Tyre rear	MT90S16
Wheels front	2.15 x 19 cast-alloy
Wheels rear	3.00 x 16 cast-alloy
Top speed	155 km/h
Fuel tank capacity	13.2 litres

Model FXE-80 1980 Series 6G
Model FXEF-80 1980 Series 6E

Motor
Type	Alternator Shovelhead
Engine	Air cooled 45° ohv V-twin
Bore and stroke	3½ x 4¼ (3.5 x 4.25) in.
Bore and stroke	88.9 x 107.95 mm
Capacity	81.78 cu. in. (1340 cc)
Power output (DIN)	60(44)/4800 rpm
	64(47)/4600 DIN
Torque (Nm)	91/3900 rpm
Compression ratio	8.0:1
Carburettor	Keihin B79D 38 mm (SAE)
Ignition	Electronic, contactless
Electrical system	12-volt
Starter	Electric starter
Exhaust system	2 into 1, 1 cylindrical silencer RHS; optional 2 into 2, 2 R/H silencers

Transmission
Gearbox	4-speed
Reduction ratios	2.45 (1)
	1.82 (2)
	1.23 (3)
	1.00 (4)
Secondary reduction ratio	2.04
Gearchange	Foot change
Final drive	Chain

Cycle parts
Frame	Twin-loop, with straight front downtubes
Front forks	Telescopic, hydraulically-damped
Rear suspension	Square-section swinging arm
Brakes	Hydraulic disc brakes
Disc diameter	254 mm front, 292 mm rear
Front brake	Hand-operated disc
Rear brake	Foot-operated disc

General data
Dry weight	290 kg
Permissible total weight	492 kg
Wheelbase	1600 mm
Tyre front	MJ90S19
Tyre rear	MT90S16
Wheels front	2.50 x 19 wire-spoked
Wheels rear	3.00 x 16 wire- or cast-alloy
Top speed	155 km/h
Fuel tank capacity	12.1 litres (FXE)
	18.9 litres (FXEF)

Model FXS 1980 Series 2F

Motor
Type	Alternator Shovelhead
Engine	Air cooled 45° ohv V-twin
Bore and stroke	3½ x 4¼ (3.5 x 4.25) in.
Bore and stroke	88.9 x 107.95 mm
Capacity	81.78 cu. in. (1340 cc)
Power output (DIN)	60(44)/4800 rpm
	64(47)/4600 DIN
Torque (Nm)	91/3900 rpm
Compression ratio	8.0:1
Carburettor	Keihin B79D 38 mm (SAE)
Ignition	Electronic, contactless
Electrical system	12-volt
Starter	Kickstarter
Exhaust system	2 into 1, 1 cylindrical silencer RHS

Transmission
Gearbox	4-speed
Reduction ratios	2.45 (1)
	1.82 (2)
	1.23 (3)
	1.00 (4)
Secondary reduction ratio	2.04
Gearchange	Foot change
Final drive	Chain

Cycle parts
Frame	Twin-loop, with straight front downtubes
Front forks	Telescopic, hydraulically-damped
Rear suspension	Square-section swinging arm
Brakes	Hydraulic disc brakes
Disc diameter	254 mm front, 292 mm rear
Front brake	Hand-operated disc
Rear brake	Foot-operated disc

General data
Dry weight	290 kg
Permissible total weight	492 kg
Wheelbase	1613 mm
Tyre front	MJ90S19
Tyre rear	MT90S16
Wheels front	2.15 x 19 cast-alloy
Wheels rear	3.00 x 16 cast-alloy
Top speed	155 km/h
Fuel tank capacity	13.2 litres

Model FXWG-80 1980 Series 9G

Motor
Type	Alternator Shovelhead
Engine	Air cooled 45° ohv V-twin
Bore and stroke	3½ x 4¼ (3.5 x 4.25) in.
Bore and stroke	88.9 x 107.95 mm
Capacity	81.78 cu. in. (1340 cc)
Power output (DIN)	68(50)/4800 rpm
	64(47)/4600 DIN
Torque (Nm)	91/3900 rpm
Compression ratio	8.0:1
Carburettor	Keihin B28D 38 mm (SAE)
Ignition	Electronic, contactless
Electrical system	12-volt
Starter	Kickstarter
Exhaust system	2 into 1, 1 cylindrical silencer RHS; optional 2 into 2, 2 R/H silencers

Transmission
Gearbox	4-speed
Reduction ratios	2.45 (1)
	1.66 (2)
	1.23 (3)
	1.00 (4)
Secondary reduction ratio	2.04
Gearchange	Foot change
Final drive	Chain

Cycle parts
Frame	Twin-loop, with straight front downtubes
Front forks	Telescopic, hydraulically-damped
Rear suspension	Square-section swinging arm
Brakes	Hydraulic disc brakes
Disc diameter	254 mm front, 292 mm rear
Front brake	Hand-operated disc
Rear brake	Foot-operated disc

General data
Dry weight	290 kg
Permissible total weight	492 kg
Wheelbase	1600 mm
Tyre front	MH90S21
Tyre rear	MT90S16
Wheels front	1.85 x 21 wire-spoked
Wheels rear	3.00 x 16 wire-spoked
Top speed	155 km/h
Fuel tank capacity	18.9 litres

Model FXB-80 1980 Series 1H

Motor
Type	Alternator Shovelhead
Engine	Air cooled 45° ohv V-twin
Bore and stroke	3½ x 4¼ (3.5 x 4.25) in.
Bore and stroke	88.9 x 107.95 mm
Capacity	81.78 cu. in. (1340 cc)
Power output (DIN)	60(44)/4800 rpm
	64(47)/4600 DIN
Torque (Nm)	91/3900 rpm
Compression ratio	8.0:1
Carburettor	Keihin B79D 38 mm (SAE)
Ignition	Electronic, contactless
Electrical system	12-volt
Starter	Kickstarter
Exhaust system	2 into 1, 1 cylindrical silencer RHS; optional 2 into 2, 2 R/H silencers

Transmission
Gearbox	4-speed
Reduction ratios	2.45 (1)
	1.66 (2)
	1.23 (3)
	1.00 (4)
Secondary reduction ratio	2.04
Gearchange	Foot change
Final drive	Belt/belt

Cycle parts
Frame	Twin-loop, with straight front downtubes
Front forks	Telescopic, hydraulically-damped
Rear suspension	Square-section swinging arm
Brakes	Hydraulic disc brakes
Disc diameter	254 mm front, 292 mm rear
Front brake	Hand-operated disc
Rear brake	Foot-operated disc

General data
Dry weight	290 kg
Permissible total weight	492 kg
Wheelbase	1613 mm
Tyre front	MJ90S19
Tyre rear	MT90S16
Wheels front	2.15 x 19 cast-alloy
Wheels rear	3.00 x 16 cast-alloy
Top speed	155 km/h
Fuel tank capacity	13.2 litres

Model FXE 1981 and 1982 Series BA
Model FXEF 1981 and 1982 Series BB

Motor
Type	Alternator Shovelhead
Engine	Air cooled 45° ohv V-twin
Bore and stroke	3½ x 4¼ (3.5 x 4.25) in.
Bore and stroke	88.9 x 107.95 mm
Capacity	81.78 cu. in. (1340 cc)
Power output (DIN)	65(48)/5400 rpm
	61(45)/5800 DIN
Torque (Nm)	91/3500 DIN
Compression ratio	7.4:1
Carburettor	Keihin B79F 38 mm (SAE)
	Keihin B79E 38 mm (DIN)
Ignition	Electronic, contactless
Electrical system	12-volt
Starter	Electric starter
Exhaust system	2 into 1, 1 cylindrical silencer RHS

Transmission
Gearbox	4-speed
Reduction ratios	2.45 (1)
	1.66 (2)
	1.23 (3)
	1.00 (4)
Secondary reduction ratio	2.04
Gearchange	Foot change
Final drive	Chain

Cycle parts
Frame	Twin-loop, with straight front downtubes
Front forks	Telescopic, hydraulically-damped
Rear suspension	Square-section swinging arm
Brakes	Hydraulic disc brakes
Disc diameter	254 mm front, 292 mm rear
Front brake	Hand-operated disc
Rear brake	Foot-operated disc

General data
Dry weight	290 kg
Permissible total weight	492 kg
Wheelbase	1600 mm
Tyre front	MJ90S19
Tyre rear	MT90S16
Wheels front	2.50 x 19 wire-spoked
Wheels rear	3.00 x 16 wire- or cast-alloy
Top speed	155 km/h
Fuel tank capacity	12.9 litres (FXE)
	14.8 litres (FXEF)

Model FXS 1981 and 1982 Series BC

Motor
Type	Alternator Shovelhead
Engine	Air cooled 45° ohv V-twin
Bore and stroke	3½ x 4¼ (3.5 x 4.25) in.
Bore and stroke	88.9 x 107.95 mm
Capacity	81.78 cu. in. (1340 cc)
Power output (DIN)	67(49)/5800 DIN
Torque (Nm)	95/3400 DIN
Compression ratio	7.4:1
Carburettor	Keihin B28E 38 mm (SAE)
	Keihin B79E 38 mm (DIN)
Ignition	Electronic, contactless
Electrical system	12-volt
Starter	Kickstarter
Exhaust system	2 into 2, 2 R/H silencers

Transmission
Gearbox	4-speed
Reduction ratios	2.45 (1)
	1.66 (2)
	1.23 (3)
	1.00 (4)
Secondary reduction ratio	1.91
Gearchange	Foot change
Final drive	Belt/belt

Cycle parts
Frame	Twin-loop, with straight front downtubes
Front forks	Telescopic, hydraulically-damped
Rear suspension	Square-section swinging arm
Brakes	Hydraulic disc brakes
Disc diameter	254 mm front, 292 mm rear
Front brake	Hand-operated disc
Rear brake	Foot-operated disc

General data
Dry weight	290 kg
Permissible total weight	492 kg
Wheelbase	1613 mm
Tyre front	MJ90S19
Tyre rear	MT90S16
Wheels front	2.15 x 19 cast-alloy
Wheels rear	3.00 x 16 cast-alloy
Top speed	155 km/h
Fuel tank capacity	14.8 litres

Model FXWG 1981 and 1982 Series BE

Motor
Type	Alternator Shovelhead
Engine	Air cooled 45° ohv V-twin
Bore and stroke	3½ x 4¼ (3.5 x 4.25) in.
Bore and stroke	88.9 x 107.95 mm
Capacity	81.78 cu. in. (1340 cc)
Power output (DIN)	67(49)/5800 DIN
Torque (Nm)	95/3400 DIN
Compression ratio	7.4:1
Carburettor	Keihin B28E 38 mm (SAE)
	Keihin B78E 38 mm (DIN)
Ignition	Electronic, contactless
Electrical system	12-volt
Starter	Kickstarter
Exhaust system	2 into 2, 2 R/H silencers

Transmission
Gearbox	4-speed
Reduction ratios	2.45 (1)
	1.66 (2)
	1.23 (3)
	1.00 (4)
Secondary reduction ratio	1.91
Gearchange	Foot change
Final drive	Chain

Cycle parts
Frame	Twin-loop, with straight front downtubes
Front forks	Telescopic, hydraulically-damped
Rear suspension	Square-section swinging arm
Brakes	Hydraulic disc brakes
Disc diameter	254 mm front, 292 mm rear
Front brake	Hand-operated disc
Rear brake	Foot-operated disc

General data
Dry weight	290 kg
Permissible total weight	492 kg
Wheelbase	1651 mm
Tyre front	MH90S21
Tyre rear	MT90S16
Wheels front	1.85 x 21 wire-spoked
Wheels rear	3.00 x 16 wire-spoked
Top speed	155 km/h
Fuel tank capacity	19.2 litres

Model FXB 1981 and 1982 Series BD

Motor
Type	Alternator Shovelhead
Engine	Air cooled 45° ohv V-twin
Bore and stroke	3½ x 4¼ (3.5 x 4.25) in.
Bore and stroke	88.9 x 107.95 mm
Capacity	81.78 cu. in. (1340 cc)
Power output (DIN)	67(49)/5800 DIN
Torque (Nm)	95/3400 DIN
Compression ratio	7.4:1
Carburettor	Keihin B28E 38 mm (SAE)
	Keihin B78E 38 mm (DIN)
Ignition	Electronic, contactless
Electrical system	12-volt
Starter	Kickstarter
Exhaust system	2 into 2, 2 R/H silencers

Transmission
Gearbox	4-speed
Reduction ratios	2.45 (1)
	1.66 (2)
	1.23 (3)
	1.00 (4)
Secondary reduction ratio	1.91
Gearchange	Foot change
Final drive	Belt/belt

Cycle parts
Frame	Twin-loop, with straight front downtubes
Front forks	Telescopic, hydraulically-damped
Rear suspension	Square-section swinging arm
Brakes	Hydraulic disc brakes
Disc diameter	254 mm front, 292 mm rear
Front brake	Hand-operated disc
Rear brake	Foot-operated disc

General data
Dry weight	290 kg
Permissible total weight	492 kg
Wheelbase	1613 mm
Tyre front	MJ90S19
Tyre rear	MT90S16
Wheels front	2.15 x 19 cast-alloy
Wheels rear	3.00 x 16 cast-alloy
Top speed	168 km/h
Fuel tank capacity	14.8 litres

Model FXE 1983 Series BA
Model FXEF 1983 Series BB

Motor
Type	Alternator Shovelhead
Engine	Air cooled 45° ohv V-twin
Bore and stroke	3½ x 4¼ (3.5 x 4.25) in.
Bore and stroke	88.9 x 107.95 mm
Capacity	81.78 cu. in. (1340 cc)
Power output (DIN)	57(42)/5200 DIN
	65(48)/5400 rpm
Torque (Nm)	89/4000 DIN
Compression ratio	7.4:1
Carburettor	Keihin B79F 38 mm (SAE)
	Keihin B79E 38 mm (DIN)
Ignition	Electronic, contactless
Electrical system	12-volt
Starter	Electric starter
Exhaust system	2 into 2, 2 R/H silencers

Transmission
Gearbox	4-speed
Reduction ratios	2.45 (1)
	1.66 (2)
	1.23 (3)
	1.00 (4)
Secondary reduction ratio	1.91
Gearchange	Foot change
Final drive	Chain

Cycle parts
Frame	Twin-loop, with straight front downtubes
Front forks	Telescopic, hydraulically-damped
Rear suspension	Square-section swinging arm
Brakes	Hydraulic disc brakes
Disc diameter	254 mm front, 292 mm rear
Front brake	Hand-operated disc
Rear brake	Foot-operated disc

General data
Dry weight	290 kg
Permissible total weight	492 kg
Wheelbase	1600 mm
Tyre front	MJ90S19
Tyre rear	MT90S16
Wheels front	2.50 x 19 wire-spoked
Wheels rear	3.00 x 16 wire- or cast-alloy
Top speed	155 km/h
Fuel tank capacity	12.9 litres (FXE)
	14.8 litres (FXEF)

Model FXSB 1983 Series BF

Motor
Type	Alternator Shovelhead
Engine	Air cooled 45° ohv V-twin
Bore and stroke	3½ x 4¼ (3.5 x 4.25) in.
Bore and stroke	88.9 x 107.95 mm
Capacity	81.78 cu. in. (1340 cc)
Power output (DIN)	57(42)/5200 DIN
	65(48)/5400 rpm
Torque (Nm)	89/4000 DIN
Compression ratio	7.4:1
Carburettor	Keihin B28E 38 mm (SAE)
	Keihin B78E 38 mm (DIN)
Ignition	Electronic, contactless
Electrical system	12-volt
Starter	Kickstarter
Exhaust system	2 into 2, 2 R/H silencers

Transmission
Gearbox	4-speed
Reduction ratios	2.45 (1)
	1.66 (2)
	1.23 (3)
	1.00 (4)
Secondary reduction ratio	2.12
Gearchange	Foot change
Final drive	Belt/belt

Cycle parts
Frame	Twin-loop, with straight front downtubes
Front forks	Telescopic, hydraulically-damped
Rear suspension	Square-section swinging arm
Brakes	Hydraulic disc brakes
Disc diameter	254 mm front, 292 mm rear
Front brake	Hand-operated disc
Rear brake	Foot-operated disc

General data
Dry weight	290 kg
Permissible total weight	492 kg
Wheelbase	1613 mm
Tyre front	MJ90S19
Tyre rear	MT90S16
Wheels front	2.15 x 19 cast-alloy
Wheels rear	3.00 x 16 cast-alloy
Top speed	168 km/h
Fuel tank capacity	14.8 litres

Model FXWG 1983 Series BE

Motor
Type	Alternator Shovelhead
Engine	Air cooled 45° ohv V-twin
Bore and stroke	3½ x 4¼ (3.5 x 4.25) in.
Bore and stroke	88.9 x 107.95 mm
Capacity	81.78 cu. in. (1340 cc)
Power output (DIN)	57(42)/5200 DIN
	65(48)/5400 rpm
Torque (Nm)	89/4000 DIN
Compression ratio	7.4:1
Carburettor	Keihin B28E 38 mm (SAE)
	Keihin B78E 38 mm (DIN)
Ignition	Electronic, contactless
Electrical system	12-volt
Starter	Kickstarter
Exhaust system	2 into 2, 2 R/H silencers

Transmission
Gearbox	4-speed
Reduction ratios	2.45 (1)
	1.66 (2)
	1.23 (3)
	1.00 (4)
Secondary reduction ratio	1.91
Gearchange	Foot change
Final drive	Chain

Cycle parts
Frame	Twin-loop, with straight front downtubes
Front forks	Telescopic, hydraulically-damped
Rear suspension	Square-section swinging arm
Brakes	Hydraulic disc brakes
Disc diameter	254 mm front, 292 mm rear
Front brake	Hand-operated disc
Rear brake	Foot-operated disc

General data
Dry weight	290 kg
Permissible total weight	492 kg
Wheelbase	1651 mm
Tyre front	MH90S21
Tyre rear	MT90S16
Wheels front	1.85 x 21 wire-spoked
Wheels rear	3.00 x 16 wire-spoked
Top speed	155 km/h
Fuel tank capacity	19.2 litres

Model FXDG 1983 Series BG

Motor
Type	Alternator Shovelhead
Engine	Air cooled 45° ohv V-twin
Bore and stroke	3½ x 4¼ (3.5 x 4.25) in.
Bore and stroke	88.9 x 107.95 mm
Capacity	81.78 cu. in. (1340 cc)
Power output (DIN)	57(43)/5200 DIN
	65(48)/5400 rpm
Torque (Nm)	89/4000 DIN
Compression ratio	7.4:1
Carburettor	Keihin B78E 38 mm (DIN)
	Keihin B28E 38 mm (SAE)
Ignition	Electronic, contactless
Electrical system	12-volt
Starter	Electric starter
Exhaust system	2 into 2, 2 R/H silencers

Transmission
Gearbox	4-speed
Reduction ratios	2.45 (1)
	1.66 (2)
	1.23 (3)
	1.00 (4)
Secondary reduction ratio	1.91
Gearchange	Foot change
Final drive	Chain/belt

Cycle parts
Frame	Twin-loop, with straight front downtubes
Front forks	Telescopic, hydraulically-damped
Rear suspension	Square-section swinging arm
Brakes	Hydraulic disc brakes
Disc diameter	254 mm front, 292 mm rear
Front brake	Hand-operated disc
Rear brake	Foot-operated disc

General data
Dry weight	290 kg
Permissible total weight	492 kg
Wheelbase	1613 mm
Tyre front	MJ90S19
Tyre rear	MT90S16
Wheels front	2.15 x 19 cast-alloy
Wheels rear	3.00 x 16 cast-alloy wheel
Top speed	155 km/h
Fuel tank capacity	14.8 litres

Model FXE 1984 Series BA
Model FXEF 1984 Series BB

Motor
Type	Alternator Shovelhead
Engine	Air cooled 45° ohv V-twin
Bore and stroke	3½ x 4¼ (3.5 x 4.25) in.
Bore and stroke	88.9 x 107.95 mm
Capacity	81.78 cu. in. (1340 cc)
Power output (DIN)	57(42)/5200 DIN
	65(48)/5400 rpm
Torque (Nm)	89/4000 DIN
Compression ratio	7.4:1
Carburettor	Keihin B28H 38 mm (SAE)
	Keihin B28E 38 mm (DIN) optional
	Keihin B28G 38 mm (DIN)
Ignition	Electronic, contactless
Electrical system	12-volt
Starter	Electric starter
Exhaust system	2 into 2, 2 R/H silencers

Transmission
Gearbox	4-speed
Reduction ratios	2.45 (1)
	1.66 (2)
	1.23 (3)
	1.00 (4)
Secondary reduction ratio	1.91
Gearchange	Foot change
Final drive	Chain

Cycle parts
Frame	Twin-loop, with straight front downtubes
Front forks	Telescopic, hydraulically-damped
Rear suspension	Square-section swinging arm
Brakes	Hydraulic disc brakes
Disc diameter	292 mm
Front brake	Hand-operated disc
Rear brake	Foot-operated disc

General data
Dry weight	290 kg
Permissible total weight	492 kg
Wheelbase	1600 mm
Tyre front	MJ90S19
Tyre rear	MT90S16
Wheels front	2.50 x 19 wire-spoked
Wheels rear	3.00 x 16 wire- or cast-alloy
Top speed	155 km/h
Fuel tank capacity	12.9 litres (FXE)
	14.8 litres (FXEF)

Model FXSB 1984 Series BF

Motor
Type	Alternator Shovelhead
Engine	Air cooled 45° ohv V-twin
Bore and stroke	3½ x 4¼ (3.5 x 4.25) in.
Bore and stroke	88.9 x 107.95 mm
Capacity	81.78 cu. in. (1340 cc)
Power output (DIN)	57(42)/5200 DIN
	65(48)/5400 rpm
Torque (Nm)	89/4000 DIN
Compression ratio	7.4:1
Carburettor	Keihin B28H 38 mm (SAE)
	Keihin B28E 38 mm (DIN) optional
	Keihin B28G 38 mm (DIN)
Ignition	Electronic, contactless
Electrical system	12-volt
Starter	Kickstarter
Exhaust system	2 into 2, 2 R/H silencers

Transmission
Gearbox	4-speed
Reduction ratios	2.45 (1)
	1.66 (2)
	1.23 (3)
	1.00 (4)
Secondary reduction ratio	2.12
Gearchange	Foot change
Final drive	Belt/belt

Cycle parts
Frame	Twin-loop, with straight front downtubes
Front forks	Telescopic, hydraulically-damped
Rear suspension	Square-section swinging arm
Brakes	Hydraulic disc brakes
Disc diameter	254 mm front, 292 mm rear
Front brake	Hand-operated disc
Rear brake	Foot-operated disc

General data
Dry weight	290 kg
Permissible total weight	492 kg
Wheelbase	1613 mm
Tyre front	MJ90S19
Tyre rear	MT90S16
Wheels front	2.15 x 19 cast-alloy
Wheels rear	3.00 x 16 cast-alloy
Top speed	155 km/h
Fuel tank capacity	14.8 litres

Model FXWG 1984 Series BE

Motor
Type	Alternator Shovelhead
Engine	Air cooled 45° ohv V-twin
Bore and stroke	3½ x 4¼ (3.5 x 4.25) in.
Bore and stroke	88.9 x 107.95 mm
Capacity	81.78 cu. in. (1340 cc)
Power output (DIN)	57(42)/5200 DIN
	65(48)/5400 rpm
Torque (Nm)	89/4000 DIN
Compression ratio	7.4:1
Carburettor	Keihin B28H 38 mm (USA)
	Keihin B28E 38 mm (DIN), later Keihin B28G, dann B28K 38 mm (DIN)
Ignition	Electronic, contactless
Electrical system	12-volt
Starter	Kickstarter
Exhaust system	2 into 2, 2 R/H silencers; optional 2 into 2, 2 R/H silencers

Transmission
Gearbox	4-speed
Reduction ratios	2.45 (1)
	1.66 (2)
	1.23 (3)
	1.00 (4)
Secondary reduction ratio	2.12
Gearchange	Foot change
Final drive	Chain/belt

Cycle parts
Frame	Twin-loop, with straight front downtubes
Front forks	Telescopic, hydraulically-damped
Rear suspension	Square-section swinging arm
Brakes	Hydraulic disc brakes
Disc diameter	292 mm
Front brake	Hand-operated disc
Rear brake	Foot-operated disc

General data
Dry weight	290 kg
Permissible total weight	492 kg
Wheelbase	1651 mm
Tyre front	MH90S21
Tyre rear	MT90S16
Wheels front	1.85 x 21 wire-spoked
Wheels rear	3.00 x 16 wire-spoked
Top speed	155 km/h
Fuel tank capacity	18.9 litres

Model FXEF 1984 Series BB

Motor
Type	Evolution
Engine	Air cooled 45° ohv V-twin
Bore and stroke	3½ x 4¼ (3.5 x 4.25) in.
Bore and stroke	88.9 x 107.95 mm
Capacity	81.78 cu. in. (1340 cc)
Power output (DIN)	64(47)/5000 DIN
	64(47)/5000 rpm
Torque (Nm)	102/3600 DIN
Compression ratio	8.5:1
Carburettor	Keihin B28H 38 mm (SAE)
	Keihin 99BA 38 mm (DIN)
Ignition	Electronic, contactless
Electrical system	12-volt
Starter	Electric starter
Exhaust system	2 into 2, 2 R/H silencers

Transmission
Gearbox	4-speed
Reduction ratios	2.45 (1)
	1.66 (2)
	1.23 (3)
	1.00 (4)
Secondary reduction ratio	2.08
Gearchange	Foot change
Final drive	Chain/belt

Cycle parts
Frame	Twin-loop, with straight front downtubes
Front forks	Telescopic, hydraulically-damped
Rear suspension	Square-section swinging arm
Brakes	Hydraulic disc brakes
Disc diameter	292 mm
Front brake	Hand-operated disc
Rear brake	Foot-operated disc

General data
Dry weight	290 kg
Permissible total weight	492 kg
Wheelbase	1600 mm
Tyre front	MJ90S19
Tyre rear	MT90S16
Wheels front	2.50 x 19 wire-spoked
Wheels rear	3.00 x 16 wire- or cast-alloy
Top speed	160 km/h
Fuel tank capacity	15.8 litres

Model FXWG 1985 and 1986 Series BE

Motor
Type	Evolution
Engine	Air cooled 45° ohv V-twin
Bore and stroke	3½ x 4¼ (3.5 x 4.25) in.
Bore and stroke	88.9 x 107.95 mm
Capacity	81.78 cu. in. (1340 cc)
Power output (DIN)	64(47)/5000 DIN
	64(47)/5000 rpm
Torque (Nm)	102/3600 DIN
Compression ratio	8.5:1
Carburettor	Keihin B28H 38 mm (USA)
	Keihin 99BA 38 mm (DIN)
Ignition	Electronic, contactless
Electrical system	12-volt
Starter	Kickstarter
Exhaust system	2 into 2, 2 R/H silencers

Transmission
Gearbox	4-speed
Reduction ratios	2.45 (1)
	1.66 (2)
	1.23 (3)
	1.00 (4)
Secondary reduction ratio	2.12
Gearchange	Foot change
Final drive	Chain/belt

Cycle parts
Frame	Twin-loop, with straight front downtubes
Front forks	Telescopic, hydraulically-damped
Rear suspension	Square-section swinging arm
Brakes	Hydraulic disc brakes
Disc diameter	292 mm
Front brake	Hand-operated disc
Rear brake	Foot-operated disc

General data
Dry weight	290 kg
Permissible total weight	492 kg
Wheelbase	1651 mm
Tyre front	MH90S21
Tyre rear	MT90S16
Wheels front	1.85 x 21 wire-spoked
Wheels rear	3.00 x 16 wire-spoked
Top speed	170 km/h
Fuel tank capacity	18.9 litres

Model FXSB 1985 Series BF

Motor
Type	Evolution
Engine	Air cooled 45° ohv V-twin
Bore and stroke	3½ x 4¼ (3.5 x 4.25) in.
Bore and stroke	88.9 x 107.95 mm
Capacity	81.78 cu. in. (1340 cc)
Power output (DIN)	64(47)/5000 DIN
	64(47)/5000 rpm
Torque (Nm)	102/3600 DIN
Compression ratio	8.5:1
Carburettor	Keihin B28H 38 mm (SAE)
	Keihin 99BA 38 mm (DIN)
Ignition	Electronic, contactless
Electrical system	12-volt
Starter	Kickstarter
Exhaust system	2 into 2, 2 R/H silencers

Transmission
Gearbox	4-speed
Reduction ratios	2.45 (1)
	1.66 (2)
	1.23 (3)
	1.00 (4)
Secondary reduction ratio	2.12
Gearchange	Foot change
Final drive	Belt/belt

Cycle parts
Frame	Twin-loop, with straight front downtubes
Front forks	Telescopic, hydraulically-damped
Rear suspension	Square-section swinging arm
Brakes	Hydraulic disc brakes
Disc diameter	292 mm
Front brake	Hand-operated disc
Rear brake	Foot-operated disc

General data
Dry weight	290 kg
Permissible total weight	492 kg
Wheelbase	1600 mm
Tyre front	MJ90S19
Tyre rear	MT90S16
Wheels front	2.50 x 19 wire-spoked
Wheels rear	3.00 x 16 wire- or cast-alloy
Top speed	160 km/h
Fuel tank capacity	15.8 litres

Military machines
1910 to 1973

From about 1910 onwards, Harley-Davidson motorcycles were ordered in steeply-rising numbers by the US Army. Since these were no different from the civilian models of the period, they are covered in those chapters.

The first motorcycle exported specifically for military use, albeit in modest quantities, was in 1912, being delivered to Japan for Army use.

Even in the USA, ever since the unleashing of unrestricted submarine warfare by the German High Command, those in the highest political and industrial circles were no longer in a position to ignore the growing threat of war. In addition, it was clear to all those involved in the USA that such a World War, a clash of nations, as this would be, was not going to be won by cavalry and horse-drawn vehicles, but by motorized units. This view was supported by the relatively successful use of cars and motorcycles in support of the 1916-7 campaign against Pancho Villa in Mexico.

So the Army ordered motorcycles from Indian and from Harley-Davidson in correspondingly large numbers to equip the new motorcycle despatch messenger units, amongst other uses. By May 1917, on the entry of the USA into the War on the side of the Allied Powers, the Army was thus splendidly prepared and sent matèriel in huge quantities to the battlefields in France, including some 70,000 motorcycles, of which 15,000 were Harleys with nearly as many sidecars.

Purpose-built sidecar outfits – the AC, or Ammunition Car of 1916 – were built; quick and manoeuvrable with some limited cross-country ability and thought capable of transporting ammunition and supplies right up to the front lines; another variation was the SC, or Stretcher Car – sidecar chassis and stretcher, complete. In this way, in the eyes of the Americans, the motorcycle began to be thought to exhibit some potential as a tactical weapon. Gun Cars (motorcycle and sidecar carrying a light gun or Colt machine gun) were built, but a prototype of an armoured motorcycle with a gun did not however make it into production; the newly-developed English Tanks were much more appropriate to the Western Front than this vehicle. However, in spite of these developments, the majority of Harleys spent their military service on despatch-riding duties.

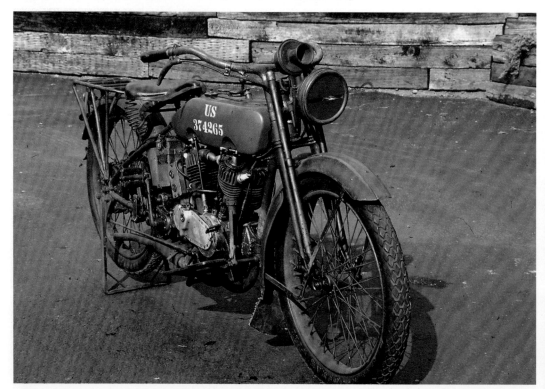

1917: A military motorcycle for the US Army. Same as the civilian models, but all in Olive Green

Shortly after the end of hostilities in November 1918 a further 4,500 vehicles were delivered to the Army. This was the last military contract for more than a decade; cars were regarded as being a better means of transport to provide mobility for the troops.

A prototype that is worthy of mention appeared in 1925; an armoured motorcycle developed for the fight against organised crime. With its thick bullet-proof glass and steel shields, it turned out however to be far too heavy and thus was only fit for limited use. A single example remained.

Only three examples of the 1920 Model W found their way into the Army. From 1920 to 1931 no H-D machines at all were ordered for the US armed forces. It wasn't until spring 1931 that an order came, for the first time in 12 years, from the Department of Defense. This contract was for the relatively small total of 110 examples of the new side-valve 45 cu. in. (742 cc) Model R, which were to be built especially to Army specifications. This preliminary order was followed in 1934 by others, this time for 15 Model RL machines with higher-compression engines and 300 of the big 74 cu. in. (1217 cc) Model VDS, with sidecars.

The Army remained faithful to its Forty-Fives and from 1937 to the end of the War in 1945 the US government ordered the brand-new WL in various specifications, according to intended use, in hitherto unequalled numbers (some 75,000 motorcycles, plus spares for another 30,000 machines!). Even after the end of the War, when new war-surplus machines were being dumped on the civilian market at rock-bottom prices, orders still came in for the good old WLA. Between 1949 and 1952, military production of the WL carrying on a year after civilian production had ended, another 1000 motorcycles were taken.

On top of this, the Canadians bought their WLC (C standing for Canada, not Civilian, as is sometimes thought) for the Canadian National Defense Forces. Other Allied armies profited by this first-class material from Milwaukee under the Lend-Lease Act; a total of 38,103 WLAs were supplied to Allied Nations – which means that more WLAs were used by foreign forces than by the United States Army. The British Empire (i.e., Australia, New Zealand, and South Africa – but some were even used by British units, notably the RAF police) took 8,686, the Free French Forces received 926, China 1,000, Brazil 45, other Latin-American republics 156, the Netherlands 420 and so on. The Red Army rolled to Berlin on, amongst other things, some 26,670 Lend-Lease Harley WLAs. There were also special models such as the WLA State Guard models delivered to Connecticut in 1951.

The big side-valve models were naturally not taken up in such quantities, but even so they successfully fulfilled various military requirements from 1938 to 1945 in the form of the low-compression 74 cu. in. (1209 cc) Models UA and US (indicating A for Army and S for Sidecar, not United States!), which were delivered with and without sidecars. A high-compression version of the ohv 60 cu. in. (989 cc) Model ELA got in on the

1918: A military motorcycle for the US Army

1934: Model VD in military trim

act and the US Navy ordered the basic Model E, while the Canadians specified the Model ELC. Other models were ordered either singly or in miniscule quantities.

A further military Harley, never produced for civilian use, saw the light of day in 1942; the Model XA (X indicating Special, or experimental, construction, A for Army). This motorcycle was a straightforward copy by H-D of one of the first German motorcycles captured from the *Wehrmacht*, a 45.49 cu. in. (745 cc.) BMW R71.

As a result it had a horizontally-opposed transverse twin-cylinder side-valve engine with a wet-sump lubrication system; a departure from the norm for Harley, whose machines were all dry-sump. Anyway, approximately 1,000 of these machines were ordered in 1942; the first deliveries were made in the same year, with the remainder following in 1943.

The Model WSR (a designation that had already been used in 1937 for an export model for Japan) was developed especially for the Red Army. Although only listed as a prototype in the factory literature, a 1943 example recently surfaced in Germany. They are supposed to have been delivered to the Russians right up to 1945.

A couple of other prototypes should not pass unmentioned by this book. One is the WRS, an offshoot of the WLA and using the same engine but with a special frame.

And the other is of course the XS, the sidecar version of the XA, the flat twin described above, with sidecar-wheel-drive. On top of that there was a further three-wheeler, the 1940/1 Model TA, which is described in its own chapter. It combined the Servi-Car concept with a Knucklehead engine which had a slightly reduced capacity of 68 cu. in. (1122 cc.). Only 16 of these curious three-wheelers were delivered before the success of the much more capable, adaptable and cheaper Jeep rendered them superfluous.

A cowled, fan-cooled version of the XA engine was built as an auxiliary powerplant for a Detroit tank manufacturer. Another attempted use of the engine was in the development, in conjunction with Willys-Overland, of the Type WAC (Willys Air-Cooled) four-wheeled light cargo carrier of which four examples were built in the winter of 1943/44. This application, which led to the creation of the 'Jeeplet', used an overhead-valve 49 cu. in. (803 cc.) version of the XA unit. The same engine was also used in the prototype of the Type XOD generator. The Army ordered a consignment of 4,975 of these in May/June 1945, but this contract was abruptly cancelled upon the end of the war in the Far East; only a single prototype survived.

WLA for the US Army

WLA for the US Army – complete with Tommy gun

Another WLA – but in anything but olive drab!

WLC for Canada

WLC for Canada – a nicely-restored example

Back to the WLAs for the USA – well, there were over 75,000 of them!

WLA taillamps – one for the blackout, and one to stop you being tailgated by tanks (some of which were faster than a WLA)!

A 120 mph speedometer for a WLA! Gives the rider something to look at, as he's ambling along ...

Model XA – 1,000 were built for the US Army, but Jeeps were cheaper, and easier to use, and could carry more ...

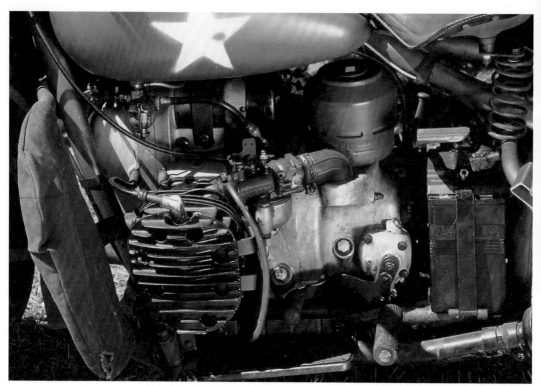

Being copies of BMWs, XAs introduced Harley Davidson to several novelties – square bore/stroke ratio, no vibration, twin ...

... carburettors, wet-sump lubrication, foot change and hand clutch, rear suspension (even if it was plunger), shaft drive ...

Some XAs escaped into civilian hands post-war. Looks like this one got out by way of the chrome shop!

Model 18FUS 1918 (sidecar-specification)
Model 19FUS 1919 (sidecar-specification)

Motor
Engine	Air-cooled 45° ioe V-twin
Bore x stroke	3³⁄₁₆ x 3½ (3.31 x 3.5) in.
Bore x stroke	84.14 x 88.9 mm
Capacity	60.33 cu. in. (989 cc)
Power output (SAE)	14 (10)/3000 rpm
Compression ratio	3.3:1
Carburettor	Schebler 1-inch
Ignition	Magneto

Transmission
Gearbox	3-speed
Gearchange	Hand change, on tank
Final drive	Chain

Cycle parts
Frame	Single-loop, straight downtube
Front forks	Springers – enclosed springs, tubular legs
Rear suspension	None (rigid)
Front brake	None
Rear brake	Drum, via rod and pedal

General data
Dry weight	approx. 160 kg
Tyre front	3 x 28 (beaded-edge)
Tyre rear	3 x 28 (beaded-edge)
Top speed	100 km/h
Fuel tank capacity	10.4 litres
Production figures	8095 (1918)
	7521 (1919)
Contract price	Not available
Sidecar	AC (Ammunition Car)
	GC (Gun Car)

Model 32R 1932

Motor
Engine	Air-cooled 45° sv V-twin
Bore x stroke	2¾ x 3¹³⁄₁₆ (2.75 x 3.81) in.
Bore x stroke	69.85 x 96.84 mm
Capacity	45.12 cu. in. (742 cc)
Power output (SAE)	17 (13)4000 rpm
Compression ratio	4.3:1
Carburettor	Schebler Deluxe 1⅛
Exhaust system	Small-fishtail silencer

Transmission
Gearbox	3-speed
Secondary reduction ratio	1.875
Gearchange	Hand change, on tank
Final drive	Chain

Cycle parts
Frame	Single-loop, full cradle
Front forks	Springers – exposed-springs, forged I-beam legs
Rear suspension	None (rigid)
Front brake	Drum, via cable and handlebar lever
Rear brake	Drum, via rod and pedal

General data
Dry weight	approx. 185 kg
Permissible total weight	330 kg
Tyres – front and rear	4.00-18 wire-edged
Wheels	18-inch wire-spoked
Top speed	approx. 110 km/h
Fuel tank capacity	21.3 litres
Production figures	110
Contract price	Not available

Model 34RL 1934
Model 35RL 1935

Motor
Engine	Air-cooled 45° sv V-twin
Bore x stroke	2¾ x 3¹³⁄₁₆ (2.75 x 3.81) in.
Bore x stroke	69.85 x 96.84 mm
Capacity	45.12 cu. in. (742 cc)
Power output (SAE)	18.5 (14)/4000 rpm
Compression ratio	5.0:1
Carburettor	Linkert 1 Type M-16
Exhaust system	High-Flo exhaust (1934) Large fishtail silencer (1935)

Transmission
Gearbox	3-speed
Secondary reduction ratio	1.875 (1934) or 2.56 (1935)
Gearchange	Hand change, on tank
Final drive	Chain

Cycle parts
Frame	Single-loop, full cradle
Front forks	Springers – exposed-springs, forged I-beam legs
Rear suspension	None (rigid)
Front brake	Drum, via cable and handlebar lever
Rear brake	Drum, via rod and pedal

General data
Dry weight	approx. 185 kg
Permissible total weight	330 kg
Tyres – front and rear	4.00-18 wire-edged
Wheels	18-inch wire-spoked
Top speed	approx. 115 km/h (1934), 95 km/h (1935)
Fuel tank capacity	21.3 litres
Production figures	9 (1934) 6 (1935)
Contract price	Not available

Model 37WL 1937

Motor
Engine	Air-cooled 45° sv V-twin
Bore x stroke	2¾ x 3¹³⁄₁₆ (2.75 x 3.81) in.
Bore x stroke	69.85 x 96.84 mm
Capacity	45.12 cu. in. (742 cc)
Power output (SAE)	22 (16)/4600 rpm
Compression ratio	5.0:1
Carburettor	Linkert Type M-41
Exhaust system	Large fishtail silencer

Transmission
Gearbox	3-speed
Secondary reduction ratio	2.56
Gearchange	Hand change, on tank
Final drive	Chain

Cycle parts
Frame	Single-loop, full cradle
Front forks	Springers – exposed-springs, forged I-beam legs
Rear suspension	None (rigid)
Front brake	Drum, via cable and handlebar lever
Rear brake	Drum, via rod and pedal

General data
Dry weight	approx. 240 kg
Permissible total weight	410 kg
Tyres – front and rear	4.00-18
Wheels	2.15 x 18 wire-spoked
Top speed	90 km/h
Fuel tank capacity	12.9 litres
Production figures	51 to 58
Contract price	Not available

Model 39W 1939 National Guard

Motor
Engine	Air-cooled 45° sv V-twin
Bore x stroke	2¾ x 3¹³⁄₁₆ (2.75 x 3.81) in.
Bore x stroke	69.85 x 96.84 mm
Capacity	45.12 cu. in. (742 cc)
Power output (SAE)	19 (14)/4400 rpm
Compression ratio	4.3:1
Carburettor	Linkert Type M-41
Exhaust system	Large fishtail silencer

Transmission
Gearbox	3-speed
Secondary reduction ratio	2.56
Gearchange	Hand change, on tank
Final drive	Chain

Cycle parts
Frame	Single-loop, full cradle
Front forks	Springers – exposed-springs, forged I-beam legs
Rear suspension	None (rigid)
Front brake	Drum, via cable and handlebar lever
Rear brake	Drum, via rod and pedal

General data
Dry weight	approx. 240 kg
Permissible total weight	410 kg
Tyres – front and rear	4.00-18
Wheels	2.15 x 18 wire-spoked
Top speed	90 km/h
Fuel tank capacity	12.9 litres
Production figures	46
Contract price	Not available

Model 40WLA 1940

Motor
Engine	Air-cooled 45° sv V-twin
Bore x stroke	2¾ x 3¹³⁄₁₆ (2.75 x 3.81) in.
Bore x stroke	69.85 x 96.84 mm
Capacity	45.12 cu. in. (742 cc)
Power output (SAE)	19 (14)/4400 rpm
Compression ratio	4.3:1
Carburettor	Linkert Type M-64
Exhaust system	Large fishtail silencer

Transmission
Gearbox	3-speed
Secondary reduction ratio	2.56
Gearchange	Hand change, on tank
Final drive	Chain

Cycle parts
Frame	Single-loop, full cradle
Front forks	Springers – exposed-springs, oval-tube legs
Rear suspension	None (rigid)
Front brake	Drum, via cable and handlebar lever
Rear brake	Drum, via rod and pedal

General data
Dry weight	approx. 240 kg
Permissible total weight	410 kg
Tyres – front and rear	4.00-18
Wheels	2.15 x 18 wire-spoked
Top speed	90 km/h
Fuel tank capacity	12.9 litres
Production figures	422
Contract price	$337

Model 40WLA Type I and II 1941

Motor
Engine	Air-cooled 45° sv V-twin
Bore x stroke	2¾ x 3¹³⁄₁₆ (2.75 x 3.81) in.
Bore x stroke	69.85 x 96.84 mm
Capacity	45.12 cu. in. (742 cc)
Power output (SAE)	19 (14)/4400 SAE (Type I)
	21 (15)/4600 SAE (Type II)
Compression ratio	4.3:1 (Type I)
	4.75:1 (Type II)
Carburettor	Linkert Type M-84 (Type I)
	Linkert Type M-88 (Type II)
Exhaust system	Large fishtail silencer

Transmission
Gearbox	3-speed
Secondary reduction ratio	2.41
Gearchange	Hand change, on tank
Final drive	Chain

Cycle parts
Frame	Single-loop, full cradle
Front forks	Springers – exposed-springs, oval-tube legs
Rear suspension	None (rigid)
Front brake	Drum, via cable and handlebar lever
Rear brake	Drum, via rod and pedal

General data
Dry weight	approx. 240 kg
Permissible total weight	410 kg
Tyres – front and rear	4.00-18
Wheels	2.15 x 18 wire-spoked
Top speed	90 km/h
Fuel tank capacity	12.9 litres
Production figures	651 (Type I)
	1800 (Type II)
Contract price	$367
Similar models	41WLA for South Africa (2350 machines) with Type M-41 carburettor and 5.00-16 tyres

Model 42WLA Type IA 1941
Model 42WLA Type IB 1941
Model 42WLA Type IC 1941

Motor

Engine	Air-cooled 45° sv V-twin
Bore x stroke	2¾ x 3¹³⁄₁₆ (2.75 x 3.81) in.
Bore x stroke	69.85 x 96.84 mm
Capacity	45.12 cu. in. (742 cc)
Power output (SAE)	22 (16)/4600 rpm
Compression ratio	5.0:1
Carburettor	Linkert Type M-88
Exhaust system	Large fishtail silencer

Transmission

Gearbox	3-speed
Secondary reduction ratio	2.56
Gearchange	Hand change, on tank
Final drive	Chain

Cycle parts

Frame	Single-loop, full cradle
Front forks	Springers – exposed-springs, oval-tube legs
Rear suspension	None (rigid)
Front brake	Drum, via cable and handlebar lever
Rear brake	Drum, via rod and pedal

General data

Dry weight	approx. 240 kg
Permissible total weight	410 kg
Wheelbase	1461 mm
Tyres – front and rear	4.00-18
Wheels	2.15 x 18 wire-spoked
Top speed	95 km/h
Fuel tank capacity	12.9 litres
Production figures	912 (Type IA)
	400 (Type IB)
	404 (Type IC)
Contract price	$395

Model 42WLA Type II 1941 to 1942

Motor

Engine	Air-cooled 45° sv V-twin
Bore x stroke	2¾ x 3¹³⁄₁₆ (2.75 x 3.81) in.
Bore x stroke	69.85 x 96.84 mm
Capacity	45.12 cu. in. (742 cc)
Power output (SAE)	22 (16)/4600 rpm
Compression ratio	5.0:1
Carburettor	Linkert Type M-88
Exhaust system	Large fishtail silencer

Transmission

Gearbox	3-speed
Secondary reduction ratio	2.56
Gearchange	Hand change, on tank
Final drive	Chain

Cycle parts

Frame	Single-loop, full cradle
Front forks	Springers – exposed-springs, oval-tube legs
Rear suspension	None (rigid)
Front brake	Drum, via cable and handlebar lever
Rear brake	Drum, via rod and pedal

General data

Dry weight	approx. 240 kg
Permissible total weight	410 kg
Wheelbase	1461 mm
Tyres – front and rear	4.00-18
Wheels	2.15 x 18 wire-spoked
Top speed	95 km/h
Fuel tank capacity	12.9 litres
Production figures	3054
	of which 350 (1941)
	of which 2704 (1942)
	in addition 830 42WLs produced for the Army
Contract price	$395
Canadian-specification	42WLC (3290 machines)

Model 42WLA Type III 1942 to 1943

Motor
Engine	Air-cooled 45° sv V-twin
Bore x stroke	2¾ x 3¹³⁄₁₆ (2.75 x 3.81) in.
Bore x stroke	69.85 x 96.84 mm
Capacity	45.12 cu. in. (742 cc)
Power output (SAE)	22 (16)/4600 rpm
Compression ratio	5.0:1
Carburettor	Linkert Type M-88
Exhaust system	Large fishtail silencer

Transmission
Gearbox	3-speed
Secondary reduction ratio	2.56
Gearchange	Hand change, on tank
Final drive	Chain

Cycle parts
Frame	Single-loop, full cradle
Front forks	Springers – exposed-springs, oval-tube legs
Rear suspension	None (rigid)
Front brake	Drum, via cable and handlebar lever
Rear brake	Drum, via rod and pedal

General data
Dry weight	approx. 240 kg
Permissible total weight	410 kg
Wheelbase	1461 mm
Tyres – front and rear	4.00-18
Wheels	2.15 x 18 wire-spoked
Top speed	95 km/h
Fuel tank capacity	12.9 litres
Production figures	20313
Contract price	$377
Canadian-specification	43-WLC (8000 motorcycles) produced from the end of 1942 to early 1944.

Model 42WLA Type IV 1943

Motor
Engine	Air-cooled 45° sv V-twin
Bore x stroke	2¾ x 3¹³⁄₁₆ (2.75 x 3.81) in.
Bore x stroke	69.85 x 96.84 mm
Capacity	45.12 cu. in. (742 cc)
Power output (SAE)	22 (16)/4600 rpm
Compression ratio	5.0:1
Carburettor	Linkert Type M-88
Exhaust system	Large fishtail silencer

Transmission
Gearbox	3-speed
Secondary reduction ratio	2.56
Gearchange	Hand change, on tank
Final drive	Chain

Cycle parts
Frame	Single-loop, full cradle
Front forks	Springers – exposed-springs, oval-tube legs
Rear suspension	None (rigid)
Front brake	Drum, via cable and handlebar lever
Rear brake	Drum, via rod and pedal

General data
Dry weight	approx. 240 kg
Permissible total weight	410 kg
Wheelbase	1461 mm
Tyres – front and rear	4.00-18
Wheels	2.15 x 18 wire-spoked
Top speed	95 km/h
Fuel tank capacity	12.9 litres
Production figures	11,600
Contract price	$377

Model 37WSR 1937
Export to Japan

Motor
Engine	Air-cooled 45° sv V-twin
Bore x stroke	2¾ x 3¹³⁄₁₆ (2.75 x 3.81) in.
Bore x stroke	69.85 x 96.84 mm
Capacity	45.12 cu. in. (742 cc)
Power output (SAE)	19 (14)/4400 rpm
Compression ratio	4.3:1
Carburettor	Linkert Type M-21
Exhaust system	Large fishtail silencer

Transmission
Gearbox	3-speed
Gearchange	Hand change, on tank
Final drive	Chain

Cycle parts
Frame	Single-loop, full cradle
Front forks	Springers – exposed-springs, forged I-beam legs
Rear suspension	None (rigid)
Front brake	Drum, via cable and handlebar lever
Rear brake	Drum, via rod and pedal

General data
Dry weight	approx. 177 kg
Permissible total weight	approx. 400 kg
Tyres – front and rear	4.00-18
Wheels	2.15 x 18 wire-spoked
Top speed	90 km/h
Fuel tank capacity	12.9 litres
Production figures	5
Contract price	Not available

Model 49WLA 1948 to 1949
Model 50WLA 1950

Motor
Engine	Air-cooled 45° sv V-twin
Bore x stroke	2¾ x 3¹³⁄₁₆ (2.75 x 3.81) in.
Bore x stroke	69.85 x 96.84 mm
Capacity	45.12 cu. in. (742 cc)
Power output (SAE)	22 (16)/4600 rpm
Compression ratio	5.0:1
Carburettor	Linkert Type M-88
Exhaust system	Large fishtail silencer

Transmission
Gearbox	3-speed
Secondary reduction ratio	2.56
Gearchange	Hand change, on tank
Final drive	Chain

Cycle parts
Frame	Single-loop, full cradle
Front forks	Springers – exposed-springs, oval-tube legs
Rear suspension	None (rigid)
Front brake	Drum, via cable and handlebar lever
Rear brake	Drum, via rod and pedal

General data
Dry weight	approx. 240 kg
Permissible total weight	410 kg
Tyres – front and rear	4.00-18
Wheels	2.15 x 18 wire-spoked
Top speed	95 km/h
Fuel tank capacity	12.9 litres
Production figures	379 (?) (1949)
	57 (?) (1950)
Contract price	Not available

Model 51WLA Connecticut State Guard 1951
Model 51WLA Type I and II 1951
Model 52WLA 1951 to 1952

Motor
Engine	Air-cooled 45° sv V-twin
Bore x stroke	2¾ x 3¹³⁄₁₆ (2.75 x 3.81) in.
Bore x stroke	69.85 x 96.84 mm
Capacity	45.12 cu. in. (742 cc)
Power output (SAE)	22 (16)/4600 rpm
Compression ratio	5.0:1
Carburettor	Linkert Type M-88 (to 1951 Type I) Linkert Type M-54B (from 1951 Type II)
Exhaust system	Large fishtail silencer

Transmission
Gearbox	3-speed
Secondary reduction ratio	2.56
Gearchange	Hand change, on tank
Final drive	Chain

Cycle parts
Frame	Single-loop, full cradle
Front forks	Springers – exposed-springs, oval-tube legs
Rear suspension	None (rigid)
Front brake	Drum, via cable and handlebar lever
Rear brake	Drum, via rod and pedal

General data
Dry weight	approx. 240 kg
Permissible total weight	410 kg
Tyres – front and rear	4.00-18
Wheels	2.15 x 18 wire-spoked
Top speed	95 km/h
Fuel tank capacity	12.9 litres
Production figures	16 (1951 Type I) 10 (1951 Type II) 549 (1952)
Contract price	Not available

Model 34VDS 1934
Model 35VDS 1935
with sidecar

Motor
Engine	Air-cooled 45° sv V-twin
Bore x stroke	3⁷⁄₁₆ x 4 (3.44 x 4) in.
Bore x stroke	87.31 x 101.6 mm
Capacity	74.24 cu. in. (1217 cc)
Power output (SAE)	25(18)/4000 rpm
Compression ratio	5.0:1
Carburettor	Linkert Type M-31
Ignition	Battery-and-coil
Exhaust system	High-Flo exhaust (1934) Large fishtail silencer (1935)

Transmission
Gearbox	3-speed without reverse
Secondary reduction ratio	1.82
Gearchange	Hand change, on tank
Final drive	Chain

Cycle parts
Frame	Single-loop, curved downtube
Front forks	Springers – exposed-springs, forged I-beam legs
Rear suspension	None (rigid)
Front brake	Drum, via cable and handlebar lever
Rear brake	Drum, via rod and pedal

General data
Dry weight	approx. 320 kg
Permissible total weight	approx. 500 kg
Tyres – front and rear	4.00-19 (1934-35) 4.50-19 (1934-35) optional 4.00-18 (1935)
Wheels	2.15 x 19 wire-spoked or 2.15 x 18 wire-spoked
Top speed	approx. 105 km/h
Fuel tank capacity	21.3 litres
Production figures	312 (1934) 12 (1935)
Contract price	Not available

Model 38US 1938
Model 39UA 1939
Model 40UA 1940
and sidecar

Motor
Engine	Air-cooled 45° sv V-twin
Bore x stroke	3⁵⁄₁₆ x 4⁹⁄₃₂ (3.31 x 4.28) in.
Bore x stroke	84.14 x 108.74 mm
Capacity	73.79 cu. in. (1209 cc)
Power output (SAE)	32.5 (24)/4200 rpm
Compression ratio	5.0:1
Carburettor	Linkert 1¼-inch Type M-51 (1938/39)
	Linkert 1¼-inch Type M-65 (1940)
Ignition	Battery-and-coil
Exhaust system	Large fishtail silencer

Transmission
Gearbox	3-speed without reverse
Secondary reduction ratio	2.32
Gearchange	Hand change, on tank
Final drive	Chain

Cycle parts
Frame	Twin-loop, with straight front downtubes
Front forks	Springers – exposed-springs, oval-tube legs
Rear suspension	None (rigid)
Front brake	Drum, via cable and handlebar lever
Rear brake	Drum, via rod and pedal

General data
Dry weight	approx. 330 kg
Permissible total weight	approx. 520 kg
Tyres – front and rear	4.50-18
Wheels	2.50 x 18 wire-spoked
Top speed	115 km/h
Fuel tank capacity	15.1 litres
Production figures	71 (1938)
	335 (1939)
	509 (1940)
Contract price	$470 to 496 (1940)

Model 41US 1941
Model 42US 1942
Model 43US 1943
Model 44US 1944
Model 45US 1945
US-Marine Corps
with sidecar

Motor
Engine	Air-cooled 45° sv V-twin
Bore x stroke	3⁵⁄₁₆ x 4⁹⁄₃₂ (3.31 x 4.28) in.
Bore x stroke	84.14 x 108.74 mm
Capacity	73.79 cu. in. (1209 cc)
Power output (SAE)	32.5 (24)/4200 rpm
Compression ratio	5.0:1
Carburettor	Linkert 1¼-inch Type M-51 (1941/42)
	Linkert 1¼-inch Type M-58 (1943/45)
Ignition	Battery-and-coil
Exhaust system	Large fishtail silencer

Transmission
Gearbox	3-speed, with reverse
Secondary reduction ratio	2.32
Gearchange	Hand change, on tank
Final drive	Chain

Cycle parts
Frame	Twin-loop, with straight front downtubes
Front forks	Springers – exposed-springs, oval-tube legs
Rear suspension	None (rigid)
Front brake	Drum, via cable and handlebar lever
Rear brake	Drum, via rod and pedal

General data
Dry weight	approx. 330 kg
Permissible total weight	approx. 520 kg
Tyres – front and rear	5.00-16
Wheels	3.00 x 16 wire-spoked
Top speed	115 km/h
Fuel tank capacity	15.1 litres
Production figures	1941 Not available
	1942 Not available
	1943 approx. 1025
	1944 approx. 240
	1945 approx. 260
Contract price	Not available

Model 41US 1941
Model 42US 1942
Model 44US 1944
South African Union Defense Forces with sidecar

Motor
Engine	Air-cooled 45° sv V-twin
Bore x stroke	3⁵⁄₁₆ x 4⁹⁄₃₂ (3.31 x 4.28) in.
Bore x stroke	84.14 x 108.74 mm
Capacity	73.79 cu. in. (1209 cc)
Power output (SAE)	30 (22)/4000 rpm
Compression ratio	4.5:1
Carburettor	Linkert 1¼-inch Type M-51 (1941) Linkert 1¼-inch Type M-58 (1942/44)
Ignition	Battery-and-coil
Exhaust system	Large fishtail silencer

Transmission
Gearbox	4-speed
Secondary reduction ratio	2.32
Gearchange	Hand change, on tank
Final drive	Chain

Cycle parts
Frame	Twin-loop, with straight front downtubes
Front forks	Springers – exposed-springs, oval-tube legs
Rear suspension	None (rigid)
Front brake	Drum, via cable and handlebar lever
Rear brake	Drum, via rod and pedal

General data
Dry weight	approx. 330 kg
Permissible total weight	approx. 520 kg
Tyres – front and rear	5.00-16
Wheels	3.00 x 16 wire-spoked
Top speed	115 km/h
Fuel tank capacity	15.1 litres
Production figures	1941 156 1942 1622 1944 44
Contract price	Not available
Similar models	44US for Australia (100 machines) None sidecar-equipped

---

Model 42E 1942
Model 43E 1943
Model 44E 1944
Model 45E 1945
US Navy

Motor
Type	Knucklehead
Engine	Air cooled 45° ohv V-twin
Bore and stroke	3⁵⁄₁₆ x 3½ (3.31 x 3.5) in.
Bore and stroke	84.14 x 88.9 mm
Capacity	60.33 cu. in. (989 cc)
Power output (SAE)	40 hp (29 kW)/4800 rpm
Compression ratio	6.5:1
Carburettor	Linkert 1½-inch Type M-35
Ignition	Battery-and-coil
Exhaust system	Rocket-Fin silencer

Transmission
Gearbox Secondary reduction ratio	4-speed (optional 3-speed) 3-speed: 3.90:1; 4-speed: 3.73:1
Gearchange	Hand change, on tank
Final drive	Chain

Cycle parts
Frame	Twin-loop, with straight front downtubes
Front forks	Springers – exposed-springs, oval-tube legs
Rear suspension	None (rigid)
Front brake	Drum, via cable and handlebar lever
Rear brake	Drum, via rod and pedal

General data
Dry weight	240 kg
Permissible total weight	440 kg
Tyres – front and rear	5.00-16
Wheels	3.00 x 16 wire-spoked
Top speed	140 km/h
Fuel tank capacity	14.2 litres
Production figures	Not available (could only have been a very few)
Contract price	Not available

Model 42ELA 1942 US Army
Model 42ELC 1942 Canadian National Defense

Motor
Type	Knucklehead
Engine	Air cooled 45° ohv V-twin
Bore and stroke	3⁵⁄₁₆ x 3½ (3.31 x 3.5) in.
Bore and stroke	84.14 x 88.9 mm
Capacity	60.33 cu. in. (989 cc)
Power output (SAE)	42(31)/4800 rpm
Compression ratio	7.0:1
Carburettor	Linkert 1½-inch Type M-90
Ignition	Battery-and-coil
Exhaust system	Rocket-Fin silencer

Transmission
Gearbox	3-speed with reverse
Secondary reduction ratio	3.90:1
Gearchange	Hand change, on tank
Final drive	Chain

Cycle parts
Frame	Twin-loop, with straight front downtubes
Front forks	Springers – exposed-springs, oval-tube legs
Rear suspension	None (rigid)
Front brake	Drum, via cable and handlebar lever
Rear brake	Drum, via rod and pedal

General data
Dry weight	240 kg
Permissible total weight	440 kg
Tyres – front and rear	5.00-16 (ELA)
	4.00-18 (ELC)
Wheels	3.00 x 16 wire-spoked (ELA)
	2.15 x 18 wire-spoked (ELC)
Top speed	140 km/h
Fuel tank capacity	14.2 litres
Production figures	8 (42-ELA)
	44 (42-ELC)
Contract price	Not available

Model 42WLC 1942 Canadian National Defense

Motor
Engine	Air-cooled 45° sv V-twin
Bore x stroke	2¾ x 3¹³⁄₁₆ (2.75 x 3.81) in.
Bore x stroke	69.85 x 96.84 mm
Capacity	45.12 cu. in. (742 cc)
Power output (SAE)	22(16)/4600 rpm
Compression ratio	5.0:1
Carburettor	Linkert Type M-88
Exhaust system	Large fishtail silencer

Transmission
Gearbox	3-speed
Secondary reduction ratio	2.56
Gearchange	Hand change, on tank
Final drive	Chain

Cycle parts
Frame	Single-loop, full cradle
Front forks	Springers – exposed-springs, oval-tube legs
Rear suspension	None (rigid)
Front brake	Drum, via cable and handlebar lever
Rear brake	Drum, via rod and pedal

General data
Dry weight	approx. 240 kg
Permissible total weight	410 kg
Tyres – front and rear	4.00-18
Wheels	2.15 x 18 wire-spoked
Top speed	95 km/h
Fuel tank capacity	12.9 litres
Production figures	Not available
Contract price	Not available

Model 42XA 1942
Model 43XA Type I and II 1943

Motor
Engine	Air cooled sv flat twin
Bore and stroke	3¹/₁₆ x 3¹/₁₆ (3.06 x 3.06) in.
Bore and stroke	77.79 x 77.79 mm
Capacity	45.12 cu. in. (739 cc)
Power output (SAE)	23(17)/4800 rpm
Compression ratio	5.7:1
Carburettor	Linkert Type M-17L/M17-R (LHS/RHS)
Exhaust system	2 Rocket-Fin silencers

Transmission
Gearbox	4-speed
Gearchange	Hand change, on tank and with foot pedal
Final drive	Shaft

Cycle parts
Frame	Single-loop, full cradle
Front forks	Springers – exposed-springs, oval-tube legs
Rear suspension	Plunger
Front brake	Drum, via cable and handlebar lever
Rear brake	Drum, via rod and pedal

General data
Dry weight	approx. 240 kg
Permissible total weight	410 kg
Tyres – front and rear	4.00-18 (Desert: 6.00-15)
Wheels	2.15 x 18 wire-spoked
Top speed	95 km/h
Fuel tank capacity	15.5 litres
Production figures	1090
Contract price	$859
Sidecars	Model 42XS Model 43XS

Model 57XLA 1957
Military Police and Shore Patrol
Model 62XLA 1962
Model 63XLA 1963

Motor
Engine	Air cooled 45° ohv V-twin
Bore and stroke	3 x 3¹³/₁₆ (3 x 3.81) in
Bore and stroke	76.2 x 96.84 mm
Capacity	53.90 cu. in. (883 cc)
Power output (SAE)	42(31)/5500 SAE (1957) 55(40)/6300 SAE (1962/63)
Torque (Nm)	35/3600 SAE (1957) 38/3800 SAE (1962/63)
Compression ratio	7.5:1 (1957) 9.0:1 (1962/63)
Carburettor	Linkert M-Type
Electrical system	6-volt
Starter	Kickstarter
Exhaust system	2 into 1, 1 RHS silencer; optional 2 into 2, 2 RHS silencers

Transmission
Gearbox	4-speed
Reduction ratios	2.52 (1) 1.83 (2) 1.38 (3) 1.00 (4)
Secondary reduction ratio	2.43
Gearchange	Foot change
Final drive	Chain

Cycle parts
Frame	Twin-loop frame
Front forks	Telescopic, hydraulically-damped
Rear suspension	Swinging-arm
Front brake	Drum, via cable and handlebar lever
Rear brake	Drum, via rod and pedal

General data
Dry weight	225 kg
Permissible total weight	400 kg
Wheelbase	1435 mm
Tyres – front and rear	3.50S18
Wheels	2.15 x 18 wire-spoked
Top speed	162 km/h (1957) 170 km/h (1962/63)
Fuel tank capacity	16.5 litres
Production figures	1957: 418 1962/63: 1000
Contract price	Not available

Model 73XLA 1973
Low compression Army Model

Motor

Engine	Air cooled 45° ohv V-twin
Bore and stroke	3³⁄₁₆ x 3¹³⁄₁₆ (3.19 x 3.81) in.
Bore and stroke	80.96 x 96.84 mm
Capacity	60.85 cu. in. (997 cc)
Power output (SAE)	approx. 55(40)/6200 rpm
Torque (Nm)	Not available
Compression ratio	approx. 7.5:1
Carburettor	Bendix-Zenith 16P12
Electrical system	12-volt
Starter	Electric starter
Exhaust system	2 into 2, 2 Standard-silencers RHS

Transmission

Gearbox	4-speed
Reduction ratios	2.52 (1)
	1.83 (2)
	1.38 (3)
	1.00 (4)
Secondary reduction ratio	2.43
Gearchange	Foot change
Final drive	Chain

Cycle parts

Frame	Twin-loop frame
Front forks	Telescopic, hydraulically-damped
Rear suspension	Swinging-arm
Front brake	Hydraulic disc (292 mm dia.)
Rear brake	203 mm dia. drum, via rod and pedal

General data

Dry weight	238 kg
Permissible total weight	408 kg
Wheelbase	1486 m
Tyres – front and rear	3.75S19 front, 4.25S18 rear
Wheels	2.15 x 19 wire-spoked optional
	2.50 x 19 wire-spoked front, 2.50 x 18 wire-spoked rear
Top speed	approx. 165 km/h
Fuel tank capacity	approx. 15 litres
Production figures	1973: 130 (some sources mention only a single prototype)
Contract price	Not available

Appendix 1: Harley-Davidson's non-motorcycle products

Pedal cycles, 1917 to 1923

To capture a wider market, Harley began, even before the end of the First World War, to offer bicycles. Depending on standard of finish, these cost between 30 and 45 bucks and were built for H-D by the Davis Sewing Machine Company. Unfortunately the number sold is unknown. Manufacture was probably nevertheless more expensive than had been envisioned and the project was shut down at the end of 1923 to free resources for more profitable products.

Stationary engines (for industry), 1930 to 1933

For four years only, during the worst part of the Depression, Harley manufactured stationary engines in many and various forms. There was a single-cylinder engine, that could even be fitted to a cultivator, and then more singles, V-twins and even a flat-twin-engined unit. Another avenue explored by Harley was the use of their engines in air compressors; for example in the road-marking VCR outfit shown on page 87, as well as the unit shown on page 377.

Snowmobiles, 1971 to 1975

At first a competitor to AMF's Ski-Daddler, in 1971 the Harley powerplant was installed in the chassis of the AMF product and the result was sold as the Harley Snowmobile. The two-stroke engine normally had a hand (recoil) starter but could be specified with electric start, and with automatic or manual transmission. The table overleaf shows the range of four models, production of which had finished by 1975.

Golf Caddy, 1963 to 1980

At the beginning of the 60s, when the sport of golf, already well-established in the USA, was becoming more than just the domain of the well-heeled, Harley-Davidson discovered the market for golf carts.

The first model appeared in 1963, a 15 cu. in./250 cc. two-stroke powered three-wheeler with a swivelling handlebar – the Tiller Steering which soon became all the rage. However, this was far from being a new in-house development but was simply the Par Car, a product of the Columbia company, which had been taken over by H-D.

Soon thereafter the same three-wheeled vehicle appeared also with a 36-volt (six 6-volt batteries linked together) electric motor (and therefore not so foul-smelling). In 1972 a four-wheeled counterpart arrived on the market. The body was of fibre-glass and the entire fibre-glass tailpiece could be simply raised for ease of maintenance. The three-wheeler could later be ordered, if so desired, with a normal steering wheel instead of the steering tiller.

As one could hardly confuse the Caddys with the early 45 cu. in./742 cc. V-twins of 1929 to 1932, the model code letter D was brought into use again. That identified the standard two-stroke three-wheeler, so that the electric three-wheeler logically became the DE. The petrol-driven four-wheeler then became the D4, so that the electric four-wheeler became the DE4.

Production of the DE4 ended completely as early as 1975, while the DE carried on until 1978, as did the petrol-driven D and D4 models. Actually that should have been the case, except that in 1980 a small final series of D, D4 (including a particularly attractive Classic version of the D4) and DE3 models was released, probably made up of left-over stocks of parts, before the axe finally fell. The table overleaf gives an overview.

Utilicar, 1965 to 1973

From something like the Golf Caddy one could of course make an outstanding small utility vehicle and H-D did just that from about 1965 with the DC model (C standing for Commercial). There were various bodystyles, from the completely open Pick-Up to a mini-utility vehicle comparable to the Vespa Ape. Naturally, one could choose between a two-stroke petrol-driven version (DC) and an electrically-driven model (DEC). Anyway, these Caddy spin-off models continued to be built up to 1973.

Minitank, 1944

Two linked Knucklehead engines with a combined capacity of 122 cu. in./2 litres were considered for the powerplant of a Canadian mini-tank. Other sources speak of it as a remote-controlled mine detector.

Snowmobile 1971 to 1975

Type	Code	Year	cc	cu. in.	Notes
Y-398/Y-400	1B	1971-75	398	24.29	Recoil starter
Y-440	9C	1972-75	433	26.42	Recoil starter
YE-398	1E	1971-75	398	24.29	Electric starter
YE-440	2E	1973-75	433	26.42	Electric starter

Golf Caddy 1963 to 1980

Type	Code	Year	cc	Notes
D	-	1963-69	250	2-stroke, three-wheeler
DE	-	1963-69	-	36-volt electric motor (6 x 6v batteries)
D	3B	1970-78 and 80	250	2-stroke, three-wheeler
DE	4B	1970-77 and 80	-	Electric, three-wheeler
D4	8C	1972-78	250	2-stroke, four-wheeler
D4 Classic	3K	1980	250	2-stroke, four-wheeler
DE4	9C	1972-75	-	Electric, four-wheeler
DE3	4K	1980	-	Electric, three-wheeler

The Harley-Davidson pedal cycle: Built for H-D between 1917 and 1922

A more recent recreation – a pedal cycle with dummy 'Harley' tank

A flat-twin stationary engine from H-D

A single-cylinder stationary engine from Harley-Davidson

Successor to the AMF Ski-Daddler …

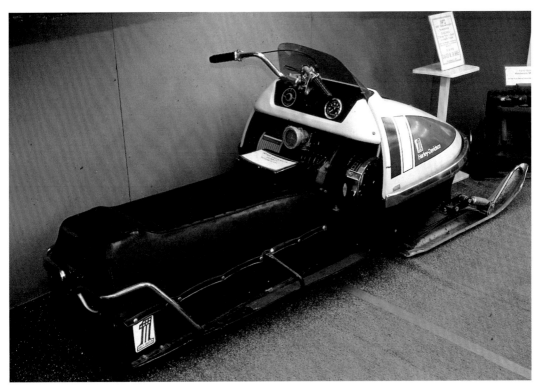

… the Harley-Davidson Snowmobile

A three-wheeler Harley-Davidson Golf Caddy with Tiller steering ...

... and a four-wheeler with steering wheel

... a three-wheeler with roof

Lots of work for one of these: an air compressor with Harley-Davidson powerplant

Harley engines have even been used as the basis for light aircraft engines

Harley-Davidson engines can even be used in cultivators

An antique Harley-Davidson oil container

Appendix 2:
80 years of Harley-Davidson

Listing of production models (less racers, military machines, prototypes and some export models) of the 1903 to 1983 model years. Sidecars are listed under their respective year headings.

1903 and 1904 model years
Model 0   25 cu. in./405 cc. ioe single   Battery-and-coil ignition   Direct drive

1905 model year
Model 1   25 cu. in./405 cc. ioe single   Battery-and-coil ignition   Direct drive

1906 model year
Model 2   27 cu. in./440 cc. ioe single   Battery-and-coil ignition   Direct drive

1907 model year
Model 3   27 cu. in./440 cc. ioe single   Battery-and-coil ignition   Direct drive

1908 model year
Model 4   27 cu. in./440 cc. ioe single   Battery-and-coil ignition   Direct drive

1909 model year
Model 5   30 cu. in./494 cc. ioe single   Battery-and-coil ignition   Direct drive, 28-in. wheels
Model 5A   30 cu. in./494 cc. ioe single   Magneto ignition   Direct drive, 28-in. wheels
Model 5B   30 cu. in./494 cc. ioe single   Battery-and-coil ignition   Direct drive, 26-in. wheels
Model 5C   30 cu. in./494 cc. ioe single   Magneto ignition   Direct drive, 26-in. wheels
Model 5D   54 cu. in./880 cc. ioe V-twin   Magneto ignition   Direct drive, 28-in. wheels

1910 model year
Model 6   30 cu. in./494 cc. ioe single   Battery-and-coil ignition   Direct drive, 28-in. wheels
Model 6A   30 cu. in./494 cc. ioe single   Magneto ignition   Direct drive, 28-in. wheels
Model 6B   30 cu. in./494 cc. ioe single   Battery-and-coil ignition   Direct drive, 26-in. wheels
Model 6C   30 cu. in./494 cc. ioe single   Magneto ignition   Direct drive, 26-in. wheels

1911 model year
Model 7   30 cu. in./494 cc. ioe single   Battery-and-coil ignition   Direct drive, 28-in. wheels
Model 7A   30 cu. in./494 cc. ioe single   Magneto ignition   Direct drive, 28-in. wheels
Model 7B   30 cu. in./494 cc. ioe single   Battery-and-coil ignition   Direct drive, 26-in. wheels
Model 7C   30 cu. in./494 cc. ioe single   Magneto ignition   Direct drive, 26-in. wheels
Model 7D   49 cu. in./811 cc. ioe V-twin   Magneto ignition   Direct drive, 28-in. wheels

1912 model year
Model 8   30 cu. in./494 cc. ioe single   Battery-and-coil ignition   Belt drive
Model X8   30 cu. in./494 cc. ioe single   Battery-and-coil ignition   Belt drive
Model 8A   30 cu. in./494 cc. ioe single   Magneto ignition   Belt drive
Model X8A   30 cu. in./494 cc. ioe single   Magneto ignition   Belt drive
Model 8D   49 cu. in./811 cc. ioe V-twin   Magneto ignition   Belt drive
Model X8D   49 cu. in./811 cc. ioe V-twin   Magneto ignition   Belt drive
Model X8E   60 cu. in./989 cc. ioe V-twin   Magneto ignition   Chain drive
The X models have a clutch in the rear hub.

1913 model year
Model 9A   34 cu. in./565 cc. ioe single   Magneto ignition   Belt drive
Model 9B   34 cu. in./565 cc. ioe single   Magneto ignition   Chain drive
Model 9E   60 cu. in./989 cc. ioe V-twin   Magneto ignition   Chain drive
Model 9G   60 cu. in./989 cc. ioe V-twin   Magneto ignition   Forecar
All models have a clutch in the rear hub.

1914 model year
Model 10A	34 cu. in./565 cc. ioe single	Magneto ignition	Belt drive
Model 10B	34 cu. in./565 cc. ioe single	Magneto ignition	Chain drive
Model 10C	34 cu. in./565 cc. ioe single	Magneto ignition	Chain drive
Model 10E	60 cu. in./989 cc. ioe V-twin	Magneto ignition	Chain drive
Model 10F	60 cu. in./989 cc. ioe V-twin	Magneto ignition	Chain drive
Model 10G	60 cu. in./989 cc. ioe V-twin	Magneto ignition	Forecar

All models have a clutch in the rear hub.

1915 model year
Model 11B	34 cu. in./565 cc. ioe single	Magneto ignition	Direct drive
Model 11C	34 cu. in./565 cc. ioe single	Magneto ignition	Two-speed gearbox
Model 11E	60 cu. in./989 cc. ioe V-twin	Magneto ignition	Direct drive
Model 11F	60 cu. in./989 cc. ioe V-twin	Magneto ignition	Two-speed gearbox
Model 11G	60 cu. in./989 cc. ioe V-twin	Magneto ignition	Forecar
Model 11H	60 cu. in./989 cc. ioe V-twin	Battery-and-coil ignition	Direct drive
Model 11J	60 cu. in./989 cc. ioe V-twin	Battery-and-coil ignition	Three-speed gearbox

All models have chain drive.

Note: Model codes 12 to 15 were bypassed in the change to using the model year codes.

1916 model year
Model 16B	34 cu. in./565 cc. ioe single	Magneto ignition	Direct drive
Model 16C	34 cu. in./565 cc. ioe single	Magneto ignition	Three-speed gearbox
Model 16E	60 cu. in./989 cc. ioe V-twin	Magneto ignition	Direct drive
Model 16F	60 cu. in./989 cc. ioe V-twin	Magneto ignition	Three-speed gearbox
Model 16J	60 cu. in./989 cc. ioe V-twin	Battery-and-coil ignition	Three-speed gearbox

All models have chain drive.

1917 model year
Model 17B	34 cu. in./565 cc. ioe single	Magneto ignition	Direct drive
Model 17C	34 cu. in./565 cc. ioe single	Magneto ignition	Three-speed gearbox
Model 17E	60 cu. in./989 cc. ioe V-twin	Magneto ignition	Direct drive
Model 17F	60 cu. in./989 cc. ioe V-twin	Magneto ignition	Three-speed gearbox
Model 17J	60 cu. in./989 cc. ioe V-twin	Battery-and-coil ignition	Three-speed gearbox

All models have chain drive.

1918 model year
Model 18B	34 cu. in./565 cc. ioe single	Magneto ignition	Direct drive
Model 18C	34 cu. in./565 cc. ioe single	Magneto ignition	Three-speed gearbox
Model 18E	60 cu. in./989 cc. ioe V-twin	Magneto ignition	Direct drive
Model 18F	60 cu. in./989 cc. ioe V-twin	Magneto ignition	Three-speed gearbox
Model 18FA	60 cu. in./989 cc. ioe V-twin	Magneto ignition	Three-speed, A Motor
Model 18FB	60 cu. in./989 cc. ioe V-twin	Magneto ignition	Three-speed, B Motor
Model 18FCA *	60 cu. in./989 cc. ioe V-twin	Magneto ignition	Three-speed, CA Motor
Model 18FCAB *	69 cu. in./1130 cc. ioe V-twin	Magneto ignition	Three-speed, CAB Motor
Model 18FF	60 cu. in./989 cc. ioe V-twin	Magneto ignition	Three-speed, 500 Motor
Model 18FFA *	60 cu. in./989 cc. ioe V-twin	Magneto ignition	Three-speed, 500 Motor
Model 18J	60 cu. in./989 cc. ioe V-twin	Battery-and-coil ignition	Three-speed
Model 18JA	60 cu. in./989 cc. ioe V-twin	Battery-and-coil ignition	Three-speed, A Motor
Model 18JB	60 cu. in./989 cc. ioe V-twin	Battery-and-coil ignition	Three-speed, B Motor
Model 18JF	60 cu. in./989 cc. ioe V-twin	Battery-and-coil ignition	Three-speed

All models have chain drive. * = with aluminium-alloy pistons.

1919 model year
Model 19F	60 cu. in./989 cc. ioe V-twin	Magneto ignition	
Model 19FA	60 cu. in./989 cc. ioe V-twin	Magneto ignition	A Motor
Model 19FB	60 cu. in./989 cc. ioe V-twin	Magneto ignition	B Motor
Model 19FS	60 cu. in./989 cc. ioe V-twin	Magneto ignition	Sidecar-specification
Model 19FCA *	60 cu. in./989 cc. ioe V-twin	Magneto ignition	CA Motor
Model 19FCAB *	69 cu. in./1130 cc. ioe V-twin	Magneto ignition	CAB Motor
Model 19FF	60 cu. in./989 cc. ioe V-twin	Magneto ignition	500 Motor
Model 19FFA *	60 cu. in./989 cc. ioe V-twin	Magneto ignition	500A Motor
Model 19J	60 cu. in./989 cc. ioe V-twin	Battery-and-coil ignition	

Model	Displacement	Ignition	Notes
Model 19JS	60 cu. in./989 cc. ioe V-twin	Battery-and-coil ignition	Sidecar-specification
Model 19JA	60 cu. in./989 cc. ioe V-twin	Battery-and-coil ignition	A Motor
Model 19JB	60 cu. in./989 cc. ioe V-twin	Battery-and-coil ignition	B Motor
Model 19JF	60 cu. in./989 cc. ioe V-twin	Battery-and-coil ignition	500 Motor
Model 19W	36 cu. in./584 cc. sv flat twin	Magneto ignition	Sport Twin

All models have chain drive and a three-speed gearbox. *= with aluminium-alloy pistons.

1920 model year

Model	Displacement	Ignition	Notes
Model 20F	60 cu. in./989 cc. ioe V-twin	Magneto ignition	
Model 20FS	60 cu. in./989 cc. ioe V-twin	Magneto ignition	Sidecar-specification
Model 20FA	60 cu. in./989 cc. ioe V-twin	Magneto ignition	A Motor
Model 20FB	60 cu. in./989 cc. ioe V-twin	Magneto ignition	B Motor
Model 20FE	60 cu. in./989 cc. ioe V-twin	Magneto ignition	E Motor
Model 20FCA *	60 cu. in./989 cc. ioe V-twin	Magneto ignition	CA Motor
Model 20FCAB *	69 cu. in./1130 cc. ioe V-twin	Magneto ignition	CAB Motor
Model 20FF	60 cu. in./989 cc. ioe V-twin	Magneto ignition	500 Motor
Model 20FFA *	60 cu. in./989 cc. ioe V-twin	Magneto ignition	500A Motor
Model 20J	60 cu. in./989 cc. ioe V-twin	Battery-and-coil ignition	
Model 20JS	60 cu. in./989 cc. ioe V-twin	Battery-and-coil ignition	Sidecar-specification
Model 20JA	60 cu. in./989 cc. ioe V-twin	Battery-and-coil ignition	A Motor
Model 20JB	60 cu. in./989 cc. ioe V-twin	Battery-and-coil ignition	B Motor
Model 20JE	60 cu. in./989 cc. ioe V-twin	Battery-and-coil ignition	E Motor
Model 20JF	60 cu. in./989 cc. ioe V-twin	Battery-and-coil ignition	500 Motor
Model 20WF	36 cu. in./584 cc. sv flat twin	Magneto ignition	Sport Twin
Model 20WJ	36 cu. in./584 cc. sv flat twin	Battery-and-coil ignition	Sport Twin

All models have chain drive and a three-speed gearbox. *= with aluminium-alloy pistons.

1921 model year

Model	Displacement	Ignition	Notes
Model 21CD	37 cu. in./608 cc. ioe single	Magneto ignition	
Model 21F	60 cu. in./989 cc. ioe V-twin	Magneto ignition	
Model 21FS	60 cu. in./989 cc. ioe V-twin	Magneto ignition	Sidecar-specification
Model 21FA	60 cu. in./989 cc. ioe V-twin	Magneto ignition	A Motor
Model 21FE	60 cu. in./989 cc. ioe V-twin	Magneto ignition	E Motor
Model 21FCA *	60 cu. in./989 cc. ioe V-twin	Magneto ignition	CA Motor
Model 21FD	74 cu. in./1217 cc. ioe V-twin	Magneto ignition	
Model 21FDS	74 cu. in./1217 cc. ioe V-twin	Magneto ignition	Sidecar-specification
Model 21FDA	74 cu. in./1217 cc. ioe V-twin	Magneto ignition	A Motor
Model 21J	60 cu. in./989 cc. ioe V-twin	Battery-and-coil ignition	
Model 21JS	60 cu. in./989 cc. ioe V-twin	Battery-and-coil ignition	Sidecar-specification
Model 21JA	60 cu. in./989 cc. ioe V-twin	Battery-and-coil ignition	A Motor
Model 21JE	60 cu. in./989 cc. ioe V-twin	Battery-and-coil ignition	E Motor
Model 21JD	74 cu. in./1217 cc. ioe V-twin	Battery-and-coil ignition	
Model 21JDS	74 cu. in./1217 cc. ioe V-twin	Battery-and-coil ignition	Sidecar-specification
Model 21JDA	74 cu. in./1217 cc. ioe V-twin	Battery-and-coil ignition	A Motor
Model 21WF	36 cu. in./584 cc. sv flat twin	Magneto ignition	Sport Twin
Model 21WJ	36 cu. in./584 cc. sv flat twin	Battery-and-coil ignition	Sport Twin

All models have chain drive and a three-speed gearbox. *= with aluminium-alloy pistons.

1922 model year

Model	Displacement	Ignition	Notes
Model 22CD	37 cu. in./608 cc. ioe single	Magneto ignition	
Model 22F	60 cu. in./989 cc. ioe V-twin	Magneto ignition	
Model 22FS	60 cu. in./989 cc. ioe V-twin	Magneto ignition	Sidecar-specification
Model 22FA	60 cu. in./989 cc. ioe V-twin	Magneto ignition	A Motor
Model 22FE	60 cu. in./989 cc. ioe V-twin	Magneto ignition	E Motor
Model 22FCA *	60 cu. in./989 cc. ioe V-twin	Magneto ignition	CA Motor
Model 22FD	74 cu. in./1217 cc. ioe V-twin	Magneto ignition	
Model 22FDS	74 cu. in./1217 cc. ioe V-twin	Magneto ignition	Sidecar-specification
Model 22FDA	74 cu. in./1217 cc. ioe V-twin	Magneto ignition	A Motor
Model 22J	60 cu. in./989 cc. ioe V-twin	Battery-and-coil ignition	
Model 22JS	60 cu. in./989 cc. ioe V-twin	Battery-and-coil ignition	Sidecar-specification
Model 22JA	60 cu. in./989 cc. ioe V-twin	Battery-and-coil ignition	A Motor
Model 22JE	60 cu. in./989 cc. ioe V-twin	Battery-and-coil ignition	E Motor
Model 22JD	74 cu. in./1217 cc. ioe V-twin	Battery-and-coil ignition	

Model 22JDS	74 cu. in./1217 cc. ioe V-twin	Battery-and-coil ignition	Sidecar-specification
Model 22JDA	74 cu. in./1217 cc. ioe V-twin	Battery-and-coil ignition	A Motor
Model 22WF	36 cu. in./584 cc. sv flat twin,	Magneto ignition	Sport Twin
Model 22WJ	36 cu. in./584 cc. sv flat twin,	Battery-and-coil ignition	Sport Twin

All models have chain drive and a three-speed gearbox. *= with aluminium-alloy pistons.

1923 model year

Model 23F	60 cu. in./989 cc. ioe V-twin	Magneto ignition	
Model 23FS	60 cu. in./989 cc. ioe V-twin	Magneto ignition	Sidecar-specification
Model 23FA	60 cu. in./989 cc. ioe V-twin	Magneto ignition	A Motor
Model 23FE	60 cu. in./989 cc. ioe V-twin	Magneto ignition	E Motor
Model 23FCA *	60 cu. in./989 cc. ioe V-twin	Magneto ignition	CA Motor
Model 23FD	74 cu. in./1217 cc. ioe V-twin	Magneto ignition	
Model 23FDS	74 cu. in./1217 cc. ioe V-twin	Magneto ignition	Sidecar-specification
Model 23FDA	74 cu. in./1217 cc. ioe V-twin	Magneto ignition	A Motor
Model 23FDCA	74 cu. in./1217 cc. ioe V-twin	Magneto ignition	CA Motor
Model 23J	60 cu. in./989 cc. ioe V-twin	Battery-and-coil ignition	
Model 23JS	60 cu. in./989 cc. ioe V-twin	Battery-and-coil ignition	Sidecar-specification
Model 23JE	60 cu. in./989 cc. ioe V-twin	Battery-and-coil ignition	E Motor
Model 23JD	74 cu. in./1217 cc. ioe V-twin	Battery-and-coil ignition	
Model 23JDS	74 cu. in./1217 cc. ioe V-twin	Battery-and-coil ignition	Sidecar-specification
Model 23JDA	74 cu. in./1217 cc. ioe V-twin	Battery-and-coil ignition	A Motor
Model 23JDCA	74 cu. in./1217 cc. ioe V-twin	Battery-and-coil ignition	CA Motor
Model 23WF	36 cu. in./584 cc. sv flat twin,	Magneto ignition	Sport Twin
Model 23WJ	36 cu. in./584 cc. sv flat twin,	Battery-and-coil ignition	Sport Twin

All models have chain drive and a three-speed gearbox. *= with aluminium-alloy pistons.

1924 model year

Model 24FE *	60 cu. in./989 cc. ioe V-twin	Magneto ignition	
Model 24FES *	60 cu. in./989 cc. ioe V-twin	Magneto ignition	Sidecar-specification
Model 24FD	74 cu. in./1217 cc. ioe V-twin	Magneto ignition	
Model 24FDS	74 cu. in./1217 cc. ioe V-twin	Magneto ignition	Sidecar-specification
Model 24FDCA *	74 cu. in./1217 cc. ioe V-twin	Magneto ignition	CA Motor
Model 24FDSCA *	74 cu. in./1217 cc. ioe V-twin	Magneto ignition	CA Motor, sidecar-spec.
Model 24JE *	60 cu. in./989 cc. ioe V-twin	Battery-and-coil ignition	
Model 24JES *	60 cu. in./989 cc. ioe V-twin	Battery-and-coil ignition	Sidecar-specification
Model 24JD	74 cu. in./1217 cc. ioe V-twin	Battery-and-coil ignition	
Model 24JDS	74 cu. in./1217 cc. ioe V-twin	Battery-and-coil ignition	Sidecar-specification
Model 24JDCA *	74 cu. in./1217 cc. ioe V-twin	Battery-and-coil ignition	CA Motor
Model 24JDSCA *	74 cu. in./1217 cc. ioe V-twin	Battery-and-coil ignition	CA Motor, sidecar-specification

All models have chain drive and a three-speed gearbox. *= with aluminium-alloy pistons.

1925 model year

Model 25FE	60 cu. in./989 cc. ioe V-twin	Magneto ignition	
Model 25FES	60 cu. in./989 cc. ioe V-twin	Magneto ignition	Sidecar-specification
Model 25FDCB	74 cu. in./1217 cc. ioe V-twin	Magneto ignition	
Model 25FDCBS	74 cu. in./1217 cc. ioe V-twin	Magneto ignition	Sidecar-specification
Model 25JE	60 cu. in./989 cc. ioe V-twin	Battery-and-coil ignition	
Model 25JES	60 cu. in./989 cc. ioe V-twin	Battery-and-coil ignition	Sidecar-specification
Model 25JDCB	74 cu. in./1217 cc. ioe V-twin	Battery-and-coil ignition	
Model 25JDCBS	74 cu. in./1217 cc. ioe V-twin	Battery-and-coil ignition	Sidecar-specification

All models have chain drive and a three-speed gearbox.

1926 model year

Model 26A	21 cu. in./346 cc. sv single	Magneto ignition	
Model 26AA	21 cu. in./346 cc. ohv single	Magneto ignition	
Model 26AAE	21 cu. in./346 cc. ohv single	Magneto ignition	Export
Model 26B	21 cu. in./346 cc. sv single	Battery-and-coil ignition	
Model 26BA	21 cu. in./346 cc. ohv single	Battery-and-coil ignition	
Model 26BAE	21 cu. in./346 cc. ohv single	Battery-and-coil ignition	Export
Model 26F	60 cu. in./989 cc. ioe V-twin	Magneto ignition	
Model 26FS	60 cu. in./989 cc. ioe V-twin	Magneto ignition	Sidecar-specification

Model 26FD	74 cu. in./1217 cc. ioe V-twin	Magneto ignition	
Model 26FDS	74 cu. in./1217 cc. ioe V-twin	Magneto ignition	Sidecar-specification
Model 26J	60 cu. in./989 cc. ioe V-twin	Battery-and-coil ignition	
Model 26JS	60 cu. in./989 cc. ioe V-twin	Battery-and-coil ignition	Sidecar-specification
Model 26JD	74 cu. in./1217 cc. ioe V-twin	Battery-and-coil ignition	
Model 26JDS	74 cu. in./1217 cc. ioe V-twin	Battery-and-coil ignition	Sidecar-specification

All models have chain drive, a three-speed gearbox and aluminium-alloy pistons.

1927 model year

Model 27A	21 cu. in./346 cc. sv single	Magneto ignition	
Model 27AA	21 cu. in./346 cc. ohv single	Magneto ignition	
Model 27AAE	21 cu. in./346 cc. ohv single	Magneto ignition	Export
Model 27B	21 cu. in./346 cc. sv single	Battery-and-coil ignition	
Model 27BA	21 cu. in./346 cc. ohv single	Battery-and-coil ignition	
Model 27BAE	21 cu. in./346 cc. ohv single	Battery-and-coil ignition	Export
Model 27F	60 cu. in./989 cc. ioe V-twin	Magneto ignition	
Model 27FS	60 cu. in./989 cc. ioe V-twin	Magneto ignition	Sidecar-specification
Model 27FK	60 cu. in./989 cc. ioe V-twin	Magneto ignition	Special
Model 27FD	74 cu. in./1217 cc. ioe V-twin	Magneto ignition	
Model 27FDL	74 cu. in./1217 cc. ioe V-twin	Magneto ignition	Special
Model 27FDS	74 cu. in./1217 cc. ioe V-twin	Magneto ignition	Sidecar-specification
Model 27J	60 cu. in./989 cc. ioe V-twin	Battery-and-coil ignition	
Model 27JS	60 cu. in./989 cc. ioe V-twin	Battery-and-coil ignition	Sidecar-specification
Model 27JK	60 cu. in./989 cc. ioe V-twin	Battery-and-coil ignition	Special
Model 27JD	74 cu. in./1217 cc. ioe V-twin	Battery-and-coil ignition	
Model 27JDL	74 cu. in./1217 cc. ioe V-twin	Battery-and-coil ignition	Special
Model 27JDS	74 cu. in./1217 cc. ioe V-twin	Battery-and-coil ignition	Sidecar-specification

All models have chain drive and a three-speed gearbox.

1928 model year

Model 28A	21 cu. in./346 cc. sv single	Magneto ignition	
Model 28AA	21 cu. in./346 cc. ohv single	Magneto ignition	
Model 28AAE	21 cu. in./346 cc. ohv single	Magneto ignition	Export
Model 28B	21 cu. in./346 cc. sv single	Battery-and-coil ignition	
Model 28BA	21 cu. in./346 cc. ohv single	Battery-and-coil ignition	
Model 28BAE	21 cu. in./346 cc. ohv single	Battery-and-coil ignition	Export
Model 28F	60 cu. in./989 cc. ioe V-twin	Magneto ignition	
Model 28FS	60 cu. in./989 cc. ioe V-twin	Magneto ignition	Sidecar-specification
Model 28FH	60 cu. in./989 cc. ioe V-twin	Magneto ignition	Special
Model 28FD	74 cu. in./1217 cc. ioe V-twin	Magneto ignition	
Model 28FDH	74 cu. in./1217 cc. ioe V-twin	Magneto ignition	Special
Model 28FDS	74 cu. in./1217 cc. ioe V-twin	Magneto ignition	Sidecar-specification
Model 28J	60 cu. in./989 cc. ioe V-twin	Battery-and-coil ignition	
Model 28JS	60 cu. in./989 cc. ioe V-twin	Battery-and-coil ignition	Sidecar-specification
Model 28JX	60 cu. in./989 cc. ioe V-twin	Battery-and-coil ignition	Sport
Model 28JXL	60 cu. in./989 cc. ioe V-twin	Battery-and-coil ignition	Special Sport
Model 28JH	60 cu. in./989 cc. ioe V-twin	Battery-and-coil ignition	Two Cam
Model 28JD	74 cu. in./1217 cc. ioe V-twin	Battery-and-coil ignition	
Model 28JDS	74 cu. in./1217 cc. ioe V-twin	Battery-and-coil ignition	Sidecar-specification
Model 28JDX	74 cu. in./1217 cc. ioe V-twin	Battery-and-coil ignition	Sport
Model 28JDXL	74 cu. in./1217 cc. ioe V-twin	Battery-and-coil ignition	Special Sport
Model 28JDH	74 cu. in./1217 cc. ioe V-twin	Battery-and-coil ignition	Two Cam

All models have chain drive and a three-speed gearbox.

1929 model year

Model 29A	21 cu. in./346 cc. sv single	Magneto ignition	
Model 29AA	21 cu. in./346 cc. ohv single	Magneto ignitionv	
Model 29AAF	21 cu. in./346 cc. ohv single	Magneto ignition	Export
Model 29B	21 cu. in./346 cc. sv single	Battery-and-coil ignition	
Model 29BA	21 cu. in./346 cc. ohv single	Battery-and-coil ignition	
Model 29BAF	21 cu. in./346 cc. ohv single	Battery-and-coil ignition	Export
Model 29C	30 cu. in./493 cc. sv single	Battery-and-coil ignition	
Model 29D	45 cu. in./742 cc. sv V-twin	Battery-and-coil ignition	Low compression

Model	Displacement	Ignition	Notes
Model 29DL	45 cu. in./742 cc. sv V-twin	Battery-and-coil ignition	Raised compression
Model 29F	60 cu. in./989 cc. ioe V-twin	Magneto ignition	
Model 29FS	60 cu. in./989 cc. ioe V-twin	Magneto ignition	Sidecar-specification
Model 29FL	60 cu. in./989 cc. ioe V-twin	Magneto ignition	L Motor
Model 29FD	74 cu. in./1217 cc. ioe V-twin	Magneto ignition	
Model 29FDH	74 cu. in./1217 cc. ioe V-twin	Magneto ignition	Special
Model 29FDL	74 cu. in./1217 cc. ioe V-twin	Magneto ignition	DL Motor
Model 29FDS	74 cu. in./1217 cc. ioe V-twin	Magneto ignition	Sidecar-specification
Model 29J	60 cu. in./989 cc. ioe V-twin	Battery-and-coil ignition	
Model 29JS	60 cu. in./989 cc. ioe V-twin	Battery-and-coil ignition	Sidecar-specification
Model 29JXL	60 cu. in./989 cc. ioe V-twin	Battery-and-coil ignition	Special Sport L
Model 29JH	60 cu. in./989 cc. ioe V-twin	Battery-and-coil ignition	Two Cam
Model 29JD	74 cu. in./1217 cc. ioe V-twin	Battery-and-coil ignition	
Model 29JDS	74 cu. in./1217 cc. ioe V-twin	Battery-and-coil ignition	Sidecar-specification
Model 29JDF	74 cu. in./1217 cc. ioe V-twin	Battery-and-coil ignition	Two Unit System
Model 29JDXL	74 cu. in./1217 cc. ioe V-twin	Battery-and-coil ignition	Special Sport DL
Model 29JDH	74 cu. in./1217 cc. ioe V-twin	Battery-and-coil ignition	Two Cam

All models have chain drive and a three-speed gearbox.

1930 model year

Model	Displacement	Ignition	Notes
Model 30A	21 cu. in./346 cc. sv single	Magneto ignition	
Model 30AA	21 cu. in./346 cc. ohv single	Magneto ignition	
Model 30B	21 cu. in./346 cc. sv single	Battery-and-coil ignition	
Model 30BA	21 cu. in./346 cc. ohv single	Battery-and-coil ignition	
Model 30BAF	21 cu. in./346 cc. ohv single	Battery-and-coil ignition	Export
Model 30C	30 cu. in./493 cc. sv single	Battery-and-coil ignition	
Model 30D	45 cu. in./742 cc. sv V-twin	Battery-and-coil ignition	Low compression
Model 30DS	45 cu. in./742 cc. sv V-twin	Battery-and-coil ignition	Low compression, sidecar-specification
Model 30DL	45 cu. in./742 cc. sv V-twin	Battery-and-coil ignition	Raised compression
Model 30DLD	45 cu. in./742 cc. sv V-twin	Battery-and-coil ignition	Special Sport
Model 30V	74 cu. in./1217 cc. sv V-twin	Battery-and-coil ignition	Low compression
Model 30VS	74 cu. in./1217 cc. sv V-twin	Battery-and-coil ignition	Low compression, sidecar-specification
Model 30VL	74 cu. in./1217 cc. sv V-twin	Battery-and-coil ignition	High compression
Model 30VC	74 cu. in./1217 cc. sv V-twin	Battery-and-coil ignition	Commercial
Model 30VCM	74 cu. in./1217 cc. sv V-twin	Magneto ignition	Commercial
Model 30VM	74 cu. in./1217 cc. sv V-twin	Magneto ignition	Low compression
Model 30VMS	74 cu. in./1217 cc. sv V-twin	Magneto ignition	Low compression, sidecar-specification
Model 30VMG	74 cu. in./1217 cc. sv V-twin	Bosch magneto ignition	Low compression
Model 30VLM	74 cu. in./1217 cc. sv V-twin	Magneto ignition	High compression

All models have chain drive and a three-speed gearbox.

1931 model year

Model	Displacement	Ignition	Notes
Model 31B	21 cu. in./346 cc. sv single	Battery-and-coil ignition	
Model 31BA	21 cu. in./346 cc. ohv single	Battery-and-coil ignition	
Model 31C	30 cu. in./493 cc. sv single	Battery-and-coil ignition	
Model 31CH	30 cu. in./493 cc. sv single	Battery-and-coil ignition	High compression
Model 31D	45 cu. in./742 cc. sv V-twin	Battery-and-coil ignition	Low compression
Model 31DS	45 cu. in./742 cc. sv V-twin	Battery-and-coil ignition	Low compression, sidecar-specification
Model 31DL	45 cu. in./742 cc. sv V-twin	Battery-and-coil ignition	Raised compression
Model 31DLD	45 cu. in./742 cc. sv V-twin	Battery-and-coil ignition	Special Sport
Model 31V	74 cu. in./1217 cc. sv V-twin	Battery-and-coil ignition	Low compression
Model 31VS	74 cu. in./1217 cc. sv V-twin	Battery-and-coil ignition	Low compression, sidecar-specification
Model 31VL	74 cu. in./1217 cc. sv V-twin	Battery-and-coil ignition	High compression
Model 31VC	74 cu. in./1217 cc. sv V-twin	Battery-and-coil ignition	Commercial
Model 31VMG	74 cu. in./1217 cc. sv V-twin	Bosch magneto ignition	Low compression

All models have chain drive and a three-speed gearbox.

1932 model year

Model	Displacement	Ignition	Notes
Model 32B	21 cu. in./346 cc. sv single	Battery-and-coil ignition	
Model 32C	30 cu. in./493 cc. sv single	Battery-and-coil ignition	
Model 32CS	30 cu. in./493 cc. sv single	Battery-and-coil ignition	Japanese-export model
Model 32R	45 cu. in./742 cc. sv V-twin	Battery-and-coil ignition	Low compression
Model 32RS	45 cu. in./742 cc. sv V-twin	Battery-and-coil ignition	Low compression, sidecar-specification
Model 32RL	45 cu. in./742 cc. sv V-twin	Battery-and-coil ignition	Raised compression
Model 32RLD	45 cu. in./742 cc. sv V-twin	Battery-and-coil ignition	Special Sport
Model 32G	45 cu. in./742 cc. sv V-twin	Battery-and-coil ignition	Servi-Car
Model 32GA	45 cu. in./742 cc. sv V-twin	Battery-and-coil ignition	Servi-Car
Model 32GD	45 cu. in./742 cc. sv V-twin	Battery-and-coil ignition	Servi-Car
Model 32GE	45 cu. in./742 cc. sv V-twin	Battery-and-coil ignition	Servi-Car
Model 32V	74 cu. in./1217 cc. sv V-twin	Battery-and-coil ignition	Low compression
Model 32VS	74 cu. in./1217 cc. sv V-twin	Battery-and-coil ignition	Low compression, sidecar-specification
Model 32VL	74 cu. in./1217 cc. sv V-twin	Battery-and-coil ignition	High compression
Model 32VC	74 cu. in./1217 cc. sv V-twin	Battery-and-coil ignition	Commercial

All models have chain drive and a three-speed gearbox.

1933 model year

Model	Displacement	Ignition	Notes
Model 33B	21 cu. in./346 cc. sv single	Battery-and-coil ignition	
Model 33C	30 cu. in./493 cc. sv single	Battery-and-coil ignition	
Model 33CB	30 cu. in./493 cc. sv single	Battery-and-coil ignition	Economy model
Model 33R	45 cu. in./742 cc. sv V-twin	Battery-and-coil ignition	Low compression
Model 33RE	45 cu. in./742 cc. sv V-twin	Battery-and-coil ignition	Low compression
Model 33RS	45 cu. in./742 cc. sv V-twin	Battery-and-coil ignition	Low compression, sidecar-specification
Model 33RL	45 cu. in./742 cc. sv V-twin	Battery-and-coil ignition	Raised compression
Model 33RLE	45 cu. in./742 cc. sv V-twin	Battery-and-coil ignition	Raised compression
Model 33RLD	45 cu. in./742 cc. sv V-twin	Battery-and-coil ignition	Special Sport
Model 33RLDE	45 cu. in./742 cc. sv V-twin	Battery-and-coil ignition	Special Sport
Model 33G	45 cu. in./742 cc. sv V-twin	Battery-and-coil ignition	Servi-Car
Model 33GA	45 cu. in./742 cc. sv V-twin	Battery-and-coil ignition	Servi-Car
Model 33GD	45 cu. in./742 cc. sv V-twin	Battery-and-coil ignition	Servi-Car
Model 33GDT	45 cu. in./742 cc. sv V-twin	Battery-and-coil ignition	Servi-Car
Model 33GE	45 cu. in./742 cc. sv V-twin	Battery-and-coil ignition	Servi-Car
Model 33V	74 cu. in./1217 cc. sv V-twin	Battery-and-coil ignition	Low compression
Model 33VS	74 cu. in./1217 cc. sv V-twin	Battery-and-coil ignition	Low compression, sidecar-specification
Model 33VE	74 cu. in./1217 cc. sv V-twin	Battery-and-coil ignition	Low compression
Model 33VSE	74 cu. in./1217 cc. sv V-twin	Battery-and-coil ignition	Low compression, sidecar-specification
Model 33VF	74 cu. in./1217 cc. sv V-twin	Battery-and-coil ignition	Low compression
Model 33VFS	74 cu. in./1217 cc. sv V-twin	Battery-and-coil ignition	Low compression, sidecar-specification
Model 33VL	74 cu. in./1217 cc. sv V-twin	Battery-and-coil ignition	High compression
Model 33VLD	74 cu. in./1217 cc. sv V-twin	Battery-and-coil ignition	High compression
Model 33VLE	74 cu. in./1217 cc. sv V-twin	Battery-and-coil ignition	High compression
Model 33VC	74 cu. in./1217 cc. sv V-twin	Battery-and-coil ignition	Commercial
Model 33VCE	74 cu. in./1217 cc. sv V-twin	Battery-and-coil ignition	Commercial

All models have chain drive and a three-speed gearbox.

1934 model year

Model	Displacement	Ignition	Notes
Model 34B	21 cu. in./346 cc. sv single	Battery-and-coil ignition	
Model 34C	30 cu. in./493 cc. sv single	Battery-and-coil ignition	Standard
Model 34CB	30 cu. in./493 cc. sv single	Battery-and-coil ignition	Economy model
Model 34R	45 cu. in./742 cc. sv V-twin	Battery-and-coil ignition	Low compression
Model 34RS	45 cu. in./742 cc. sv V-twin	Battery-and-coil ignition	Low compression, sidecar-specification
Model 34RSR	45 cu. in./742 cc. sv V-twin	Battery-and-coil ignition	Low compression, sidecar-specification, Japanese-export model

Model	Displacement	Ignition	Notes
Model 34RL	45 cu. in./742 cc. sv V-twin	Battery-and-coil ignition	Raised compression
Model 34RLD	45 cu. in./742 cc. sv V-twin	Battery-and-coil ignition	Special Sport
Model 34G	45 cu. in./742 cc. sv V-twin	Battery-and-coil ignition	Servi-Car
Model 34GA	45 cu. in./742 cc. sv V-twin	Battery-and-coil ignition	Servi-Car
Model 34GD	45 cu. in./742 cc. sv V-twin	Battery-and-coil ignition	Servi-Car
Model 34GDT	45 cu. in./742 cc. sv V-twin	Battery-and-coil ignition	Servi-Car
Model 34GE	45 cu. in./742 cc. sv V-twin	Battery-and-coil ignition	Servi-Car
Model 34VFD	74 cu. in./1217 cc. sv V-twin	Battery-and-coil ignition	Low compression
Model 34VFDS	74 cu. in./1217 cc. sv V-twin	Battery-and-coil ignition	Low compression, sidecar-specification
Model 34VD	74 cu. in./1217 cc. sv V-twin	Battery-and-coil ignition	High compression
Model 34VDS	74 cu. in./1217 cc. sv V-twin	Battery-and-coil ignition	High compression, sidecar-specification
Model 34VLD	74 cu. in./1217 cc. sv V-twin	Battery-and-coil ignition	Special Sport

All models have chain drive and a three-speed gearbox.

1935 model year

Model	Displacement	Ignition	Notes
Model 35R	45 cu. in./742 cc. sv V-twin	Battery-and-coil ignition	Low compression
Model 35RS	45 cu. in./742 cc. sv V-twin	Battery-and-coil ignition	Low compression, sidecar-specification
Model 35RSR	45 cu. in./742 cc. sv V-twin	Battery-and-coil ignition	Low compression, sidecar-specification, Japanese-export model
Model 35RL	45 cu. in./742 cc. sv V-twin	Battery-and-coil ignition	Raised compression
Model 35RLD	45 cu. in./742 cc. sv V-twin	Battery-and-coil ignition	Special Sport
Model 35G	45 cu. in./742 cc. sv V-twin	Battery-and-coil ignition	Servi-Car
Model 35GA	45 cu. in./742 cc. sv V-twin	Battery-and-coil ignition	Servi-Car
Model 35GD	45 cu. in./742 cc. sv V-twin	Battery-and-coil ignition	Servi-Car
Model 35GDT	45 cu. in./742 cc. sv V-twin	Battery-and-coil ignition	Servi-Car
Model 35GE	45 cu. in./742 cc. sv V-twin	Battery-and-coil ignition	Servi-Car
Model 35VFD	74 cu. in./1217 cc. sv V-twin	Battery-and-coil ignition	Low compression
Model 35VFDS	74 cu. in./1217 cc. sv V-twin	Battery-and-coil ignition	Low compression, sidecar-specification
Model 35VD	74 cu. in./1217 cc. sv V-twin	Battery-and-coil ignition	High compression
Model 35VDS	74 cu. in./1217 cc. sv V-twin	Battery-and-coil ignition	High compression, sidecar-specification
Model 35VLD	74 cu. in./1217 cc. sv V-twin	Battery-and-coil ignition	Special Sport
Model 35VLDD	79 cu. in./1293 cc. sv V-twin	Battery-and-coil ignition	
Model 35VDDS	79 cu. in./1293 cc. sv V-twin	Battery-and-coil ignition	Sidecar-specification

All models have chain drive and a three-speed gearbox.

1936 model year

Model 36R	45 cu. in./742 cc. sv V-twin	Three-speed, chain drive	Low compression
Model 36RS	45 cu. in./742 cc. sv V-twin	Three-speed, chain drive	Low compression, sidecar-specification
Model 36RSR	45 cu. in./742 cc. sv V-twin	Three-speed, chain drive	Low compression, sidecar-specification, Japanese-export model
Model 36RL	45 cu. in./742 cc. sv V-twin	Three-speed, chain drive	Raised compression
Model 36RLD	45 cu. in./742 cc. sv V-twin	Three-speed, chain drive	Special Sport
Model 36G	45 cu. in./742 cc. sv V-twin	Three-speed, chain drive	Servi-Car
Model 36GA	45 cu. in./742 cc. sv V-twin	Three-speed, chain drive	Servi-Car
Model 36GD	45 cu. in./742 cc. sv V-twin	Three-speed, chain drive	Servi-Car
Model 36GDT	45 cu. in./742 cc. sv V-twin	Three-speed, chain drive	Servi-Car
Model 36GE	45 cu. in./742 cc. sv V-twin	Three-speed, chain drive	Servi-Car
Model 36VFD	74 cu. in./1217 cc. sv V-twin	Three-speed, chain drive	Low compression
Model 36VFDS	74 cu. in./1217 cc. sv V-twin	Three-speed, chain drive	Low compression, sidecar-specification
Model 36VD	74 cu. in./1217 cc. sv V-twin	Three-speed, chain drive	High compression
Model 36VDS	74 cu. in./1217 cc. sv V-twin	Three-speed, chain drive	High compression, sidecar-specification
Model 36VLD	74 cu. in./1217 cc. sv V-twin	Three-speed, chain drive	Special Sport
Model 36VFH	79 cu. in./1293 cc. sv V-twin	Three-speed, chain drive	Low compression

Model 36VFHS	79 cu. in./1293 cc. sv V-twin	Three-speed, chain drive	Low compression, sidecar-specification
Model 36VLH	79 cu. in./1293 cc. sv V-twin	Three-speed, chain drive	High compression
Model 36VHS	79 cu. in./1293 cc. sv V-twin	Three-speed, chain drive	Sidecar-specification
Model 36E	60 cu. in./989 cc. ohv V-twin	Four-speed, chain drive	Low compression
Model 36ES	60 cu. in./989 cc. ohv V-twin	Four-speed, chain drive	Low compression, sidecar-specification
Model 36EL	60 cu. in./989 cc. ohv V-twin	Four-speed, chain drive	High compression

1937 model year

Model 37W	45 cu. in./742 cc. sv V-twin	Three-speed, chain drive	Low compression
Model 37WS	45 cu. in./742 cc. sv V-twin	Three-speed, chain drive	Low compression, sidecar-specification
Model 37WL	45 cu. in./742 cc. sv V-twin	Three-speed, chain drive	Raised compression
Model 37WLD	45 cu. in./742 cc. sv V-twin	Three-speed, chain drive	Special Sport
Model 37G	45 cu. in./742 cc. sv V-twin	Three-speed, chain drive	Servi-Car
Model 37GA	45 cu. in./742 cc. sv V-twin	Three-speed, chain drive	Servi-Car
Model 37GD	45 cu. in./742 cc. sv V-twin	Three-speed, chain drive	Servi-Car
Model 37GDT	45 cu. in./742 cc. sv V-twin	Three-speed, chain drive	Servi-Car
Model 37GE	45 cu. in./742 cc. sv V-twin	Three-speed, chain drive	Servi-Car
Model 37U	74 cu. in./1209 cc. sv V-twin	Three-speed, chain drive	Low compression
Model 37UMG	74 cu. in./1209 cc. sv V-twin	Three-speed, chain drive	Low compression, magneto ignition
Model 37US	74 cu. in./1209 cc. sv V-twin	Three-speed, chain drive	Low compression, sidecar-specification
Model 37UL	74 cu. in./1209 cc. sv V-twin	Three-speed, chain drive	High compression
Model 37UH	79 cu. in./1302 cc. sv V-twin	Three-speed, chain drive	Low compression
Model 37UHS	79 cu. in./1302 cc. sv V-twin	Three-speed, chain drive	Low compression, sidecar-specification
Model 37ULH	79 cu. in./1302 cc. sv V-twin	Three-speed, chain drive	High compression
Model 37E	60 cu. in./989 cc. ohv V-twin	Four-speed, chain drive	Low compression
Model 37ES	60 cu. in./989 cc. ohv V-twin	Four-speed, chain drive	Low compression, sidecar-specification
Model 37EL	60 cu. in./989 cc. ohv V-twin	Four-speed, chain drive	High compression

1938 model year

Model 38W	45 cu. in./742 cc. sv V-twin	Three-speed, chain drive	Low compression
Model 38WS	45 cu. in./742 cc. sv V-twin	Three-speed, chain drive	Low compression, sidecar-specification
Model 38WL	45 cu. in./742 cc. sv V-twin	Three-speed, chain drive	Raised compression
Model 38WLD	45 cu. in./742 cc. sv V-twin	Three-speed, chain drive	Special Sport
Model 38G	45 cu. in./742 cc. sv V-twin	Three-speed, chain drive	Servi-Car
Model 38GA	45 cu. in./742 cc. sv V-twin	Three-speed, chain drive	Servi-Car
Model 38GD	45 cu. in./742 cc. sv V-twin	Three-speed, chain drive	Servi-Car
Model 38GDT	45 cu. in./742 cc. sv V-twin	Three-speed, chain drive	Servi-Car
Model 38U	74 cu. in./1209 cc. sv V-twin	Three-speed, chain drive	Low compression
Model 38UMG	74 cu. in./1209 cc. sv V-twin	Three-speed, chain drive	Low compression, magneto ignition
Model 38US	74 cu. in./1209 cc. sv V-twin	Three-speed, chain drive	Low compression, sidecar-specification
Model 38UL	74 cu. in./1209 cc. sv V-twin	Three-speed, chain drive	High compression
Model 38UH	79 cu. in./1302 cc. sv V-twin	Three-speed, chain drive	Low compression
Model 38UHS	79 cu. in./1302 cc. sv V-twin	Three-speed, chain drive	Low compression, sidecar-specification
Model 38ULH	79 cu. in./1302 cc. sv V-twin	Three-speed, chain drive	High compression
Model 38ES	60 cu. in./989 cc. ohv V-twin	Four-speed, chain drive	Low compression, sidecar-specification
Model 38EL	60 cu. in./989 cc. ohv V-twin	Four-speed, chain drive	High compression

1939 model year

| Model 39W | 45 cu. in./742 cc. sv V-twin | Three-speed, chain drive | Low compression |
| Model 39WS | 45 cu. in./742 cc. sv V-twin | Three-speed, chain drive | Low compression, sidecar-specification |

Model	Displacement	Transmission	Notes
Model 39WL	45 cu. in./742 cc. sv V-twin	Three-speed, chain drive	Raised compression
Model 39WLD	45 cu. in./742 cc. sv V-twin	Three-speed, chain drive	Special Sport
Model 39G	45 cu. in./742 cc. sv V-twin	Three-speed, chain drive	Servi-Car
Model 39GA	45 cu. in./742 cc. sv V-twin	Three-speed, chain drive	Servi-Car
Model 39GD	45 cu. in./742 cc. sv V-twin	Three-speed, chain drive	Servi-Car
Model 39GDT	45 cu. in./742 cc. sv V-twin	Three-speed, chain drive	Servi-Car
Model 39U	74 cu. in./1209 cc. sv V-twin	Three-speed, chain drive	Low compression
Model 39UMG	74 cu. in./1209 cc. sv V-twin	Three-speed, chain drive	Low compression, magneto ignition
Model 39US	74 cu. in./1209 cc. sv V-twin	Three-speed, chain drive	Low compression, sidecar-specification
Model 39UL	74 cu. in./1209 cc. sv V-twin	Three-speed, chain drive	High compression
Model 39UH	79 cu. in./1302 cc. sv V-twin	Three-speed, chain drive	Low compression
Model 39UHS	79 cu. in./1302 cc. sv V-twin	Three-speed, chain drive	Low compression, sidecar-specification
Model 39ULH	79 cu. in./1302 cc. sv V-twin	Three-speed, chain drive	High compression
Model 39ES	60 cu. in./989 cc. ohv V-twin	Four-speed, chain drive	Low compression, sidecar-specification
Model 39EL	60 cu. in./989 cc. ohv V-twin	Four-speed, chain drive	High compression

1940 model year

Model	Displacement	Transmission	Notes
Model 40W	45 cu. in./742 cc. sv V-twin	Three-speed, chain drive	Low compression
Model 40WS	45 cu. in./742 cc. sv V-twin	Three-speed, chain drive	Low compression, sidecar-specification
Model 40WL	45 cu. in./742 cc. sv V-twin	Three-speed, chain drive	Raised compression
Model 40WLD	45 cu. in./742 cc. sv V-twin	Three-speed, chain drive	Special Sport
Model 40G	45 cu. in./742 cc. sv V-twin	Three-speed, chain drive	Servi-Car
Model 40GA	45 cu. in./742 cc. sv V-twin	Three-speed, chain drive	Servi-Car
Model 40GD	45 cu. in./742 cc. sv V-twin	Three-speed, chain drive	Servi-Car
Model 40GDT	45 cu. in./742 cc. sv V-twin	Three-speed, chain drive	Servi-Car
Model 40U	74 cu. in./1209 cc. sv V-twin	Three-speed, chain drive	Low compression
Model 40UMG	74 cu. in./1209 cc. sv V-twin	Three-speed, chain drive	Low compression, magneto ignition
Model 40US	74 cu. in./1209 cc. sv V-twin	Three-speed, chain drive	Low compression, sidecar-specification
Model 40UL	74 cu. in./1209 cc. sv V-twin	Three-speed, chain drive	High compression
Model 40UH	79 cu. in./1302 cc. sv V-twin	Three-speed, chain drive	Low compression
Model 40UHS	79 cu. in./1302 cc. sv V-twin	Three-speed, chain drive	Low compression, sidecar-specification
Model 40ULH	79 cu. in./1302 cc. sv V-twin	Three-speed, chain drive	High compression
Model 40ES	60 cu. in./989 cc. ohv V-twin	Four-speed, chain drive	Low compression, sidecar-specification
Model 40EL	60 cu. in./989 cc. ohv V-twin	Four-speed, chain drive	High compression

1941 model year

Model	Displacement	Transmission	Notes
Model 41WL	45 cu. in./742 cc. sv V-twin	Three-speed, chain drive	Raised compression
Model 41WLS	45 cu. in./742 cc. sv V-twin	Three-speed, chain drive	Raised compression, sidecar-specification
Model 41WLD	45 cu. in./742 cc. sv V-twin	Three-speed, chain drive	Special Sport
Model 41WLDR	45 cu. in./742 cc. sv V-twin	Three-speed, chain drive	Road-legal version of racer
Model 41G	45 cu. in./742 cc. sv V-twin	Three-speed, chain drive	Servi-Car
Model 41GA	45 cu. in./742 cc. sv V-twin	Three-speed, chain drive	Servi-Car
Model 41GD	45 cu. in./742 cc. sv V-twin	Three-speed, chain drive	Servi-Car
Model 41GDT	45 cu. in./742 cc. sv V-twin	Three-speed, chain drive	Servi-Car
Model 41U	74 cu. in./1209 cc. sv V-twin	Three-speed, chain drive	Low compression
Model 41US	74 cu. in./1209 cc. sv V-twin	Three-speed, chain drive	Low compression, sidecar-specification
Model 41UL	74 cu. in./1209 cc. sv V-twin	Three-speed, chain drive	High compression
Model 41UH	79 cu. in./1302 cc. sv V-twin	Three-speed, chain drive	Low compression
Model 41UHS	79 cu. in./1302 cc. sv V-twin	Three-speed, chain drive	Low compression, sidecar-specification
Model 41ULH	79 cu. in./1302 cc. sv V-twin	Three-speed, chain drive	High compression
Model 41ES	60 cu. in./989 cc. ohv V-twin	Four-speed, chain drive	Low compression, sidecar-specification

Model 41EL	60 cu. in./989 cc. ohv V-twin	Four-speed, chain drive	High compression
Model 41FS	74 cu. in./1207 cc. ohv V-twin	Four-speed, chain drive	Low compression, sidecar-specification
Model 41FL	74 cu. in./1207 cc. ohv V-twin	Four-speed, chain drive	High compression

1942 model year

Model 42WL	45 cu. in./742 cc. sv V-twin	Three-speed, chain drive	Raised compression
Model 42WLS	45 cu. in./742 cc. sv V-twin	Three-speed, chain drive	Raised compression, sidecar-specification
Model 42WLD	45 cu. in./742 cc. sv V-twin	Three-speed, chain drive	Special Sport
Model 42G	45 cu. in./742 cc. sv V-twin	Three-speed, chain drive	Servi-Car
Model 42GA	45 cu. in./742 cc. sv V-twin	Three-speed, chain drive	Servi-Car
Model 42U	74 cu. in./1209 cc. sv V-twin	Three-speed, chain drive	Low compression
Model 42US	74 cu. in./1209 cc. sv V-twin	Three-speed, chain drive	Low compression, sidecar-specification
Model 42UL	74 cu. in./1209 cc. sv V-twin	Three-speed, chain drive	High compression
Model 42E	60 cu. in./989 cc. ohv V-twin	Four-speed, chain drive	Low compression
Model 42EL	60 cu. in./989 cc. ohv V-twin	Four-speed, chain drive	High compression
Model 42F	74 cu. in./1207 cc. ohv V-twin	Four-speed, chain drive	Low compression
Model 42FL	74 cu. in./1207 cc. ohv V-twin	Four-speed, chain drive	High compression

From February 1942 all civilian production was reduced to a minimum. From the beginning of 1942 to the summer of 1945, a civilian motorcycle could only be ordered with the special permission of the authorities.

1943 model year

Model 43G	45 cu. in./742 cc. sv V-twin	Three-speed, chain drive	Servi-Car
Model 43GA	45 cu. in./742 cc. sv V-twin	Three-speed, chain drive	Servi-Car
Model 43U	74 cu. in./1209 cc. sv V-twin	Three-speed, chain drive	Low compression
Model 43UL	74 cu. in./1209 cc. sv V-twin	Three-speed, chain drive	High compression
Model 43E	60 cu. in./989 cc. ohv V-twin	Four-speed, chain drive	Low compression
Model 43EL	60 cu. in./989 cc. ohv V-twin	Four-speed, chain drive	High compression
Model 43F	74 cu. in./1207 cc. ohv V-twin	Four-speed, chain drive	Low compression
Model 43FL	74 cu. in./1207 cc. ohv V-twin	Four-speed, chain drive	High compression

1944 model year

Model 44G	45 cu. in./742 cc. sv V-twin	Three-speed, chain drive	Servi-Car
Model 44GA	45 cu. in./742 cc. sv V-twin	Three-speed, chain drive	Servi-Car
Model 44U	74 cu. in./1209 cc. sv V-twin	Three-speed, chain drive	Low compression
Model 44UL	74 cu. in./1209 cc. sv V-twin	Three-speed, chain drive	High compression
Model 44E	60 cu. in./989 cc. ohv V-twin	Four-speed, chain drive	Low compression
Model 44EL	60 cu. in./989 cc. ohv V-twin	Four-speed, chain drive	High compression
Model 44F	74 cu. in./1207 cc. ohv V-twin	Four-speed, chain drive	Low compression
Model 44FL	74 cu. in./1207 cc. ohv V-twin	Four-speed, chain drive	High compression

1945 model year

Model 45WL	45 cu. in./742 cc. sv V-twin	Three-speed, chain drive	
Model 45G	45 cu. in./742 cc. sv V-twin	Three-speed, chain drive	Servi-Car
Model 45GA	45 cu. in./742 cc. sv V-twin	Three-speed, chain drive	Servi-Car
Model 45U	74 cu. in./1209 cc. sv V-twin	Three-speed, chain drive	Low compression
Model 45US	74 cu. in./1209 cc. sv V-twin	Three-speed, chain drive	Low compression, sidecar-specification
Model 45UL	74 cu. in./1209 cc. sv V-twin	Three-speed, chain drive	High compression
Model 45E	60 cu. in./989 cc. ohv V-twin	Four-speed, chain drive	Low compression
Model 45ES	60 cu. in./989 cc. ohv V-twin	Four-speed, chain drive	Low compression, sidecar-specification
Model 45EL	60 cu. in./989 cc. ohv V-twin	Four-speed, chain drive	High compression
Model 45F	74 cu. in./1207 cc. ohv V-twin	Four-speed, chain drive	Low compression
Model 45FS	74 cu. in./1207 cc. ohv V-twin	Four-speed, chain drive	Low compression, sidecar-specification
Model 45FL	74 cu. in./1207 cc. ohv V-twin	Four-speed, chain drive	High compression

1946 model year

Model 46WL	45 cu. in./742 cc. sv V-twin	Three-speed, chain drive	

Model 46WL-SP	45 cu. in./742 cc. sv V-twin	Three-speed, chain drive	Aluminium-alloy cylinder heads
Model 46G	45 cu. in./742 cc. sv V-twin	Three-speed, chain drive	Servi-Car
Model 46GA	45 cu. in./742 cc. sv V-twin	Three-speed, chain drive	Servi-Car
Model 46U	74 cu. in./1209 cc. sv V-twin	Three-speed, chain drive	Low compression
Model 46US	74 cu. in./1209 cc. sv V-twin	Three-speed, chain drive	Low compression, sidecar-specification
Model 46UL	74 cu. in./1209 cc. sv V-twin	Three-speed, chain drive	High compression
Model 46E	60 cu. in./989 cc. ohv V-twin	Four-speed, chain drive	Low compression
Model 46ES	60 cu. in./989 cc. ohv V-twin	Four-speed, chain drive	Low compression, sidecar-specification
Model 46EL	60 cu. in./989 cc. ohv V-twin	Four-speed, chain drive	High compression
Model 46F	74 cu. in./1207 cc. ohv V-twin	Four-speed, chain drive	Low compression
Model 46FS	74 cu. in./1207 cc. ohv V-twin	Four-speed, chain drive	Low compression, sidecar-specification
Model 46FL	74 cu. in./1207 cc. ohv V-twin	Four-speed, chain drive	High compression

1947 model year

Model 47WL	45 cu. in./742 cc. sv V-twin	Three-speed, chain drive	
Model 47WL-SP	45 cu. in./742 cc. sv V-twin	Three-speed, chain drive	Aluminium-alloy cylinder heads
Model 47G	45 cu. in./742 cc. sv V-twin	Three-speed, chain drive	Servi-Car
Model 47GA	45 cu. in./742 cc. sv V-twin	Three-speed, chain drive	Servi-Car
Model 47U	74 cu. in./1209 cc. sv V-twin	Three-speed, chain drive	Low compression
Model 47US	74 cu. in./1209 cc. sv V-twin	Three-speed, chain drive	Low compression, sidecar-specification
Model 47UL	74 cu. in./1209 cc. sv V-twin	Three-speed, chain drive	High compression
Model 47E	60 cu. in./989 cc. ohv V-twin	Four-speed, chain drive	Low compression
Model 47ES	60 cu. in./989 cc. ohv V-twin	Four-speed, chain drive	Low compression, sidecar-specification
Model 47EL	60 cu. in./989 cc. ohv V-twin	Four-speed, chain drive	High compression
Model 47F	74 cu. in./1207 cc. ohv V-twin	Four-speed, chain drive	Low compression
Model 47FS	74 cu. in./1207 cc. ohv V-twin	Four-speed, chain drive	Low compression, sidecar-specification
Model 47FL	74 cu. in./1207 cc. ohv V-twin	Four-speed, chain drive	High compression

1948 model year

Model 48S	7.6 cu. in./125 cc. 2-stroke single	Three-speed, chain drive	Model 125, DKW-derived
Model 48WL	45 cu. in./742 cc. sv V-twin	Three-speed, chain drive	
Model 48WLS	45 cu. in./742 cc. sv V-twin	Three-speed, chain drive	Sidecar-specification
Model 48WL-SP	45 cu. in./742 cc. sv V-twin	Three-speed, chain drive	Aluminium-alloy cylinder heads
Model 48G	45 cu. in./742 cc. sv V-twin	Three-speed, chain drive	Servi-Car
Model 48GA	45 cu. in./742 cc. sv V-twin	Three-speed, chain drive	Servi-Car
Model 48U	74 cu. in./1209 cc. sv V-twin	Three-speed, chain drive	Low compression
Model 48US	74 cu. in./1209 cc. sv V-twin	Three-speed, chain drive	Low compression, sidecar-specification
Model 48UL	74 cu. in./1209 cc. sv V-twin	Three-speed, chain drive	High compression
Model 48E	60 cu. in./989 cc. ohv V-twin	Four-speed, chain drive	Low compression
Model 48ES	60 cu. in./989 cc. ohv V-twin	Four-speed, chain drive	Low compression, sidecar-specification
Model 48EL	60 cu. in./989 cc. ohv V-twin	Four-speed, chain drive	High compression
Model 48F	74 cu. in./1207 cc. ohv V-twin	Four-speed, chain drive	Low compression
Model 48FS	74 cu. in./1207 cc. ohv V-twin	Four-speed, chain drive	Low compression, sidecar-specification
Model 48FL	74 cu. in./1207 cc. ohv V-twin	Four-speed, chain drive	High compression

1949 model year

Model 49S	7.6 cu. in./125 cc. 2-stroke single	Three-speed, chain drive	Model 125, DKW-derived
Model 49WL	45 cu. in./742 cc. sv V-twin	Three-speed, chain drive	
Model 49WLS	45 cu. in./742 cc. sv V-twin	Three-speed, chain drive	Sidecar-specification
Model 49WL-SP	45 cu. in./742 cc. sv V-twin	Three-speed, chain drive	Aluminium-alloy cylinder heads

Model 49G	45 cu. in./742 cc. sv V-twin	Three-speed, chain drive	Servi-Car
Model 49GA	45 cu. in./742 cc. sv V-twin	Three-speed, chain drive	Servi-Car
Model 49E	60 cu. in./989 cc. ohv V-twin	Four-speed, chain drive	Low compression
Model 49ES	60 cu. in./989 cc. ohv V-twin	Four-speed, chain drive	Low compression, sidecar-specification
Model 49EL	60 cu. in./989 cc. ohv V-twin	Four-speed, chain drive	High compression
Model 49F	74 cu. in./1207 cc. ohv V-twin	Four-speed, chain drive	Low compression
Model 49FS	74 cu. in./1207 cc. ohv V-twin	Four-speed, chain drive	Low compression, sidecar-specification
Model 49FL	74 cu. in./1207 cc. ohv V-twin	Four-speed, chain drive	High compression

1950 model year

Model 50S	7.6 cu. in./125 cc. 2-stroke single	Three-speed, chain drive	Model 125, DKW-derived
Model 50WL	45 cu. in./742 cc. sv V-twin	Three-speed, chain drive	
Model 50WLS	45 cu. in./742 cc. sv V-twin	Three-speed, chain drive	Sidecar-specification
Model 50WL-SP	45 cu. in./742 cc. sv V-twin	Three-speed, chain drive	Aluminium-alloy cyl heads
Model 50G	45 cu. in./742 cc. sv V-twin	Three-speed, chain drive	Servi-Car
Model 50GA	45 cu. in./742 cc. sv V-twin	Three-speed, chain drive	Servi-Car
Model 50E	60 cu. in./989 cc. ohv V-twin	Four-speed, chain drive	Low compression
Model 50ES	60 cu. in./989 cc. ohv V-twin	Four-speed, chain drive	Low compression, sidecar-specification
Model 50EL	60 cu. in./989 cc. ohv V-twin	Four-speed, chain drive	Hig-----h compression
Model 50F	74 cu. in./1207 cc. ohv V-twin	Four-speed, chain drive	Low compression
Model 50FS	74 cu. in./1207 cc. ohv V-twin	Four-speed, chain drive	Low compression, sidecar-specification
Model 50FL	74 cu. in./1207 cc. ohv V-twin	Four-speed, chain drive	High compression

1951 model year

Model 51S	7.6 cu. in./125 cc. 2-stroke single	Three-speed, chain drive	Model 125, DKW-derived
Model 51WL	45 cu. in./742 cc. sv V-twin	Three-speed, chain drive	
Model 51WLS	45 cu. in./742 cc. sv V-twin	Three-speed, chain drive	Sidecar-specification
Model 51WL-SP	45 cu. in./742 cc. sv V-twin	Three-speed, chain drive	Aluminium-alloy cyl-heads
Model 51G	45 cu. in./742 cc. sv V-twin	Three-speed, chain drive	Servi-Car
Model 51GA	45 cu. in./742 cc. sv V-twin	Three-speed, chain drive	Servi-Car
Model 51EL	60 cu. in./989 cc. ohv V-twin	Four-speed, chain drive	
Model 51ELS	60 cu. in./989 cc. ohv V-twin	Four-speed, chain drive	Sidecar-specification
Model 51FL	74 cu. in./1207 cc. ohv V-twin	Four-speed, chain drive	
Model 51FLS	74 cu. in./1207 cc. ohv V-twin	Four-speed, chain drive	Sidecar-specification

1952 model year

Model 52S	7.6 cu. in./125 cc. 2-stroke single	Three-speed, chain drive	Model 125, DKW-derived
Model 52K	45 cu. in./742 cc. sv V-twin	Four-speed, chain drive	
Model 52G	45 cu. in./742 cc. sv V-twin	Three-speed, chain drive	Servi-Car
Model 52GA	45 cu. in./742 cc. sv V-twin	Three-speed, chain drive	Servi-Car
Model 52EL	60 cu. in./989 cc. ohv V-twin	Four-speed, chain drive	
Model 52ELS	60 cu. in./989 cc. ohv V-twin	Four-speed, chain drive	Sidecar-specification
Model 52FL	74 cu. in./1207 cc. ohv V-twin	Four-speed, chain drive	
Model 52FLS	74 cu. in./1207 cc. ohv V-twin	Four-speed, chain drive	Sidecar-specification

The E and F models were available with foot or hand gearchange.

1953 model year

Model 53ST	10 cu. in./166 cc. 2-stroke single	Three-speed, chain drive	Model 165, DKW-derived
Model 53K	45 cu. in./742 cc. sv V-twin	Four-speed, chain drive	
Model 53KK	45 cu. in./742 cc. sv V-twin	Four-speed, chain drive	Special speed kit
Model 53G	45 cu. in./742 cc. sv V-twin	Three-speed, chain drive	Servi-Car
Model 53GA	45 cu. in./742 cc. sv V-twin	Three-speed, chain drive	Servi-Car
Model 53FL	74 cu. in./1207 cc. ohv V-twin	Four-speed, chain drive	
Model 53FLE	74 cu. in./1207 cc. ohv V-twin	Four-speed, chain drive	

The F models were available with foot or hand gearchange, and with or without sidecar-specification.

1954 model year
Model 54ST	10 cu. in./166 cc. 2-stroke single	Three-speed, chain drive	Model 165, DKW-derived
Model 54STU	10 cu. in./166 cc. 2-stroke single	Three-speed, chain drive	Model 165, DKW-derived
Model 54KH	54 cu. in./888 cc. sv V-twin	Four-speed, chain drive	
Model 54G	45 cu. in./742 cc. sv V-twin	Three-speed, chain drive	Servi-Car
Model 54GA	45 cu. in./742 cc. sv V-twin	Three-speed, chain drive	Servi-Car
Model 54FL	74 cu. in./1207 cc. ohv V-twin	Four-speed, chain drive	
Model 54FLE	74 cu. in./1207 cc. ohv V-twin	Four-speed, chain drive	

The F models were available with foot or hand gearchange, and with or without sidecar-specification.

1955 model year
Model 55B	7.6 cu. in./125 cc. 2-stroke single	Three-speed, chain drive	Hummer, DKW-derived
Model 55ST	10 cu. in./166 cc. 2-stroke single	Three-speed, chain drive	Model 165, DKW-derived
Model 55STU	10 cu. in./166 cc. 2-stroke single	Three-speed, chain drive	Model 165, DKW-derived
Model 55KH	54 cu. in./888 cc. sv V-twin	Four-speed, chain drive	
Model 55KHK	54 cu. in./888 cc. sv V-twin	Four-speed, chain drive	
Model 55G	45 cu. in./742 cc. sv V-twin	Three-speed, chain drive	Servi-Car
Model 55GA	45 cu. in./742 cc. sv V-twin	Three-speed, chain drive	Servi-Car
Model 55FL	74 cu. in./1207 cc. ohv V-twin	Four-speed, chain drive	
Model 55FLE	74 cu. in./1207 cc. ohv V-twin	Four-speed, chain drive	
Model 55FLH	74 cu. in./1207 cc. ohv V-twin	Four-speed, chain drive	

The F models were available with foot or hand gearchange, and with or without sidecar-specification.

1956 model year
Model 56B	7.6 cu. in./125 cc. 2-stroke single	Three-speed, chain drive	Hummer, DKW-derived
Model 56ST	10 cu. in./166 cc. 2-stroke single	Three-speed, chain drive	Model 165, DKW-derived
Model 56STU	10 cu. in./166 cc. 2-stroke single	Three-speed, chain drive	Model 165, DKW-derived
Model 56KH	54 cu. in./888 cc. sv V-twin	Four-speed, chain drive	
Model 56KHK	54 cu. in./888 cc. sv V-twin	Four-speed, chain drive	Speed Kit
Model 56G	45 cu. in./742 cc. sv V-twin	Three-speed, chain drive	Servi-Car
Model 56GA	45 cu. in./742 cc. sv V-twin	Three-speed, chain drive	Servi-Car
Model 56FL	74 cu. in./1207 cc. ohv V-twin	Four-speed, chain drive	
Model 56FLE	74 cu. in./1207 cc. ohv V-twin	Four-speed, chain drive	
Model 56FLH	74 cu. in./1207 cc. ohv V-twin	Four-speed, chain drive	

The FL models were available with foot or hand gearchange, and with or without sidecar-specification.

1957 model year
Model 57B	7.6 cu. in./125 cc. 2-stroke single	Three-speed, chain drive	Hummer, DKW-derived
Model 57ST	10 cu. in./166 cc. 2-stroke single	Three-speed, chain drive	Model 165, DKW-derived
Model 57STU	10 cu. in./166 cc. 2-stroke single	Three-speed, chain drive	Model 165, DKW-derived
Model 57XL	54 cu. in./883 cc. ohv V-twin	Four-speed, chain drive	Sportster
Model 57G	45 cu. in./742 cc. sv V-twin	Three-speed, chain drive	Servi-Car
Model 57GA	45 cu. in./742 cc. sv V-twin	Three-speed, chain drive	Servi-Car
Model 57FL	74 cu. in./1207 cc. ohv V-twin	Four-speed, chain drive	
Model 57FLH	74 cu. in./1207 cc. ohv V-twin	Four-speed, chain drive	

The FL models were available with foot or hand gearchange, and with or without sidecar-specification.

1958 model year
Model 58B	7.6 cu. in./125 cc. 2-stroke single	Three-speed, chain drive	Hummer, DKW-derived
Model 58ST	10 cu. in./166 cc. 2-stroke single	Three-speed, chain drive	Model 165, DKW-derived
Model 58STU	10 cu. in./166 cc. 2-stroke single	Three-speed, chain drive	Model 165, DKW-derived
Model 58XL	54 cu. in./883 cc. ohv V-twin	Four-speed, chain drive	Sportster
Model 58XLH	54 cu. in./883 cc. ohv V-twin	Four-speed, chain drive	Sportster H
Model 58G	45 cu. in./742 cc. sv V-twin	Three-speed, chain drive	Servi-Car
Model 58GA	45 cu. in./742 cc. sv V-twin	Three-speed, chain drive	Servi-Car
Model 58FL	74 cu. in./1207 cc. ohv V-twin	Four-speed, chain drive	
Model 58FLH	74 cu. in./1207 cc. ohv V-twin	Four-speed, chain drive	

The FL models were available with foot or hand gearchange, and with or without sidecar-specification.
The Models XLC and XLCH were competition only – not road-legal.

1959 model year
Model 59A	10 cu. in./166 cc. 2-stroke single	Automatic, V-belt & chain	Topper
Model 59B	7.6 cu. in./125 cc. 2-stroke single	Three-speed, chain drive	Hummer, DKW-derived

Model	Displacement	Transmission	Name
Model 59ST	10 cu. in./166 cc. 2-stroke single	Three-speed, chain drive	Model 165, DKW-derived
Model 59STU	10 cu. in./166 cc. 2-stroke single	Three-speed, chain drive	Model 165, DKW-derived
Model 59XL	54 cu. in./883 cc. ohv V-twin	Four-speed, chain drive	Sportster
Model 59XLH	54 cu. in./883 cc. ohv V-twin	Four-speed, chain drive	Sportster H
Model 59XLCH	54 cu. in./883 cc. ohv V-twin	Four-speed, chain drive	Sportster CH
Model 59G	45 cu. in./742 cc. sv V-twin	Three-speed, chain drive	Servi-Car
Model 59GA	45 cu. in./742 cc. sv V-twin	Three-speed, chain drive	Servi-Car
Model 59FL	74 cu. in./1207 cc. ohv V-twin	Four-speed, chain drive	
Model 59FLH	74 cu. in./1207 cc. ohv V-twin	Four-speed, chain drive	

The F models were available with foot or hand gearchange, and with or without sidecar-specification.

1960 model year

Model	Displacement	Transmission	Name
Model 60A	10 cu. in./166 cc. 2-stroke single	Automatic, V-belt & chain	Topper
Model 60AU	10 cu. in./166 cc. 2-stroke single	Automatic, V-belt & chain	Topper
Model 60BT	10 cu. in./166 cc. 2-stroke single	Three-speed, chain drive	Super 10, DKW-derived
Model 60BTU	10 cu. in./166 cc. 2-stroke single	Three-speed, chain drive	Super 10, DKW-derived
Model 60XLH	54 cu. in./883 cc. ohv V-twin	Four-speed, chain drive	Sportster H
Model 60XLCH	54 cu. in./883 cc. ohv V-twin	Four-speed, chain drive	Sportster CH
Model 60G	45 cu. in./742 cc. sv V-twin	Three-speed, chain drive	Servi-Car
Model 60GA	45 cu. in./742 cc. sv V-twin	Three-speed, chain drive	Servi-Car
Model 60FL	74 cu. in./1207 cc. ohv V-twin	Four-speed, chain drive	
Model 60FLH	74 cu. in./1207 cc. ohv V-twin	Four-speed, chain drive	

The FL models were available with foot or hand gearchange, and with or without sidecar-specification.

1961 model year

Model	Displacement	Transmission	Name
Model 61AH	10 cu. in./166 cc. 2-stroke single	Automatic, V-belt & chain	Topper
Model 61AU	10 cu. in./166 cc. 2-stroke single	Automatic, V-belt & chain	Topper
Model 61BT	10 cu. in./166 cc. 2-stroke single	Three-speed, chain drive	Super 10, DKW-derived
Model 61BTU	10 cu. in./166 cc. 2-stroke single	Three-speed, chain drive	Super 10, DKW-derived
Model 61C	15 cu. in./246 cc. ohv single	Four-speed, chain drive	Aermacchi Sprint C
Model 61XLH	54 cu. in./883 cc. ohv V-twin	Four-speed, chain drive	Sportster H
Model 61XLCH	54 cu. in./883 cc. ohv V-twin	Four-speed, chain drive	Sportster CH
Model 61G	45 cu. in./742 cc. sv V-twin	Three-speed, chain drive	Servi-Car
Model 61GA	45 cu. in./742 cc. sv V-twin	Three-speed, chain drive	Servi-Car
Model 61FL	74 cu. in./1207 cc. ohv V-twin	Four-speed, chain drive	
Model 61FLH	74 cu. in./1207 cc. ohv V-twin	Four-speed, chain drive	

The FL models were available with foot or hand gearchange, and with or without sidecar-specification.

1962 model year

Model	Displacement	Transmission	Name
Model 62AH	10 cu. in./166 cc. 2-stroke single	Automatic, V-belt & chain	Topper
Model 62AU	10 cu. in./166 cc. 2-stroke single	Automatic, V-belt & chain	Topper
Model 62BT	11 cu. in./175 cc. 2-stroke single	Three-speed, chain drive	Pacer, DKW-derived
Model 62BTU	10 cu. in./166 cc. 2-stroke single	Three-speed, chain drive	Pacer, DKW-derived
Model 62BTF	10 cu. in./166 cc. 2-stroke single	Three-speed, chain drive	Ranger, DKW-derived
Model 62BTH	11 cu. in./175 cc. 2-stroke single	Three-speed, chain drive	Scat, DKW-derived
Model 62C	15 cu. in./246 cc. ohv single	Four-speed, chain drive	Aermacchi Sprint C
Model 62H	15 cu. in./246 cc. ohv single	Four-speed, chain drive	Aermacchi Sprint H
Model 62XLH	54 cu. in./883 cc. ohv V-twin	Four-speed, chain drive	Sportster H
Model 62XLCH	54 cu. in./883 cc. ohv V-twin	Four-speed, chain drive	Sportster CH
Model 62G	45 cu. in./742 cc. sv V-twin	Three-speed, chain drive	Servi-Car
Model 62GA	45 cu. in./742 cc. sv V-twin	Three-speed, chain drive	Servi-Car
Model 62FL	74 cu. in./1207 cc. ohv V-twin	Four-speed, chain drive	
Model 62FLH	74 cu. in./1207 cc. ohv V-twin	Four-speed, chain drive	

The FL models were available with foot or hand gearchange, and with or without sidecar-specification.

1963 model year

Model	Displacement	Transmission	Name
Model 63AH	10 cu. in./166 cc. 2-stroke single	Automatic, V-belt & chain	Topper
Model 63AU	10 cu. in./166 cc. 2-stroke single	Automatic, V-belt & chain	Topper
Model 63BT	11 cu. in./175 cc. 2-stroke single	Three-speed, chain drive	Pacer, DKW-derived
Model 63BTU	11 cu. in./175 cc. 2-stroke single	Three-speed, chain drive	Pacer, DKW-derived
Model 63BTH	11 cu. in./175 cc. 2-stroke single	Three-speed, chain drive	Scat, DKW-derived
Model 63C	15 cu. in./246 cc. ohv single	Four-speed, chain drive	Aermacchi Sprint C
Model 63H	15 cu. in./246 cc. ohv single	Four-speed, chain drive	Aermacchi Sprint H

Model	Engine	Transmission	Name
Model 63XLH	54 cu. in./883 cc. ohv V-twin	Four-speed, chain drive	Sportster H
Model 63XLCH	54 cu. in./883 cc. ohv V-twin	Four-speed, chain drive	Sportster CH
Model 63G	45 cu. in./742 cc. sv V-twin	Three-speed, chain drive	Servi-Car
Model 63GA	45 cu. in./742 cc. sv V-twin	Three-speed, chain drive	Servi-Car
Model 63FL	74 cu. in./1207 cc. ohv V-twin	Four-speed, chain drive	
Model 63FLH	74 cu. in./1207 cc. ohv V-twin	Four-speed, chain drive	

The FL models were available with foot or hand gearchange, and with or without sidecar-specification.

1964 model year

Model	Engine	Transmission	Name
Model 64AH	10 cu. in./166 cc. 2-stroke single	Automatic, V-belt & chain	Topper
Model 64AU	10 cu. in./166 cc. 2-stroke single	Automatic, V-belt & chain	Topper
Model 64BT	11 cu. in./175 cc. 2-stroke single	Three-speed, chain drive	Pacer, DKW-derived
Model 64BTU	11 cu. in./175 cc. 2-stroke single	Three-speed, chain drive	Pacer, DKW-derived
Model 64BTH	11 cu. in./175 cc. 2-stroke single	Three-speed, chain drive	Scat, DKW-derived
Model 64C	15 cu. in./246 cc. ohv single	Four-speed, chain drive	Aermacchi Sprint C
Model 64H	15 cu. in./246 cc. ohv single	Four-speed, chain drive	Aermacchi Scrambler
Model 64XLH	54 cu. in./883 cc. ohv V-twin	Four-speed, chain drive	Sportster H
Model 64XLCH	54 cu. in./883 cc. ohv V-twin	Four-speed, chain drive	Sportster CH
Model 64GE	45 cu. in./742 cc. sv V-twin	Three-speed, chain drive	Servi-Car
Model 64FL	74 cu. in./1207 cc. ohv V-twin	Four-speed, chain drive	
Model 64FLH	74 cu. in./1207 cc. ohv V-twin	Four-speed, chain drive	

The FL models were available with foot or hand gearchange, and with or without sidecar-specification.

1965 model year

Model	Engine	Transmission	Name
Model 65M 50	3 cu. in./47 cc. 2-stroke single	Three-speed, chain drive	Aermacchi Leggero
Model 65AH	10 cu. in./166 cc. 2-stroke single	Automatic, V-belt & chain	Topper
Model 65BT	11 cu. in./175 cc. 2-stroke single	Three-speed, chain drive	Pacer, DKW-derived
Model 65BTH	11 cu. in./175 cc. 2-stroke single	Three-speed, chain drive	Scat, DKW-derived
Model 65C	15 cu. in./246 cc. ohv single	Four-speed, chain drive	Aermacchi Sprint C
Model 65H	15 cu. in./246 cc. ohv single	Four-speed, chain drive	Aermacchi Scrambler
Model 65XLH	54 cu. in./883 cc. ohv V-twin	Four-speed, chain drive	Sportster H
Model 65XLCH	54 cu. in./883 cc. ohv V-twin	Four-speed, chain drive	Sportster CH
Model 65GE	45 cu. in./742 cc. sv V-twin	Three-speed, chain drive	Servi-Car
Model 65FL	74 cu. in./1207 cc. ohv V-twin	Four-speed, chain drive	
Model 65FLH	74 cu. in./1207 cc. ohv V-twin	Four-speed, chain drive	

The FL models were available with foot or hand gearchange, and with or without sidecar-specification.

1966 model year

Model	Engine	Transmission	Name
Model 66M-50	3 cu. in./47 cc. 2-stroke single	Three-speed, chain drive	Aermacchi Leggero
Model 66M-50S	3 cu. in./47 cc. 2-stroke single	Three-speed, chain drive	Aermacchi Leggero
Model 66BTH	11 cu. in./175 cc. 2-stroke single	Three-speed, chain drive	Bobcat, DKW-derived
Model 66C	15 cu. in./246 cc. ohv single	Four-speed, chain drive	Aermacchi Sprint C
Model 66H	15 cu. in./246 cc. ohv single	Four-speed, chain drive	Aermacchi Scrambler
Model 66XLH	54 cu. in./883 cc. ohv V-twin	Four-speed, chain drive	Sportster H
Model 66XLCH	54 cu. in./883 cc. ohv V-twin	Four-speed, chain drive	Sportster CH
Model 66GE	45 cu. in./742 cc. sv V-twin	Three-speed, chain drive	Servi-Car
Model 66FL	74 cu. in./1207 cc. ohv V-twin	Four-speed, chain drive	
Model 66FLH	74 cu. in./1207 cc. ohv V-twin	Four-speed, chain drive	

The FL models were available with foot or hand gearchange, and with or without sidecar-specification.

1967 model year

Model	Engine	Transmission	Name
Model 67M-65	4 cu. in./64 cc. 2-stroke single	Three-speed, chain drive	Aermacchi Leggero
Model 67M-65S	4 cu. in./64 cc. 2-stroke single	Three-speed, chain drive	Aermacchi Leggero
Model 67SS-250	15 cu. in./248 cc. ohv single	Four-speed, chain drive	Aermacchi Sprint SS
Model 67H	15 cu. in./248 cc. ohv single	Four-speed, chain drive	Aermacchi Sprint H
Model 67XLH	54 cu. in./883 cc. ohv V-twin	Four-speed, chain drive	Sportster H
Model 67XLCH	54 cu. in./883 cc. ohv V-twin	Four-speed, chain drive	Sportster CH
Model 67GE	45 cu. in./742 cc. sv V-twin	Three-speed, chain drive	Servi-Car
Model 67FL	74 cu. in./1207 cc. ohv V-twin	Four-speed, chain drive	
Model 67FLH	74 cu. in./1207 cc. ohv V-twin	Four-speed, chain drive	

The FL models were available with foot or hand gearchange, and with or without sidecar-specification.

1968 model year
Model 68M-65	4 cu. in./64 cc. 2-stroke single	Three-speed, chain drive	Aermacchi Leggero
Model 68M-65S	4 cu. in./64 cc. 2-stroke single	Three-speed, chain drive	Aermacchi Leggero
Model 68M-125	7.5 cu. in./123 cc. 2-stroke single	Three-speed, chain drive	Aermacchi Rapido
Model 68SS-250	15 cu. in./248 cc. ohv single	Four-speed, chain drive	Aermacchi Sprint SS
Model 68H	15 cu. in./248 cc. ohv single	Four-speed, chain drive	Aermacchi Sprint H
Model 68XLH	54 cu. in./883 cc. ohv V-twin	Four-speed, chain drive	Sportster H
Model 68XLCH	54 cu. in./883 cc. ohv V-twin	Four-speed, chain drive	Sportster CH
Model 68GE	45 cu. in./742 cc. sv V-twin	Three-speed, chain drive	Servi-Car
Model 68FL	74 cu. in./1207 cc. ohv V-twin	Four-speed, chain drive	
Model 68FLH	74 cu. in./1207 cc. ohv V-twin	Four-speed, chain drive	

The FL models were available with foot or hand gearchange, and with or without sidecar-specification.

1969 model year
Model 69M-65	4 cu. in./64 cc. 2-stroke single	Three-speed, chain drive	Aermacchi Leggero
Model 69M-65S	4 cu. in./64 cc. 2-stroke single	Three-speed, chain drive	Aermacchi Leggero
Model 69ML-125	7.5 cu. in./123 cc. 2-stroke single	Three-speed, chain drive	Aermacchi Rapido
Model 69SS-350	21 cu. in./342 cc. ohv single	Four-speed, chain drive	Aermacchi Sprint SS
Model 69XLH	54 cu. in./883 cc. ohv V-twin	Four-speed, chain drive	Sportster H
Model 69XLCH	54 cu. in./883 cc. ohv V-twin	Four-speed, chain drive	Sportster CH
Model 69GE	45 cu. in./742 cc. sv V-twin	Three-speed, chain drive	Servi-Car
Model 69FL	74 cu. in./1207 cc. ohv V-twin	Four-speed, chain drive	
Model 69FLH	74 cu. in./1207 cc. ohv V-twin	Four-speed, chain drive	

The FL models were available with foot or hand gearchange, and with or without sidecar-specification.

1970 model year
Model 70M-65S	4 cu. in./64 cc. 2-stroke single	Three-speed, chain drive	Aermacchi Leggero
Model 70MLS-125	7.5 cu. in./123 cc. 2-stroke single	Four-speed, chain drive	Aermacchi Rapido
Model 70SS-350	21 cu. in./342 cc. ohv single	Four-speed, chain drive	Aermacchi Sprint SS
Model 70XLH	54 cu. in./883 cc. ohv V-twin	Four-speed, chain drive	Sportster H
Model 70XLCH	54 cu. in./883 cc. ohv V-twin	Four-speed, chain drive	Sportster CH
Model 70GE	45 cu. in./742 cc. sv V-twin	Three-speed, chain drive	Servi-Car
Model 70FL	74 cu. in./1207 cc. ohv V-twin	Four-speed, chain drive	
Model 70FLH	74 cu. in./1207 cc. ohv V-twin	Four-speed, chain drive	

The FL models were available with foot or hand gearchange, and with or without sidecar-specification.

1971 model year
Model 71M-65S	4 cu. in./64 cc. 2-stroke single	Three-speed, chain drive	Aermacchi Leggero
Model 71MSR-100L	6 cu. in./98 cc. 2-stroke single	Five-speed, chain drive	Aermacchi Baja
Model 71MLS-125	7.5 cu. in./123 cc. 2-stroke single	Four-speed, chain drive	Aermacchi Rapido
Model 71SS-350	21 cu. in./342 cc. ohv single	Four-speed, chain drive	Aermacchi Sprint SS
Model 71SX-350	21 cu. in./342 cc. ohv single	Four-speed, chain drive	Aermacchi Sprint SX
Model 71XLH	54 cu. in./883 cc. ohv V-twin	Four-speed, chain drive	Sportster H
Model 71XLCH	54 cu. in./883 cc. ohv V-twin	Four-speed, chain drive	Sportster CH
Model 71GE	45 cu. in./742 cc. sv V-twin	Three-speed, chain drive	Servi-Car
Model 71FX	74 cu. in./1207 cc. ohv V-twin	Four-speed, chain drive	Super Glide
Model 71FL	74 cu. in./1207 cc. ohv V-twin	Four-speed, chain drive	Electra-Glide (Police)
Model 71FLH	74 cu. in./1207 cc. ohv V-twin	Four-speed, chain drive	Electra-Glide

The FL models were available with foot or hand gearchange, and with or without sidecar-specification.

1972 model year
Model 72M-65S	4 cu. in./64 cc. 2-stroke single	Three-speed, chain drive	Aermacchi Leggero
Model 72MC-65	4 cu. in./64 cc. 2-stroke single	Three-speed, chain drive	Aermacchi Shortster
Model 72MSR-100L	6 cu. in./98 cc. 2-stroke single	Five-speed, chain drive	Aermacchi Baja
Model 72MLS-125	7.5 cu. in./123 cc. 2-stroke single	Four-speed, chain drive	Aermacchi Rapido
Model 72SS-350	21 cu. in./342 cc. ohv single	Four-speed, chain drive	Aermacchi Sprint SS
Model 72SX-350	21 cu. in./342 cc. ohv single	Four-speed, chain drive	Aermacchi Sprint SX
Model 72GE	45 cu. in./742 cc. sv V-twin	Three-speed, chain drive	Servi-Car
Model 72XLH	61 cu. in./997 cc. ohv ohv V-twin	Four-speed, chain drive	Sportster H
Model 72XLCH	61 cu. in./997 cc. ohv V-twin	Four-speed, chain drive	Sportster CH
Model 72FX	74 cu. in./1207 cc. ohv V-twin	Four-speed, chain drive	Super Glide
Model 72FL	74 cu. in./1207 cc. ohv V-twin	Four-speed, chain drive	Electra-Glide (Police)
Model 72FLH	74 cu. in./1207 cc. ohv V-twin	Four-speed, chain drive	Electra-Glide

The FL models were available with foot or hand gearchange, and with or without sidecar-specification.

1973 model year
Model 73X	5.5 cu. in./90 cc. 2-stroke single	Four-speed, chain drive	Aermacchi Shortster
Model 73Z	5.5 cu. in./90 cc. 2-stroke single	Four-speed, chain drive	Aermacchi Leggero
Model 73MSR-100L	6 cu. in./98 cc. 2-stroke single	Five-speed, chain drive	Aermacchi Baja
Model 73TX-125	7.5 cu. in./123 cc. 2-stroke single	Five-speed, chain drive	Aermacchi Rapido
Model 73SS-350	21 cu. in./342 cc. ohv single	Five-speed, chain drive	Aermacchi Sprint SS
Model 73SX-350	21 cu. in./342 cc. ohv single	Five-speed, chain drive	Aermacchi Sprint SX
Model 73GE	45 cu. in./742 cc. sv V-twin	Three-speed, chain drive	Servi-Car
Model 73XLH	61 cu. in./997 cc. ohv V-twin	Four-speed, chain drive	Sportster H
Model 73XLCH	61 cu. in./997 cc. ohv V-twin	Four-speed, chain drive	Sportster CH
Model 73FX	74 cu. in./1207 cc. ohv V-twin	Four-speed, chain drive	Super Glide
Model 73FL	74 cu. in./1207 cc. ohv V-twin	Four-speed, chain drive	Electra-Glide (Police)
Model 73FLH	74 cu. in./1207 cc. ohv V-twin	Four-speed, chain drive	Electra-Glide

The FL models were available with foot or hand gearchange, and with or without sidecar-specification.

1974 model year
Model 74X	5.5 cu. in./90 cc. 2-stroke single	Four-speed, chain drive	Aermacchi Shortster
Model 74Z	5.5 cu. in./90 cc. 2-stroke single	Four-speed, chain drive	Aermacchi Leggero
Model 74SR-100	6 cu. in./98 cc. 2-stroke single	Five-speed, chain drive	Aermacchi Baja
Model 74SX-125	7.5 cu. in./123 cc. 2-stroke single	Five-speed, chain drive	Aermacchi Rapido
Model 74SX-175	11 cu. in./174 cc. 2-stroke single	Five-speed, chain drive	Aermacchi SX-175
Model 74SX-250	15 cu. in./243 cc. 2-stroke single	Five-speed, chain drive	Aermacchi SX-250
Model 74SS-350	21 cu. in./342 cc. ohv single	Five-speed, chain drive	Aermacchi Sprint SS
Model 74SX-350	21 cu. in./342 cc. ohv single	Five-speed, chain drive	Aermacchi Sprint SX
Model 74XLH	61 cu. in./997 cc. ohv V-twin	Four-speed, chain drive	Sportster H
Model 74XLCH	61 cu. in./997 cc. ohv V-twin	Four-speed, chain drive	Sportster CH
Model 74FX	74 cu. in./1207 cc. ohv V-twin	Four-speed, chain drive	Super Glide
Model 74FXE	74 cu. in./1207 cc. ohv V-twin	Four-speed, chain drive	Super Glide
Model 74FL	74 cu. in./1207 cc. ohv V-twin	Four-speed, chain drive	Electra-Glide
Model 74FLH	74 cu. in./1207 cc. ohv V-twin	Four-speed, chain drive	Electra-Glide

The FL models were available with foot or hand gearchange, and with or without sidecar-specification.

1975 model year
Model 75X	5.5 cu. in./90 cc. 2-stroke single	Four-speed, chain drive	Aermacchi Shortster
Model 75Z	5.5 cu. in./90 cc. 2-stroke single	Four-speed, chain drive	Aermacchi Leggero
Model 75SXT-125	7.5 cu. in./123 cc. 2-stroke single	Five-speed, chain drive	Aermacchi Rapido
Model 75SX-175	11 cu. in./174 cc. 2-stroke single	Five-speed, chain drive	Aermacchi SX-175
Model 75SS-250	15 cu. in./243 cc. 2-stroke single	Five-speed, chain drive	Aermacchi SS-250
Model 75SX-250	15 cu. in./243 cc. 2-stroke single	Five-speed, chain drive	Aermacchi SX-250
Model 75XLH	61 cu. in./997 cc. ohv V-twin	Four-speed, chain drive	Sportster H
Model 75XLCH	61 cu. in./997 cc. ohv V-twin	Four-speed, chain drive	Sportster CH
Model 75FX	74 cu. in./1207 cc. ohv V-twin	Four-speed, chain drive	Super Glide
Model 75FXE	74 cu. in./1207 cc. ohv V-twin	Four-speed, chain drive	Super Glide
Model 75FL	74 cu. in./1207 cc. ohv V-twin	Four-speed, chain drive	Electra-Glide (Police)
Model 75FLH	74 cu. in./1207 cc. ohv V-twin	Four-speed, chain drive	Electra-Glide

The FL models were available with foot or hand gearchange, and with or without sidecar-specification.

1976 model year
Model 76SS-125	7.5 cu. in./123 cc. 2-stroke single	Five-speed, chain drive	Aermacchi SS-125
Model 76SXT-125	7.5 cu. in./123 cc. 2-stroke single	Five-speed, chain drive	Aermacchi SXT-125
Model 76SS-175	11 cu. in./174 cc. 2-stroke single	Five-speed, chain drive	Aermacchi SS-175
Model 76SX-175	11 cu. in./174 cc. 2-stroke single	Five-speed, chain drive	Aermacchi SX-175
Model 76SS-250	15 cu. in./243 cc. 2-stroke single	Five-speed, chain drive	Aermacchi SS-250
Model 76SST-250	15 cu. in./243 cc. 2-stroke single	Five-speed, chain drive	Aermacchi SST-250
Model 76SX-250	15 cu. in./243 cc. 2-stroke single	Five-speed, chain drive	Aermacchi SX-250
Model 76XLH	61 cu. in./997 cc. ohv V-twin	Four-speed, chain drive	Sportster H
Model 76XLCH	61 cu. in./997 cc. ohv V-twin	Four-speed, chain drive	Sportster CH
Model 76FX	74 cu. in./1207 cc. ohv V-twin	Four-speed, chain drive	Super Glide
Model 76FXE	74 cu. in./1207 cc. ohv V-twin	Four-speed, chain drive	Super Glide
Model 76FL	74 cu. in./1207 cc. ohv V-twin	Four-speed, chain drive	Electra-Glide
Model 76FLH	74 cu. in./1207 cc. ohv V-twin	Four-speed, chain drive	Electra-Glide

The FL models could be ordered with or without sidecar-specification.

Special models
Model 76XLH	Liberty Edition
Model 76XLCH	Liberty Edition
Model 76FX	Liberty Edition
Model 76FXE	Liberty Edition
Model 76FLH	Liberty Edition

1977 model year

Model	Displacement	Transmission	Name
Model 77SS-125	7.5 cu. in./123 cc. 2-stroke single	Five-speed, chain drive	Aermacchi SS-125
Model 77SXT-125	7.5 cu. in./123 cc. 2-stroke single	Five-speed, chain drive	Aermacchi SXT-125
Model 77SS-175	11 cu. in./174 cc. 2-stroke single	Five-speed, chain drive	Aermacchi SS-175
Model 77SS-250	15 cu. in./243 cc. 2-stroke single	Five-speed, chain drive	Aermacchi SS-250
Model 77SST-250	15 cu. in./243 cc. 2-stroke single	Five-speed, chain drive	Aermacchi SST-250
Model 77SX-250	15 cu. in./243 cc. 2-stroke single	Five-speed, chain drive	Aermacchi SX-250
Model 77XLH	61 cu. in./997 cc. ohv V-twin	Four-speed, chain drive	Sportster H
Model 77XLCH	61 cu. in./997 cc. ohv V-twin	Four-speed, chain drive	Sportster CH
Model 77XLT	61 cu. in./997 cc. ohv V-twin	Four-speed, chain drive	Sportster Touring
Model 77XLCR	61 cu. in./997 cc. ohv V-twin	Four-speed, chain drive	Sportster Café Racer
Model 77FX	74 cu. in./1207 cc. ohv V-twin	Four-speed, chain drive	Super Glide
Model 77FXE	74 cu. in./1207 cc. ohv V-twin	Four-speed, chain drive	Super Glide
Model 77$^{1}/_{2}$-FXS	74 cu. in./1207 cc. ohv V-twin	Four-speed, chain drive	Super Glide Low Rider
Model 77FL	74 cu. in./1207 cc. ohv V-twin	Four-speed, chain drive	Electra-Glide (Police)
Model 77FLH	74 cu. in./1207 cc. ohv V-twin	Four-speed, chain drive	Electra-Glide
Model 77FLHS	74 cu. in./1207 cc. ohv V-twin	Four-speed, chain drive	Electra-Glide Sport

The FL models could be ordered with or without sidecar-specification.

1978 model year

Model	Displacement	Transmission	Name
Model 78SS-175	11 cu. in./174 cc. 2-stroke single	Five-speed, chain drive	Aermacchi SS-175
Model 78SX-175	11 cu. in./174 cc. 2-stroke single	Five-speed, chain drive	Aermacchi SX-175
Model 78SS-250	15 cu. in./243 cc. 2-stroke single	Five-speed, chain drive	Aermacchi SS-250
Model 78SST-250	15 cu. in./243 cc. 2-stroke single	Five-speed, chain drive	Aermacchi SST-250
Model 78XLH	61 cu. in./997 cc. ohv V-twin	Four-speed, chain drive	Sportster H
Model 78XLCH	61 cu. in./997 cc. ohv V-twin	Four-speed, chain drive	Sportster CH
Model 78XLT	61 cu. in./997 cc. ohv V-twin	Four-speed, chain drive	Sportster Touring
Model 78XLCR	61 cu. in./997 cc. ohv V-twin	Four-speed, chain drive	Sportster Café Racer
Model 78FX	74 cu. in./1207 cc. ohv V-twin	Four-speed, chain drive	Super Glide
Model 78FXE	74 cu. in./1207 cc. ohv V-twin	Four-speed, chain drive	Super Glide
Model 78FXS	74 cu. in./1207 cc. ohv V-twin	Four-speed, chain drive	Super Glide Low Rider
Model 78FL	74 cu. in./1207 cc. ohv V-twin	Four-speed, chain drive	Electra-Glide (Police)
Model 78FLH	74 cu. in./1207 cc. ohv V-twin	Four-speed, chain drive	Electra-Glide
Model 78FL-80	82 cu. in./1340 cc. ohv V-twin	Four-speed, chain drive	Electra-Glide (Police)
Model 78FLH-80	82 cu. in./1340 cc. ohv V-twin	Four-speed, chain drive	Electra-Glide

The FL models could be ordered with or without sidecar-specification.

Special models
Model 78XLH	Anniversary Edition
Model 78FLH	Anniversary Edition

1979 model year

Model	Displacement	Transmission	Name
Model 79XLH	61 cu. in./997 cc. ohv V-twin	Five-speed, chain drive	Sportster H
Model 79XLCH	61 cu. in./997 cc. ohv V-twin	Four-speed, chain drive	Sportster CH
Model 79XLS	61 cu. in./997 cc. ohv V-twin	Four-speed, chain drive	Sportster Roadster
Model 79FXE	74 cu. in./1207 cc. ohv V-twin	Four-speed, chain drive	Super Glide
Model 79FXS	74 cu. in./1207 cc. ohv V-twin	Four-speed, chain drive	Low Rider
Model 79FXS-80	82 cu. in./1340 cc. ohv V-twin	Four-speed, chain drive	Low Rider
Model 79FXEF	74 cu. in./1207 cc. ohv V-twin	Four-speed, chain drive	Fat Bob
Model 79FXEF-80	82 cu. in./1340 cc. ohv V-twin	Four-speed, chain drive	Fat Bob
Model 79FLH	74 cu. in./1207 cc. ohv V-twin	Four-speed, chain drive	Electra-Glide
Model 79FL-80	82 cu. in./1340 cc. ohv V-twin	Four-speed, chain drive	Electra-Glide (Police)
Model 79FLH-80	82 cu. in./1340 cc. ohv V-twin	Four-speed, chain drive	Electra-Glide
Model 79FLH-80	82 cu. in./1340 cc. ohv V-twin	Four-speed, chain drive	Electra-Glide Shrine

The FL models could be ordered with or without sidecar-specification.

1980 model year
Model 80XLH	61 cu. in./997 cc. ohv V-twin	Four-speed, chain drive	Sportster
Model 80XLS	61 cu. in./997 cc. ohv V-twin	Four-speed, chain drive	Roadster
Model 80FXE	74 cu. in./1207 cc. ohv V-twin	Four-speed, chain drive	Super Glide
Model 80FXE-80	82 cu. in./1340 cc. ohv V-twin	Four-speed, chain drive	Super Glide
Model 80FXS	74 cu. in./1207 cc. ohv V-twin	Four-speed, chain drive	Low Rider
Model 80FXS-80	82 cu. in./1340 cc. ohv V-twin	Four-speed, chain drive	Low Rider
Model 80FXEF	74 cu. in./1207 cc. ohv V-twin	Four-speed, chain drive	Fat Bob
Model 80FXEF-80	82 cu. in./1340 cc. ohv V-twin	Four-speed, chain drive	Fat Bob
Model 80FXWG-80	82 cu. in./1340 cc. ohv V-twin	Four-speed, chain drive	Wide Glide
Model 80FXB-80	82 cu. in./1340 cc. ohv V-twin	Four-speed belt drive	Sturgis
Model 80FLH	74 cu. in./1207 cc. ohv V-twin	Four-speed, chain drive	Electra-Glide
Model 80FL-80	82 cu. in./1340 cc. ohv V-twin	Four-speed, chain drive	Electra-Glide (Pol)
Model 80FLH-80	82 cu. in./1340 cc. ohv V-twin	Four-speed, chain drive	Electra-Glide
Model 80FLH-80	82 cu. in./1340 cc. ohv V-twin	Four-speed, chain drive	Electra-Glide Shrine
Model 80FLHC-80	82 cu. in./1340 cc. ohv V-twin	Four-speed, chain drive	Electra-Glide Classic
Model 80FLHS-80	82 cu. in./1340 cc. ohv V-twin	Four-speed, chain drive	Electra-Glide Sport

The FL models could be ordered with or without sidecar-specification.

1981 model year
Model 81XLH	61 cu. in./997 cc. ohv V-twin	Four-speed, chain drive	Sportster
Model 81XLS	61 cu. in./997 cc. ohv V-twin	Four-speed, chain drive	Roadster
Model 81FXE	82 cu. in./1340 cc. ohv V-twin	Four-speed, chain drive	Super Glide
Model 81FXS	82 cu. in./1340 cc. ohv V-twin	Four-speed, chain drive	Low Rider
Model 81FXEF	82 cu. in./1340 cc. ohv V-twin	Four-speed, chain drive	Fat Bob
Model 81FXWG	82 cu. in./1340 cc. ohv V-twin	Four-speed, chain drive	Wide Glide
Model 81FXB	82 cu. in./1340 cc. ohv V-twin	Four-speed, belt drive	Sturgis
Model 81FLH	82 cu. in./1340 cc. ohv V-twin	Four-speed, chain drive	Electra-Glide
Model 81FLH	82 cu. in./1340 cc. ohv V-twin	Four-speed, chain drive	Electra-Glide Shrine
Model 81FLH	82 cu. in./1340 cc. ohv V-twin	Four-speed, chain drive	Electra-Glide Heritage
Model 81FLHC	82 cu. in./1340 cc. ohv V-twin	Four-speed, chain drive	Electra-Glide Classic
Model 81FLHS	82 cu. in./1340 cc. ohv V-twin	Four-speed, chain drive	Electra-Glide Sport

The FL models could be ordered with or without sidecar-specification.

1982 model year
Model 82XLH	61 cu. in./997 cc. ohv V-twin	Four-speed, chain drive	Sportster
Model 82XLS	61 cu. in./997 cc. ohv V-twin	Four speed, chain drive	Roadster
Model 82FXE	82 cu. in./1340 cc. ohv V-twin	Four-speed, chain drive	Super Glide
Model 82FXS	82 cu. in./1340 cc. ohv V-twin	Four-speed, chain drive	Low Rider
Model 82FXE/F	82 cu. in./1340 cc. ohv V-twin	Four-speed, chain drive	Fat Bob
Model 82FXWG	82 cu. in./1340 cc. ohv V-twin	Four-speed, chain drive	Wide Glide
Model 82FXB	82 cu. in./1340 cc. ohv V-twin	Four-speed belt drive	Sturgis
Model 82FXRS	82 cu. in./1340 cc. ohv V-twin	Four-speed, chain drive	Low Glide
Model 82FLH	82 cu. in./1340 cc. ohv V-twin	Four-speed, chain drive	Electra-Glide
Model 82FLH	82 cu. in./1340 cc. ohv V-twin	Four-speed, chain drive	Electra-Glide Shrine
Model 82FLH	82 cu. in./1340 cc. ohv V-twin	Four-speed, chain drive	Electra-Glide Heritage
Model 82FLHC	82 cu. in./1340 cc. ohv V-twin	Four-speed, chain drive	Electra-Glide Classic
Model 82FLHS	82 cu. in./1340 cc. ohv V-twin	Four-speed, chain drive	Electra-Glide Sport

The FL models could be ordered with or without sidecar-specification.

Special models
Model 82XLH	Anniversary Sportster
Model 82XLS	Anniversary Roadster

1983 model year
Model 83XLX	61 cu. in./997 cc. ohv V-twin	Four-speed, chain drive	Sportster
Model 83XLH	61 cu. in./997 cc. ohv V-twin	Four-speed, chain drive	Sportster Standard
Model 83XLS	61 cu. in./997 cc. ohv V-twin	Four-speed, chain drive	Roadster
Model 83XR-1000	61 cu. in./997 cc. ohv V-twin	Four-speed, chain drive	Sportster
Model 83FXE	82 cu. in./1340 cc. ohv V-twin	Four-speed, chain drive	Super Glide
Model 83FXE/F	82 cu. in./1340 cc. ohv V-twin	Four-speed, chain drive	Fat Bob
Model 83FXWG	82 cu. in./1340 cc. ohv V-twin	Four-speed, chain drive	Wide Glide
Model 83FXSB	82 cu. in./1340 cc. ohv V-twin	Four-speed belt drive	Low Rider

Model 83FXDG	82 cu. in./1340 cc. ohv V-twin	Four-speed, chain drive	Disc Glide
Model 83FXRS	82 cu. in./1340 cc. ohv V-twin	Four-speed, chain drive	Low Glide
Model 83FXRT	82 cu. in./1340 cc. ohv V-twin	Four-speed, chain drive	Sport Glide
Model 83FLH	82 cu. in./1340 cc. ohv V-twin	Four-speed, chain drive	Electra-Glide
Model 83FLH	82 cu. in./1340 cc. ohv V-twin	Four-speed, chain drive	Electra-Glide Shrine
Model 83FLHS	82 cu. in./1340 cc. ohv V-twin	Four-speed, chain drive	Electra-Glide Sport
Model 83FLHT	82 cu. in./1340 cc. ohv V-twin	Five-speed, chain drive	Electra-Glide
Model 83FLHT	82 cu. in./1340 cc. ohv V-twin	Five-speed, chain drive	Electra-Glide Shrine
Model 83FLHTC	82 cu. in./1340 cc. ohv V-twin	Five-speed, chain drive	Electra-Glide Classic
Model 83FLHTC	82 cu. in./1340 cc. ohv V-twin	Five-speed belt drive	Electra-Glide Classic

The FL models could be ordered with or without sidecar-specification.

Special models
Model 83FLHS   Limited Edition

Appendix 3: Harley-Davidson production totals, 1903 to 1983

Model year	Machines	Notes
1903	3	First motorcycle
1904	8	
1905	16	
1906	50	Incorporation of the Harley-Davidson Motor Company
1907	150	
1908	450	
1909	1,149	
1910	3,168	
1911	5,625	
1912	9,571	
1913	12,904	
1914	16,284	
1915	16,645	
1916	17,439	
1917	19,763	Entry of the USA into the First World War
1918	19,359	End of the First World War
1919	24,292	
1920	28,189	Figure not to be reached again for 22 years
1921	10,202	
1922	12,759	
1923	18,430	
1924	13,996	
1925	16,929	
1926	23,354	
1927	19,911	
1928	22,350	
1929	21,242	October: Black Thursday; start of the Great Depression
1930	17,422	
1931	10,500	
1932	6,841	
1933	3,703	
1934	10,231	End of the Great Depression
1935	10,368	
1936	9,812	
1937	11,674	
1938	9,934	
1939	8,355	Start of the Second World War
1940	10,855	
1941	18,428	7, Dec, 1941: Japanese attack on Pearl Harbor
1942	29,603	1920 production total exceeded for the first time
1943	29,243	
1944	17,006	
1945	11,978	End of the Second World War
1946	15,554	
1947	20,392	
1948	31,163	New production record
1949	23,740	

Model year	Machines	Notes
1950	18,355	Korean war
1951	14,580	
1952	17,250	
1953	14,050	
1954	12,250	
1955	9,750	
1956	11,906	
1957	13,079	
1958	12,676	
1959	12,342	
1960	15,728	
1961	10,497	Cuban Missile Crisis
1962	11,144	
1963	10,407	
1964	13,270	
1965	25,328	
1966	36,310	Breaking the 1948 record
1967	27,202	
1968	26,748	
1969	27,375	
1970	28,850	
1971	37,620	
1972	59,908	Annual production exceeded 50,000 machines for the first time
1973	70,903	
1974	68,210	
1975	75,403	End of the Vietnam war. Annual production exceeded 75,000 machines for the first time
1976	61,375	
1977	45,608	
1978	47,401	Less special export models
1979	49,578	Less special export models
1980	48,181	Less special export models
1981	41,606	Less special export models
1982	37,943	Less special export models
1983	29,573	Less special export models